Lecture Notes in Computer Science 15050

Founding Editors

Gerhard Goos
Juris Hartmanis

Editorial Board Members

Elisa Bertino, *Purdue University, West Lafayette, IN, USA*
Wen Gao, *Peking University, Beijing, China*
Bernhard Steffen ⓘ, *TU Dortmund University, Dortmund, Germany*
Moti Yung ⓘ, *Columbia University, New York, NY, USA*

The series Lecture Notes in Computer Science (LNCS), including its subseries Lecture Notes in Artificial Intelligence (LNAI) and Lecture Notes in Bioinformatics (LNBI), has established itself as a medium for the publication of new developments in computer science and information technology research, teaching, and education.

LNCS enjoys close cooperation with the computer science R & D community, the series counts many renowned academics among its volume editors and paper authors, and collaborates with prestigious societies. Its mission is to serve this international community by providing an invaluable service, mainly focused on the publication of conference and workshop proceedings and postproceedings. LNCS commenced publication in 1973.

Laura Kovács · Ana Sokolova
Editors

Reachability Problems

18th International Conference, RP 2024
Vienna, Austria, September 25–27, 2024
Proceedings

Editors
Laura Kovács
TU Wien
Vienna, Austria

Ana Sokolova
Paris Lodron University of Salzburg
Salzburg, Austria

ISSN 0302-9743 ISSN 1611-3349 (electronic)
Lecture Notes in Computer Science
ISBN 978-3-031-72620-0 ISBN 978-3-031-72621-7 (eBook)
https://doi.org/10.1007/978-3-031-72621-7

© The Editor(s) (if applicable) and The Author(s), under exclusive license
to Springer Nature Switzerland AG 2024

This work is subject to copyright. All rights are solely and exclusively licensed by the Publisher, whether the whole or part of the material is concerned, specifically the rights of translation, reprinting, reuse of illustrations, recitation, broadcasting, reproduction on microfilms or in any other physical way, and transmission or information storage and retrieval, electronic adaptation, computer software, or by similar or dissimilar methodology now known or hereafter developed.
The use of general descriptive names, registered names, trademarks, service marks, etc. in this publication does not imply, even in the absence of a specific statement, that such names are exempt from the relevant protective laws and regulations and therefore free for general use.
The publisher, the authors and the editors are safe to assume that the advice and information in this book are believed to be true and accurate at the date of publication. Neither the publisher nor the authors or the editors give a warranty, expressed or implied, with respect to the material contained herein or for any errors or omissions that may have been made. The publisher remains neutral with regard to jurisdictional claims in published maps and institutional affiliations.

This Springer imprint is published by the registered company Springer Nature Switzerland AG
The registered company address is: Gewerbestrasse 11, 6330 Cham, Switzerland

If disposing of this product, please recycle the paper.

Preface

This volume contains the papers presented at the 18th International Conference on Reachability Problems (RP 2024), co-organized by the Formal Methods in Systems Engineering (FORSYTE) research unit of the TU Wien, the Paris Lodron University of Salzburg (PLUS), and the Wolfgang Pauli Institute (WPI) Vienna, Austria.

The RP 2024 conference took place as a physical event during September 25–27, 2024 at the TU Wien, Vienna, Austria. Previous events in the RP conference series were located at the Laboratoire d'Informatique, Signaux et Systèmes de Sophia Antipolis of the Université Côte d'Azur, France (2023); the University of Kaiserslautern, Germany (2022); the University of Liverpool, UK (2021); Université Paris Cité, France (2020); Université Libre de Bruxelles, Belgium (2019); Aix-Marseille University, France (2018); Royal Holloway, University of London, UK (2017); Aalborg University, Denmark (2016); the University of Warsaw, Poland (2015); the University of Oxford, UK (2014); Uppsala University, Sweden (2013); the University of Bordeaux, France (2012); the University of Genoa, Italy (2011); Masaryk University, Czech Republic (2010); École Polytechnique, France (2009); the University of Liverpool, UK (2008); and Turku University, Finland (2007).

The RP conference aims to gather together scholars from diverse disciplines and backgrounds interested in reachability problems that appear, among others, in algebraic structures, automata theory and formal languages, computational game theory, concurrency and distributed computation, decision procedures in computational models, hybrid dynamical systems, logic and model checking, and verification of finite- and infinite-state systems. In addition, the conference promotes the exploration and combination of new approaches for the modeling and analysis of computational processes by combining mathematical, algorithmic, and computational techniques. As such, the RP conference provides an active forum for discussion and networking for researchers interested in diverse fields of reachability analysis.

RP 2024 received 37 submissions, consisting of 19 regular research papers, 2 invited papers, and 16 presentation-only submissions. All regular paper submissions to RP 2024 were reviewed using a single-blind reviewing process: each paper was assigned for review to at least three Program Committee (PC) members, who made use of sub-reviewers. Using the review reports, the PC had a thorough discussion on each regular paper, which, in some cases, changed the PC support on the respective paper. Submissions with predominantly supportive reviews were agreed upon to be accepted by the PC: in total, the PC accepted 13 regular papers and 16 presentation-only submissions, in addition to the 2 invited paper contributions. This volume contains the 13 regular papers accepted at RP 2024, lists the 16 presentation-only submissions, and also includes the 2 invited papers of RP 2024.

The RP 2024 conference also featured invited talks by

- **Ezio Bartocci** (TU Wien, Austria), delivering a talk about *"Quantifying Uncertainty in Probabilistic Loops without Sampling: a Fully Automated Approach"*,

- **Joost-Pieter Katoen** (RWTH Aachen, Germany), discussing *"Programmatic Synthesis for Infinite MDPs Using Program Refinement"*,
- **Antonín Kučera** (Masaryk University, Czech Republic), detailing results on *"The Satisfiability and Validity Problems for Probabilistic CTL"*,
- **Ruzica Piskac** (Yale University, USA), presenting *"Proofs as Polynomials"*.

In addition, we are also very grateful to our RP 2024 invited tutorial speaker, **K. S. Thejaswini** (IST Austria, Austria), who surveyed results and challanges in *"Solving Parity and Rabin Games by Constructing Universal Trees"*. Abstracts of the RP 2024 invited talks and our invited tutorial are also included in this volume.

We would like to thank everyone who helped to make RP 2024 successful. We thank the authors for submitting their papers to RP 2024. The PC members and additional reviewers did an excellent job in reviewing papers: they provided detailed reports and engaged in the PC discussions. We thank the RP steering committee, and especially Igor Potapov for his valuable advice. We are grateful to members of our local organizing committee, and in particular to Norbert Mauser (co-organizer, WPI) and Beatrix Buhl (responsible for local arrangements). We acknowledge the financial support provided by Amazon Web Services and Springer Nature. We are grateful for the institutional support RP 2024 received from the Paris Lodron University of Salzburg (PLUS), the TU Wien, and WPI. We also would like to thank the EasyChair developers for their help and assistance. Finally, a big thank you goes to all our RP 2024 authors, invited speakers, and invited tutorial speaker for their high-quality contributions, and the participants for making RP 2024 a success!

September 2024

Laura Kovács
Ana Sokolova

Organization

Program Committee Chairs

Laura Kovács — TU Wien, Austria
Ana Sokolova — Paris Lodron University of Salzburg, Austria

Program Committee

Parosh Aziz Abdulla — Uppsala University, Sweden
Luca Aceto — Reykjavik University, Iceland
Christel Baier — TU Dresden, Germany
Valerie Berthe — CNRS IRIF, France
Valentina Castiglioni — Eindhoven University of Technology, The Netherlands
Michele Chiari — TU Wien, Austria
Laure Daviaud — University of East Anglia, UK
Jim de Groot — Australian National University, Australia
Christoph Haase — University of Oxford, UK
Vesa Halava — University of Turku, Finland
Ichiro Hasuo — National Institute of Informatics, Japan
Jarkko Kari — University of Turku, Finland
George Kenison — Liverpool John Moores University, UK
Sandra Kiefer — University of Oxford, UK
Laura Kovács — TU Wien, Austria
Jérôme Leroux — CNRS, Université de Bordeaux, France
Rupak Majumdar — MPI-SWS, Germany
Kaushik Mallik — Institute of Science and Technology Austria, Austria
Tobias Meggendorfer — Lancaster University Leipzig, Germany
Anca Muscholl — LaBRI, Universite Bordeaux, France
Markus Müller-Olm — Universität Münster, Germany
Dave Parker — University of Oxford, UK
Igor Potapov — University of Liverpool, UK
Amaury Pouly — IRIF/CNRS - Université Paris Cité, France
Jurriaan Rot — Radboud University, The Netherlands
Helmut Seidl — Technische Universität München, Germany
Martina Seidl — Johannes Kepler University Linz, Austria
Mahsa Shirmohammadi — CNRS, France

Ana Sokolova Paris Lodron University of Salzburg, Austria
Maximilian Weininger Institute of Science and Technology Austria, Austria
Thorsten Wißmann Friedrich-Alexander-Universität Erlangen-Nürnberg, Germany
James Worrell University of Oxford, UK
Dmitry Zaitsev University of Derby, UK
Florian Zuleger TU Wien, Austria

Steering Committee

Parosh Aziz Abdulla Uppsala University, Sweden
Olivier Bournez École Polytechnique, LIX, IPP, France
Vesa Halava University of Turku, Finland
Alain Finkel École normale supérieure Paris-Saclay, France
Oscar Ibarra University of California Santa Barbara, USA
Juhani Karhumaki University of Turku, Finland
Jérôme Leroux Université de Bordeaux, France
Joël Ouaknine Max Planck Institute for Software Systems, Germany
Igor Potapov University of Liverpool, UK
James Worrell University of Oxford, UK

Local Organizing Committee

Laura Kovács (co-organizer) TU Wien c/o WPI, Austria
Norbert Mauser (co-organizer) WPI, Austria
Ana Sokolova (co-organizer) Paris Lodron University of Salzburg, Austria
Beatrix Buhl (local arrangements) TU Wien, Austria
Robin Coutelier TU Wien, Austria
Benjamin Deutsch TU Wien, Austria
Clemens Eisenhofer TU Wien, Austria
Katalin Fazekas TU Wien, Austria
Thomas Hader TU Wien, Austria
Florian Sextl TU Wien, Austria
Anton Varonka TU Wien, Austria

Additional Reviewers

Ait El Manssour, Rida
Jajcayova, Tatiana

Abstracts of Invited Speakers

Quantifying Uncertainty in Probabilistic Loops without Sampling: a Fully Automated Approach

Ezio Bartocci

TU Wien, Austria
ezio.bartocci@tuwien.ac.at

Abstract. A probabilistic loop is a programming control flow structure whose behavior depends on random variables' assignments and probabilistic conditions. One challenging problem is quantifying automatically the uncertainty of the probabilistic loop behavior for a potentially unbounded number of iterations. Although this problem is generally highly undecidable, we have explored the necessary restrictions enabling the automated analysis of probabilistic loops without user intervention. Our symbolic approach leverages algebraic methods and the statistical properties of well-defined probability distributions to derive closed-form expressions of the higher-order statistical moments for the program's random variables at each loop iteration. In this talk, we demonstrate the application of our methodology through a series of examples.

Programmatic Synthesis for Infinite MDPs Using Program Refinement

Joost-Pieter Katoen

RWTH Aachen, Germany
katoen@cs.rwth-aachen.de

Abstract. We consider imperative programs that involve both randomization and pure nondeterminism. The central question is how to find a strategy resolving the pure nondeterminism such that the so-obtained determinized program satisfies a given quantitative specification, i.e., bounds on expected outcomes such as the expected final value of a program variable or the probability to terminate in a given set of states.

We show how memoryless and deterministic (MD) strategies can be obtained in a semi-automatic fashion using deductive verification. For loop-free programs, the MD strategies resulting from our weakest precondition-style framework are correct by construction. This extends to loopy programs, provided the loops are equipped with suitable loop invariants - just like in program verification.

We show how our technique relates to the well-studied problem of obtaining strategies in countably infinite Markov decision processes with reachability-reward objectives.

I will show the applicability of this approach by means of some case studies.

This talk is based on joint work with Kevin Batz, Tom Biskup, and Tobias Winkler.

The Satisfiability and Validity Problems for Probabilistic CTL

Antonín Kučera

Masaryk University, Czech Republic
tony@fi.muni.cz

Abstract. Probabilistic CTL is obtained from the standard CTL (Computational Tree Logic) by replacing the existential and universal path quantifiers with the probabilistic operator where the probability of runs satisfying a given path formula is bounded by a rational constant. We survey the existing results about the satisfiability and validity problems for probabilistic CTL, and we also present some of the underlying proof techniques.

Proofs as Polynomials

Ruzica Piskac

Yale University, USA
ruzica.piskac@yale.edu

Abstract. Zero-knowledge (ZK) protocols are well-known cryptographic primitives that allow one party to prove to another party a statement without revealing anything beyond the statement. A ZK protocol consists of two parties: a "prover" and a "verifier". In our work, the prover holds a secret formula and its proof of validity and needs to convince the verifier about the correctness of the proof. The verifier validates the prover's claims, by checking every step of the proof. To be able to do that without revealing any details about the formula, we use so-called commitment schemes. Commitment schemes are a fundamental part of zero-knowledge protocols as they allow a prover to commit to a value while keeping it hidden, ensuring the value cannot be altered later. A polynomial commitment scheme can be used to commit polynomials and prove the properties of the polynomials. Our work encodes proofs as polynomials and in this way transforms checking the proof steps into checking relations between polynomials. By doing this, we are able to verify the proof without revealing the formulae (and the proof itself).

In this talk, we focus on proofs for formulas produced in the verification process and we explain how to encode them as polynomials. Initially, we develop a protocol for validating the unsatisfiability of Boolean formulas in privacy-preserving settings. We use the resolution calculus to produce a proof of unsatisfiability: we encode each clause appearing in the proof as a polynomial and we reduce checking the correctness of the resolution rule to checking the divisibility of two polynomials.

A natural extension of this technique is to consider more expressive logics, such as those supported by SMT (Satisfiability Modulo Theories) solvers. To this end, we extend our initial work and develop a virtual machine for validating general unsatisfiability proofs. This virtual machine can support the majority of popular theories when proving program safety while being complete and sound. To demonstrate this, we use theories of equality and linear integer arithmetic as examples. These theories require non-trivial checking procedures and we propose optimized arithmetizations based on multiset interpretation and polynomial encodings.

Finally, we conclude the talk by outlining how this approach benefits and empowers the verification process: we can now obtain privacy while preserving correctness.

This talk is based on the following papers:

[1] Ning Luo, Timos Antonopoulos, William Harris, Ruzica Piskac, Eran Tromer, Xiao Wang: *Proving UNSAT in Zero Knowledge, CCS 2022*
[2] Daniel Luick, John C. Kolesar, Timos Antonopoulos, William R. Harris, James Parker, Ruzica Piskac, Eran Tromer, Xiao Wang, Ning Luo: *ZKSMT: A VM for Proving SMT Theorems in Zero Knowledge.* USENIX'24 Security.

Abstract of Invited Tutorial

Solving Parity and Rabin Games by Constructing Universal Trees

K. S. Thejaswini[ID]

IST Austria, Austria
thejaswini.k.s@ist.ac.at

Abstract. The computational problem of deciding the winner in two-player games on finite graphs with parity and Rabin objectives is a fundamental problem with applications to program verification and synthesis, and it is closely linked to problems in automata theory and logic. We focus on understanding the structural properties of these games and devising algorithms that harness these properties.

At the core of these games is the concept of an attractor decomposition, a structured representation of a parity game that serves as a witness of winning for a player and that naturally corresponds to a tree. It has been established that universal trees—trees capable of embedding all possible trees emerging from an input parity game—play a pivotal role in serving as a search space for all known algorithms designed to solve parity games. We survey these techniques and further, we also discuss the extensions to these universal trees that help efficiently solve Rabin games, as well as Strahler universal trees that solve parity games efficiently for certain parameter settings.

List of Presentation-Only Papers

- Julia Klein, Alberto D'Onofrio and Tatjana Petrov. *Exploring Consensus Robustness in Swarms with Disruptive Individuals*
- George Kenison. *The Threshold Problem for Hypergeometric Sequences with Quadratic Parameters*
- Tobias Meggendorfer and Maximilian Weininger. *Playing Games with your PET: Solving Stochastic Games, Reliably*
- Ali Asadi, Krishnendu Chatterjee, Raimundo Saona Urmeneta and Jakub Svoboda. *Deterministic Sub-exponential Algorithm for Discounted-sum Games with Unary Weights*
- Udi Boker, Thomas Henzinger, Nicolas Mazzocchi and N. Ege Saraç. *Safety and Liveness of Quantitative Properties and Automata*
- Guy Avni, Ehsan Kafshdar Goharshady, Thomas A. Henzinger and Kaushik Mallik. *Bidding Games with Charging*
- Rida Ait El Manssour, George Kenison, Mahsa Shirmohammadi and Anton Varonka. *Simple Linear Loops: Algebraic Invariants and Synthesis*
- Munehiro Iwami and Takahito Aoto. *Preservation of Rationality in Infinitary Rewriting by Top-Down Tree Transducers*
- Quentin Guilmant, Engel Lefaucheux, Joël Ouaknine and James Worell. *The 2-Dimensional Constraint Loop Problem is Decidable*
- Filip Cano, Thomas A. Henzinger, Bettina Könighofer , Konstantin Kueffner and Kaushik Mallik. *Abstraction-Based Decision Making for Statistical Properties*
- Ashwani Anand, Sylvain Schmitz, Lia Schütze and Georg Zetzsche. *Verifying Unboundedness via Amalgamation*
- Florian Bruse, Martin Lange and Sören Möller. *The Calculus of Dual Influence*
- Arnd Hartmanns, Sebastian Junges, Tim Quatmann and Maximilian Weininger. *The Revised Practitioner's Guide to MDP Model Checking Algorithms*
- R Govind, S. Akshay, Paul Gastin and B Srivathsan. *MITL Model Checking via Generalized Timed Automata and a new Liveness Algorithm*
- Bruno Loff and Mateusz Skomra. *Smoothed Analysis of Deterministic Discounted and Mean-Payoff Games*
- Quentin Guilmant and Joël Ouaknine. *Inaproximability in Weighted Timed Games*

Contents

Invited Papers

Quantifying Uncertainty in Probabilistic Loops Without Sampling:
A Fully Automated Approach 3
   Ezio Bartocci

The Satisfiability and Validity Problems for Probabilistic CTL 9
   Antonín Kučera

Computability and Reachability

Computing Reachable Simulations on Transition Systems 21
   Pierre Ganty, Nicolas Manini, and Francesco Ranzato

Computing All Minimal Ways to Reach a Context-Free Language 38
   Florian Bruse and Martin Lange

On Solving All-Path Reachability Problems for Starvation Freedom
of Concurrent Rewrite Systems Under Process Fairness 54
   Misaki Kojima and Naoki Nishida

Automata and Complexity

Rollercoasters with Plateaus 73
   Duncan Adamson, Pamela Fleischmann, and Annika Huch

Quantum Automata and Languages of Finite Index 88
   Andrea Benso, Flavio D'Alessandro, and Paolo Papi

On Shortest Products for Nonnegative Matrix Mortality 104
   Andrew Ryzhikov

Hardness of Busy Beaver Value BB(15) 120
   Tristan Stérin and Damien Woods

Linear Systems and Recurrences

On the Complexity of Reachability and Mortality for Bounded Piecewise Affine Maps 141
   Olga Tveretina

Semi-linear VASR for Over-Approximate Semi-linear Transition System Reachability 154
   Nikhil Pimpalkhare and Zachary Kincaid

Reachability in Linear Recurrence Automata 167
   Mika Hirvensalo, Akitoshi Kawamura, Igor Potapov, and Takao Yuyama

Games and Abstractions

Robust Deterministic Abstractions for Supervising Discrete-Time Continuous Systems 187
   Gwendal Priser, Elena Vanneaux, and Goran Frehse

Markov Decision Processes with Sure Parity and Multiple Reachability Objectives 203
   Raphaël Berthon, Joost-Pieter Katoen, and Tobias Winkler

Modelling Dynamical Systems: Learning ODEs with No Internal ODE Resolution 221
   Johanne Cohen, Emmanuel Goutierre, Hayg Guler, Fatios Kapotos, Sida-Bastien Li, Michèle Sébag, and Bowen Zhu

Author Index 239

Invited Papers

Quantifying Uncertainty in Probabilistic Loops Without Sampling: A Fully Automated Approach

Ezio Bartocci[✉] [iD]

TU Wien, 1040 Vienna, Austria
ezio.bartocci@tuwien.ac.at

Abstract. A probabilistic loop is a programming control flow structure whose behavior depends on random variables' assignments and probabilistic conditions. One challenging problem is quantifying automatically the uncertainty of the probabilistic loop behavior for a potentially unbounded number of iterations. Although this problem is generally highly undecidable, we have explored the necessary restrictions enabling the automated analysis of probabilistic loops without user intervention. Our symbolic approach leverages algebraic methods and the statistical properties of well-defined probability distributions to derive closed-form expressions of the higher-order statistical moments for the program's random variables at each loop iteration. In this talk, we demonstrate the application of our methodology through a series of examples.

Keywords: Probabilistic Programs · Probabilistic reasoning · Moment-based invariants

1 Overview

Probabilistic loops are employed in many applications, including rejection sampling for probabilistic machine learning models [20], randomized algorithms [16], simulating stochastic dynamical systems such as biological and cyber-physical systems [13,14], or controlling cyber-physical systems [19]. Listing 1.1 shows an illustrative example of a probabilistic loop that simulates two people playing the "Rock-Paper-Scissors" game[1]. Each loop iteration represents a match among two players. For each match, a player can choose one of the following symbols: r for *rock*, p for *paper* and s for *scissors*. Players display each other the chosen symbol

---

[1] https://en.wikipedia.org/wiki/Rock_paper_scissors

This work was supported by the Vienna Science and Technology Fund (WWTF) [10.47379/ICT19018] (ProbInG). The work presented in this invited talk was previously published in several papers in collaboration with Daneshvar Amrollahi, Efstathia Bura, George Kenison, Ahmad Karimi, Andrey Kofnov, Laura Kovács, Marcel Moosbrugger, Miroslav Stankovic.

© The Author(s), under exclusive license to Springer Nature Switzerland AG 2024
L. Kovács and A. Sokolova (Eds.): RP 2024, LNCS 15080, pp. 3–8, 2024.
https://doi.org/10.1007/978-3-031-72621-7_1

simultaneously. If both players have chosen the same symbol, the result is a tie. Otherwise, the player choosing *rock* symbol loses against the player choosing *paper* symbol and wins against the player choosing *scissors* symbol; the player choosing *paper* symbol loses against the player choosing *scissors* symbol. The variables $p1w$ and $p2w$ record how often the first and the second players will win after n matches, respectively. The variable tie computes the number of matches ending in a tie after n matches. All $p1w$, $p2w$ and tie are unbounded variables generating infinite state space. The players' choices $p1$ and $p2$ are discrete random variables that can assume only three values whose probability distribution is categorical and depends on parameters describing the probability for each player to choose a particular symbol. The problem we want to solve here is to quantify automatically, without any user intervention, the expected number of matches each player can win and the number of ties.

```
1  #p1    - player1, p2 - player2
2  #p1w   - Win Player1, p2w - Win Player2
3  #tie   - number of ties
4  p1, p2, p1w, p2w, tie = 0, 0, 0, 0, 0
5
6  # r - rock, p - paper, s - scissors
7  r, p, s = 0, 1, 2
8
9  while true:
10     p1 = Categorical(r_1, p_1, s_1)
11     p2 = Categorical(r_2, p_2, s_2)
12     if p1 == r:
13         if p2 == p:
14             p2w = p2w + 1
15         elif p2 == s:
16             p1w = p1w + 1
17         elif p2 == r:
18             tie = tie + 1
19         end
20     elif p1 == p:
21         if p2 == r:
22             p1w = p1w + 1
23         elif p2 == s:
24             p2w = p2w + 1
25         elif p2 == p:
26             tie = tie + 1
27         end
28     elif p1 == s:
29         if p2 == r:
30             p2w = p2w + 1
31         elif p2 == p:
32             p1w = p1w + 1
33         elif p2 == s:
34             tie = tie + 1
35         end
36     end
37 end
```

Listing 1.1. A probabilistic loop simulating the Rock-Paper-Scissors game.

Probabilistic model checkers [8,15] cannot be employed to solve this problem because they limit their analysis to finite-state Markov chains, and they are not designed to compute invariants over expected values. Computing expected values through sampling and statistical approaches [22] is unfeasible in this case for two main reasons: the probabilities in the example are provided as symbolic parameters and sampling requires setting a particular number of iterations and parameters' values. There are symbolic approaches [7,11] to compute quantitative invariants for unbounded loops, but necessitate the user to annotate the loop with a linear expression template. Other methods like PSI [6] are fully automatic but do not support unbounded probabilistic loops.

In contrast, we have developed several automated methods [1–3,5,14,16,20] to analyze, without sampling, unbounded loops with multi-path executions, probabilistic conditions, symbolic parameters, random variables drawing their values from well-defined continuous/discrete probability distributions with bounded/unbounded support, and with (non-)polynomial updates. All these techniques are implemented in our POLAR [16] tool. POLAR computes automatically the expected values of $p1w$, $p2w$ and tie in Listing 1.1 with these closed-form solutions (n is the number of iteration):

$$\mathbb{E}[p1w] = (-p_1 \cdot p_2 - 2 \cdot p_1 \cdot s_2 + p_1 + p_2 \cdot s_1 - s_1 \cdot s_2 + s_2)n$$

$$\mathbb{E}[p2w] = (-p_1 \cdot p_2 + p_1 \cdot s_2 - 2 \cdot p_2 \cdot s_1 + p_2 - s_1 \cdot s_2 + s_1)n$$

$$\mathbb{E}[tie] = (2 \cdot p_1 \cdot p_2 + p_1 \cdot s_2 - p_1 + p_2 \cdot s_1 - p_2 + 2 \cdot s_1 \cdot s_2 - s_1 - s_2 + 1)n$$

When the probability to choose one of the three symbols is equal for all the players, $r_1 = r_2 = p_1 = p_2 = s_1 = s_2 = 1/3$ then the expected number of matches that are ties are equivalent to expected number of matches won by one of the two players $\mathbb{E}[p1w] = \mathbb{E}[p2w] = \mathbb{E}[tie] = \frac{1}{3}n$. The situation remains the same if only just one of the two players chooses each symbol with the same probability $r_i = p_i = s_i = 1/3$. However, if both players chooses different probabilities for each symbol, such as for example $r_1 = r_2 = 1/2$, $p_1 = p_2 = 1/8$ and $p_1 = p_2 = 3/8$ then the results will be different with $\mathbb{E}[p1w] = \mathbb{E}[p2w] = \frac{19}{64}n, \mathbb{E}[tie] = \frac{13}{32}n$.

First-order statistical moments of random variables alone are usually insufficient to determine the probabilistic behavior of two programs: while their expected values might be the same, their distributions could be entirely different. For this reason, our tool also provides the possibility to compute the closed-form expression of high-order statistical moments that can be used to approximate the distributions of the program random variables using Gram-Charlier series [10,16].

Our approach consists of the following steps:

1. We impose restrictions on the language of the probabilistic program that our method can analyze. The key restrictions are: (a) if-statement conditions can only be applied to variables with a finite number of values, (b) variables' updates must not depend on other variables in a non-linearly cyclical way, and (c) parameters in probability distributions and choices must remain constant throughout loop executions. Restriction (a) ensures that the class of loops is not Turing-complete, thereby keeping our procedure decidable (see [16,17] for more details about decidability results). Restriction (b) in some cases can be lifted (see our work on unsolvable loops [1,2]).

2. Our method requires as input a probabilistic loop and a higher-order statistical moment of a random variable for analysis. Once the desired statistical moment is provided as input, our approach transforms the probabilistic loop into a deterministic moment-preserving loop by replacing if-statements with polynomial assignments, as well as substituting the drawing from well-defined probability distributions with their well-defined moments.
3. By introducing new auxiliary variables representing the monomials of the polynomial updates we can build a system of linear C-finite recurrences [12,16] representing the deterministic updates obtained in the previous step. The closed-form solution for this system of recurrences has the form of an exponential polynomial function in the number of the loop iterations.

1.1 Probabilistic Loops with Non-polynomial Updates

Our initial approach [16] considered only probabilistic loops with polynomial updates. However, using probabilistic loops, we can encode the numerical simulation of stochastic dynamical systems with non-polynomial updates such as trigonometric, exponential and logarithmic functions. To address the analysis of such probabilistic loops we have introduced two different methods in [13,14]. The first, more general, consists in approximating the non-polynomial updates with a sum of orthogonal polynomials [13,14] by leveraging the theory of polynomial chaos expansion [21]. Tight approximations require large polynomial expressions with several monomials, making our general procedure computationally slower. However, in [14], we have introduced also an exact method for computing closed-form solutions of statistical moments in probabilistic loops with mixed trigonometric and exponential updates. Our results build on the work of Jasour et al. in [9] that provides the solution of statistical moment of mixed trigonometric functions when they are applied to well-defined distributions. In [14], we demonstrated how to extend such results also to the case of exponential update functions, further enriching the class of function updates that our method can handle exactly in probabilistic loops.

1.2 Analyzing Bayesian Networks via Probabilistic Loops Analysis

Bayesian networks are established probabilistic machine learning models introduced by Judea Pearl in [18]. These graphical models have a directed acyclical graph (DAG) structure in which the nodes are random variables and the edges are the conditional probability dependencies among them. In [4,20], we demonstrated how to encode the inference process (with rejection sampling) in Bayesian networks (BNs) using probabilistic loops with polynomial updates. The acyclical structure of BNs ensures that the random variables' dependency in the probabilistic loop is also acyclical, meeting one the main requirements (we support only linear cyclical dependency) in our framework. We could encode different types of BNs in our framework, including discrete BNs, linear conditional Gaussian BNs and dynamic BNs. Using our methodology, we can solve many problems without sampling, including exact inference, sensitivity analysis, filtering problems on dynamic BNs, and computing the expected number of samples necessary to observe a particular event. In [20], we successfully demonstrated our approach on different size BNs, highlighting the applicability of our approach in different scenarios.

References

1. Amrollahi, D., Bartocci, E., Kenison, G., Kovács, L., Moosbrugger, M., Stankovic, M.: Solving invariant generation for unsolvable loops. In: Singh, G., Urban, C. (eds.) SAS 2022. LNCS, vol. 13790, pp. 19–43. Springer, Cham (2022). https://doi.org/10.1007/978-3-031-22308-2_3
2. Amrollahi, D., Bartocci, E., Kenison, G., Kovács, L., Moosbrugger, M., Stankovic, M.: (Un)Solvable loop analysis. Formal Methods Syst. Des. (2024). https://doi.org/10.1007/s10703-024-00455-0
3. Bartocci, E., Kovács, L., Stankovič, M.: Automatic generation of moment-based invariants for prob-solvable loops. In: Chen, Y.-F., Cheng, C.-H., Esparza, J. (eds.) ATVA 2019. LNCS, vol. 11781, pp. 255–276. Springer, Cham (2019). https://doi.org/10.1007/978-3-030-31784-3_15
4. Bartocci, E., Kovács, L., Stankovič, M.: Analysis of Bayesian networks via prob-solvable loops. In: Pun, V.K.I., Stolz, V., Simao, A. (eds.) ICTAC 2020. LNCS, vol. 12545, pp. 221–241. Springer, Cham (2020). https://doi.org/10.1007/978-3-030-64276-1_12
5. Bartocci, E., Kovács, L., Stankovič, M.: MORA - automatic generation of moment-based invariants. In: TACAS 2020. LNCS, vol. 12078, pp. 492–498. Springer, Cham (2020). https://doi.org/10.1007/978-3-030-45190-5_28
6. Gehr, T., Misailovic, S., Vechev, M.: PSI: exact symbolic inference for probabilistic programs. In: Chaudhuri, S., Farzan, A. (eds.) CAV 2016. LNCS, vol. 9779, pp. 62–83. Springer, Cham (2016). https://doi.org/10.1007/978-3-319-41528-4_4
7. Gretz, F., Katoen, J.-P., McIver, A.: PRINSYS—on a quest for probabilistic loop invariants. In: Joshi, K., Siegle, M., Stoelinga, M., D'Argenio, P.R. (eds.) QEST 2013. LNCS, vol. 8054, pp. 193–208. Springer, Heidelberg (2013). https://doi.org/10.1007/978-3-642-40196-1_17
8. Hensel, C., Junges, S., Katoen, J., Quatmann, T., Volk, M.: The probabilistic model checker storm. Int. J. Softw. Tools Technol. Transf. **24**(4), 589–610 (2022). https://doi.org/10.1007/S10009-021-00633-Z
9. Jasour, A., Wang, A., Williams, B.C.: Moment-based exact uncertainty propagation through nonlinear stochastic autonomous systems. CoRR abs/2101.12490 (2021). https://arxiv.org/abs/2101.12490
10. Karimi, A., Moosbrugger, M., Stankovic, M., Kovács, L., Bartocci, E., Bura, E.: Distribution estimation for probabilistic loops. In: Ábrahám, E., Paolieri, M. (eds.) QEST 2022. LNCS, vol. 13479, pp. 26–42. Springer, Cham (2022). https://doi.org/10.1007/978-3-031-16336-4_2
11. Katoen, J.-P., McIver, A.K., Meinicke, L.A., Morgan, C.C.: Linear-invariant generation for probabilistic programs: automated support for proof-based methods. In: Cousot, R., Martel, M. (eds.) SAS 2010. LNCS, vol. 6337, pp. 390–406. Springer, Heidelberg (2010). https://doi.org/10.1007/978-3-642-15769-1_24
12. Kauers, M., Paule, P.: The Concrete Tetrahedron. Texts and Monographs in Symbolic Computation, Springer, Vienna (2011)

13. Kofnov, A., Moosbrugger, M., Stankovic, M., Bartocci, E., Bura, E.: Moment-based invariants for probabilistic loops with non-polynomial assignments. In: QEST 2022. LNCS, vol. 13479, pp. 3–25. Springer, Cham (2022). https://doi.org/10.1007/978-3-031-16336-4_1
14. Kofnov, A., Moosbrugger, M., Stankovič, M., Bartocci, E., Bura, E.: Exact and approximate moment derivation for probabilistic loops with non-polynomial assignments. ACM Trans. Model. Comput. Simul. **34**(3), 1–25 (2024). https://doi.org/10.1145/3641545
15. Kwiatkowska, M., Norman, G., Parker, D.: PRISM 4.0: verification of probabilistic real-time systems. In: Gopalakrishnan, G., Qadeer, S. (eds.) CAV 2011. LNCS, vol. 6806, pp. 585–591. Springer, Heidelberg (2011). https://doi.org/10.1007/978-3-642-22110-1_47
16. Moosbrugger, M., Stankovič, M., Bartocci, E., Kovács, L.: This is the moment for probabilistic loops. Proc. ACM Program. Lang. **6**(OOPSLA2), 1497–1525 (2022)
17. Müller-Olm, M., Seidl, H.: Computing polynomial program invariants. Inf. Process. Lett. **91**(5), 233–244 (2004)
18. Pearl, J.: Bayesian networks: a model of self-activated memory for evidential reasoning. In: Proceedings of the 7th Conference of the Cognitive Science Society, pp. 329–334 (1985)
19. Selyunin, K., Ratasich, D., Bartocci, E., Islam, M.A., Smolka, S.A., Grosu, R.: Neural programming: towards adaptive control in cyber-physical systems. In: Proceedings of CDC 2015, pp. 6978–6985. IEEE (2015). https://doi.org/10.1109/CDC.2015.7403319
20. Stankovic, M., Bartocci, E., Kovács, L.: Moment-based analysis of Bayesian network properties. Theor. Comput. Sci. **903**, 113–133 (2022). https://doi.org/10.1016/J.TCS.2021.12.021
21. Xiu, D., Karniadakis, G.E.: The Wiener-Askey polynomial chaos for stochastic differential equations. SIAM J. Sci. Comput. **24**(2), 619–644 (2002). https://doi.org/10.1137/S1064827501387826
22. Younes, H.L.S., Simmons, R.G.: Statistical probabilistic model checking with a focus on time-bounded properties. Inf. Comput. **204**(9), 1368–1409 (2006). https://doi.org/10.1016/j.ic.2006.05.002

The Satisfiability and Validity Problems for Probabilistic CTL

Antonín Kučera

Faculty of Informatics, Masaryk University, Botanická 68a, 60200 Brno, Czechia
tony@fi.muni.cz

Abstract. Probabilistic CTL is obtained from the standard CTL (Computational Tree Logic) by replacing the existential and universal path quantifiers with the probabilistic operator where the probability of runs satisfying a given path formula is bounded by a rational constant. We survey the existing results about the satisfiability and validity problems for probabilistic CTL, and we also present some of the underlying proof techniques.

Keywords: Satisfiability · Probabilistic temporal logics · Probabilistic CTL

1 Introduction

Probabilistic Computational Tree Logic (PCTL) [20] is obtained from the standard CTL [15] by replacing the existential and universal path quantifiers with the probabilistic operator. PCTL is interpreted over Markov chains and can express explicit constraints on the probability of all runs satisfying a given path formula. Since the probabilistic operator is bound to precisely one temporal connective, the probability constraint can be associated directly with the connective. For example, the formula

$$P(\mathbf{G}(\neg a \wedge P(\mathbf{F}\, a) > 0)) > 0$$

can be written as

$$\mathbf{G}_{>0}(\neg a \wedge \mathbf{F}_{>0}\, a) \tag{1}$$

The intuitive meaning of (1) is "with positive probability, all states visited along a run do *not* satisfy the proposition a, but each of these states can reach a state satisfying a". Also, note that the semantics of (1) is substantially different from the semantics of its non-probabilistic CTL counterpart

$$\mathbf{EG}(\neg a \wedge \mathbf{EF}\, a) \tag{2}$$

In particular, the CTL formula (2) has the following simple model:

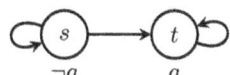

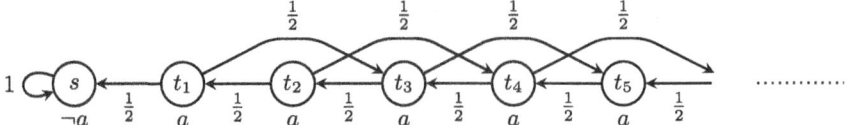

Fig. 1. An infinite-state model of the PCTL formula $\mathbf{G}_{>0}(\neg a \wedge \mathbf{F}_{>0}\, a)$.

However, when the above transition system is changed into a Markov chain by assigning probabilities to the transitions, then the state s does *not* satisfy the PCTL formula (1) (regardless of how the transition probabilities are chosen). In fact, the formula (1) does not have *any* finite-state model, although it *is* satisfiable. An infinite-state model of the formula is shown in Fig. 1. Observe that

- $t_i \models \neg a \wedge \mathbf{F}_{>0}\, a$ for every $i \geq 1$,
- the probability of all runs w initiated in t_i such that w never visits the state s is positive for every $i \geq 1$.

Hence, $t_i \models \mathbf{G}_{>0}(\neg a \wedge \mathbf{F}_{>0}\, a)$ for every $i \geq 1$.

The previous example shows that PCTL does not inherit the *small model property* of CTL [16] and other non-probabilistic temporal logics such as the modal μ-calculus [23] guaranteeing the existence of a bounded-size model for every satisfiable formula. Still, the existence of a bounded-size (perhaps even exponential-size) model for every *finite-satisfiable* PCTL formula is a plausible conjecture. At first glance, there is no apparent way how a finite-satisfiable PCTL formula φ can enforce the existence of at least $F(|\varphi|)$ states in its model, where the function F grows faster than an arbitrary computable function.

The *PCTL satisfiability problem* has been studied in two basic variants:

- *finite PCTL satisfiability*, asking about the existence of a finite-state model for a given PCTL formula;
- *general PCTL satisfiability*, asking about the existence of an unrestricted (possibly infinite-state) model.

The satisfiability problem can also be restricted to other meaningful subclasses of eligible models, as we shall see in Sect. 2.3.

Intuitively, the finite PCTL satisfiability problem appears simpler. For a given PCTL formula φ, the existence of a model with at most n states (where $n > 1$ is a constant) is decidable is space polynomial in $|\varphi|$ and n. This result is obtained by a straightforward encoding of the question into the existential fragment of first-order theory of the reals (see, e.g., [14]). Hence, the finite PCTL satisfiability problem is at least semi-decidable. The only missing ingredient for obtaining the decidability result is a computable upper bound on the size of a model for a finitely satisfiable PCTL formula. Many of the existing decidability results about the (finite or general) satisfiability for various PCTL fragments were obtained by establishing such a bound.

1.1 Existing Results

The PCTL satisfiability problem has been initially studied for probabilistic logics corresponding to the *qualitative fragment* of PCTL, where the admissible probability constraints is restricted to $\{=0, >0, =1, <1\}$ [21,24,26]. Both general and finite satisfiability for qualitative PCTL are shown decidable in these works. A precise complexity classification of general and finite satisfiability for qualitative PCTL, together with the construction of (a finite description of) a model, are given in [7]. In this paper, it is also shown that both general and finite satisfiability are *undecidable* when the class of models is restricted to Markov chains where every state has precisely k immediate successors, where $k \geq 2$ is an arbitrary constant. A variant of the bounded satisfiability problem, where transition probabilities are restricted to $\{\frac{1}{2}, 1\}$, is proven **NP**-complete in [3].

The decidability of finite satisfiability for various PCTL fragments with general probability constrains is established in [11,14,25]. More concretely, in [11], it is shown that every formula φ of the *bounded fragment* of PCTL, where the validity of φ in a state s depends only on a bounded prefix of a run initiated in s, has a bounded-size tree model. In [25], several PCTL fragments based on **F** and **G** operators are studied. For each of these fragments, it is shown that every finite satisfiable formula has a bounded-size model where every non-bottom SCC is a singleton. In [14], a more abstract decidability result based on isolating the progress achieved along a chain of visited SCCs is presented.

For unrestricted PCTL, the finite and general satisfiability problems have recently been shown *undecidable* [12,13]. The general PCTL satisfiability is even *highly undecidable*, i.e., beyond the arithmetical hierarchy. Consequently, the general and finite *validity* problems for PCTL are not even semi-decidable. Hence, there are no sound and complete deductive systems proving all valid and finitely valid PCTL formulae.

The *model-checking* problem for PCTL has been studied both for finite Markov chains (see, e.g., [1,2,4,22]) and for infinite Markov chains generated by probabilistic pushdown automata and their subclasses [10,17,18]. PCTL formulae have also been used as *objectives* in Markov decision processes (MDPs) and stochastic games, where the players controlling non-deterministic states strive to satisfy/falsify a given PCTL formula. Positive decidability results exist for finite MDPs and qualitative PCTL formulae [9]. For general PCTL and finite MDPs, the problem becomes undecidable [6].

2 Preliminaries

We use $\mathbb{N}$, $\mathbb{Q}$, and $\mathbb{R}$ to denote the sets of non-negative integers, rational numbers, and real numbers, respectively.

2.1 The Logic PCTL

The logic PCTL [20] is obtained from the standard Computational Tree Logic (CTL, see [15]) by replacing the existential and universal path quantifiers with the probabilistic operator $P(\Phi) \bowtie r$, where Φ is a path formula, $\bowtie$ is comparison such as ">" or "$\leq$", and $r \in [0,1]$ is a rational constant.

Definition 1 (PCTL). *Let AP be a set of atomic propositions. The syntax of PCTL state and path formulae is defined by the following abstract syntax equations:*

$$\varphi ::= a \mid \neg\varphi \mid \varphi_1 \wedge \varphi_2 \mid P(\Phi) \bowtie r$$
$$\Phi ::= \mathbf{X}\,\varphi \mid \varphi_1 \mathbf{U} \varphi_2 \mid \varphi_1 \mathbf{U}^k \varphi_2$$

Here, $a \in AP$, $\bowtie \in \{\geq, >, \leq, <, =\}$, $r \in [0,1]$ is a rational constant, and $k \in \mathbb{N}$.

The formulae *true*, *false*, and the other Boolean connectives are defined using $\neg$ and $\wedge$ in the standard way. We also use $\mathbf{F}\,\varphi$ and $\mathbf{F}^k\,\varphi$ to abbreviate the formulae *true* $\mathbf{U}\,\varphi$ and *true* $\mathbf{U}^k\,\varphi$, respectively. Furthermore, we often abbreviate a formula of the form $P(\Phi) \bowtie r$ by omitting P and adjoining the probability constraint directly to the topmost path operator of Φ. For example, we write $\mathbf{X}_{=1}\,\varphi$ instead of $P(\mathbf{X}\,\varphi) = 1$. We also write $\mathbf{G}_{=1}\,\varphi$ instead of $\mathbf{F}_{=0}\,\neg\varphi$, and $\mathbf{G}_{\geq r}\,\varphi$ instead of $\mathbf{F}_{<1-r}\,\neg\varphi$ where $0 < r < 1$.

The *qualitative* fragment of PCTL is obtained by restricting the set of eligible constraints r in $P(\Phi) \bowtie r$ to $\{0,1\}$. Hence, in qualitative PCTL, the probabilistic operator can only say that the probability of all runs satisfying Φ is zero, positive, less than one, or equal to one.

2.2 Markov Chains

PCTL formulae are interpreted over Markov chains where every state s is assigned a subset $v(s) \subseteq AP$ of propositions valid in s.

Definition 2 (Markov chain). *A Markov chain is a triple $M = (S, P, v)$, where S is a finite or countably infinite set of states, $P \colon S \times S \to [0,1]$ is a function such that $\sum_{t \in S} P(s,t) = 1$ for every $s \in S$, and $v \colon S \to 2^{AP}$ is a valuation.*

For $s, t \in S$, we say that t is an *immediate successor* of s if $P(s,t) > 0$. A *path* in M is a finite sequence $w = s_0, \ldots, s_n$ of states where $n \geq 0$ and $P(s_i, s_{i+1}) > 0$ for all $i < n$. We say that t is *reachable* from s if there is a path where the first and the last state is s and t, respectively.

A *run* in M is an infinite sequence $\pi = s_0, s_1, \ldots$ of states such that every finite prefix of π is a path in M. We also use $\pi(i)$ to denote the state s_i of π.

For every path $w = s_0, \ldots, s_n$, let $Run(w)$ be the set of all runs starting with w, and let $\mathbb{P}(Run(w)) = \prod_{i=0}^{n-1} P(s_i, s_{i+1})$. To every state s, we associate the probability space $(Run(s), \mathcal{F}_s, \mathbb{P}_s)$, where $\mathcal{F}_s$ is the σ-field generated by all $Run(w)$ where w starts in s, and $\mathbb{P}_s$ is the unique probability measure obtained by extending $\mathbb{P}$ in the standard way (see, e.g., [5]).

The *validity* of a PCTL state/path formula for a given state/run of M is defined inductively as follows:

$$
\begin{aligned}
s &\models a & &\text{iff} & a &\in v(s), \\
s &\models \neg\varphi & &\text{iff} & s &\not\models \varphi, \\
s &\models \varphi_1 \wedge \varphi_2 & &\text{iff} & s &\models \varphi_1 \text{ and } s \models \varphi_2, \\
s &\models P(\Phi) \bowtie r & &\text{iff} & \mathbb{P}_s&(\{\pi \in Run(s) \mid \pi \models \Phi\}) \bowtie r, \\
\pi &\models \mathbf{X}\,\varphi & &\text{iff} & \pi(1) &\models \varphi, \\
\pi &\models \varphi_1\,\mathbf{U}\,\varphi_2 & &\text{iff} & &\text{there is } j \geq 0 \text{ such that } \pi(j) \models \varphi_2 \\
& & & & &\text{and } \pi(i) \models \varphi_1 \text{ for all } 0 \leq i < j, \\
\pi &\models \varphi_1\,\mathbf{U}^k\,\varphi_2 & &\text{iff} & &\text{there is } 0 \leq j \leq k \text{ such that } \pi(j) \models \varphi_2 \\
& & & & &\text{and } \pi(i) \models \varphi_1 \text{ for all } 0 \leq i < j.
\end{aligned}
$$

We say that M is a *model* of φ if $s \models \varphi$ for some state s of M.

2.3 The PCTL Satisfiability Problem and Its Variants

The *general/finite PCTL satisfiability problem* is the question of whether a given PCTL formula has a general/finite model.

More generally, the PCTL satisfiability problem can be parameterized by a class $\mathcal{C}$ of eligible models. Apart of the classes of all Markov chains and all finite-state Markov chains, other natural classes of models include

- the class of all *finite tree-like* Markov chains, where the underlying graph is a finite tree with self-loops on all leaves;
- the class D_k of all Markov chains where every state has precisely k immediate successors, where $k \geq 2$ is a constant.

3 Positive Decidability Results

3.1 Qualitative PCTL

As we already mentioned in Sect. 1, the first positive decidability results about PCTL satisfiability have been obtained for the qualitative fragment of PCTL. Here, the underlying proof techniques are similar to those used for non-probabilistic CTL [16].

More concretely, for every satisfiable qualitative PCTL formula φ, one can construct a *pseudo-model* by filtration through a variant of Fischer-Ladner closure [19] of φ. The closure contains all subformulae of φ together with finitely many additional formulae, and the states of a pseudo-model correspond to "consistent" subsets of the closure. This gives an exponential bound on the number of states of a pseudo-model. Then, it is shown how to extract a pseudo-model from a "real" model of φ. In the non-probabilistic case, the main problem with this technique is the introduction of new cycles that may "spoil" subformulae of the form $\mathbf{AF}\,\psi$. In the probabilistic case, these new cycles do not impose a major difficulty, because they are either harmless or they are eventually left with probability one. On the other hand, it may happen that a state of the obtained pseudo-model contains a formula $\neg(\psi_1 \mathbf{U}_{=1} \psi_2)$, but it does not satisfy this formula (since we are dealing with qualitative PCTL, the exact values of edge probabilities in the pseudo-model do not matter). If this happens, then the real model allows the extraction of a special "certificate" in the pseudo-model, witnessing the invalidity of $\psi_1 \mathbf{U}_{=1} \psi_2$. Intuitively, this certificate is a subset of states of the pseudo-model B that is strongly connected and all states of B satisfy $\neg\psi_2$. Hence, all runs staying in B do *not* satisfy $\psi_1 \mathbf{U}_{=1} \psi_2$. Although it may happen that almost all runs of the pseudo-model may leave B, one can understand the pseudo-model as a *symbolic* representation of an infinite Markov chain obtained by unfolding the pseudo-model into an infinite tree where the transition probabilities are chosen so that the probability of staying in B becomes positive.

The above considerations ultimately lead to the conclusion that every satisfiable qualitative PCTL formula φ has a model representable by a finite directed graph $\mathcal{G}$ such that

- the number of vertices of $\mathcal{G}$ is at most exponential in $|\varphi|$;
- every vertex of $\mathcal{G}$ is assigned a subset of atomic propositions occurring in φ;
- every vertex of $\mathcal{G}$ has at least one outgoing edge;
- every edge is declared as *ordinary* or *special*.

If φ is finite satisfiable, then the graph $\mathcal{G}$ can be constructed so that it contains no special edges and the Markov chain obtained from $\mathcal{G}$ by assigning a positive probability to every edge is a model of φ. If φ is satisfiable but nor finitely satisfiable, then $\mathcal{G}$ represents an infinite model $M_\mathcal{G}$ of φ obtained by unfolding $\mathcal{G}$ into an infinite tree, where the probabilities of special edges "increase at an exponential rate". More precisely, $M_\mathcal{G}$ is constructed as follows:

- The states of $M_\mathcal{G}$ are finite paths in $\mathcal{G}$.
- For every finite path wv in $\mathcal{G}$ ending in a vertex v, we have that $wv \to wvv'$ is a transition of $M_\mathcal{G}$ iff (v, v') is an edge of $\mathcal{G}$. This transition is *ordinary* or *special* depending on the type of (v, v'). The state wv has no other outgoing transitions.
- The probability assignment is chosen so that for every state w of $M_\mathcal{G}$, there exist $p, p' \in [0, 1]$ so that the following conditions hold:
  - the probability of every ordinary outgoing transition of w is equal to p;
  - the probability of every special outgoing transition of w is equal to p';
  - if w has at least one special outgoing transition, then the total probability of all special outgoing transitions of w is equal to $1 - 1/4^{|w|}$, where $|w|$ is the length of w.

As an example, consider the PCTL formula

$$\mathbf{G}_{>0}(\neg a \wedge \mathbf{F}_{>0}\, a)$$

introduced in Sect. 1. The formula is satisfiable but not finitely satisfiable, and the following graph $\mathcal{G}$ represents an infinite model of the formula (special edges are in dashed):

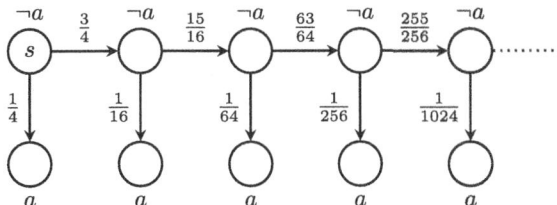

Fig. 2. An infinite-state model $M_{\mathcal{G}}$ of the formula $\mathbf{G}_{>0}(\neg a \wedge \mathbf{F}_{>0}\, a)$.

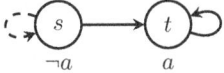

An initial part of $M_{\mathcal{G}}$ is shown in Fig. 2.

Recall that the formula $\mathbf{G}_{>0}(\neg a \wedge \mathbf{F}_{>0}\, a)$ has another infinite-state model where the transition probabilities are bounded away from zero (see Fig. 1). However, using arbitrarily small probabilities is unavoidable in general. A simple example is the qualitative PCTL formula

$$\mathbf{G}_{=1}(\mathbf{X}_{>0}\, a) \;\wedge\; \mathbf{G}_{>0}\, \neg a \tag{3}$$

Observe that the Markov chain of Fig. 2 is a model of (3), but the formula does not have any model where the transition probabilities are bounded away from zero by some constant.

3.2 Non-qualitative PCTL Fragments

The existing works on the satisfiability of non-qualitative PCTL [11,14,25] identify fragments where all satisfiable formulae have a model of specific shape and bounded size. In [11], it is shown that every formula φ of the PCTL fragment without the **U** operator has a bounded-size tree-like model. In [25], several PCTL fragments based on **F** and **G** operators are studied. For each fragment, it is shown that every finitely satisfiable formula has a bounded-size model where every non-bottom SCC is a singleton. It is also shown that there are finitely satisfiable PCTL formulae without a model of this shape. An example is the following formula:

$$\mathbf{G}_{=1}\left(\mathbf{F}_{\geq 0.5}(a \wedge \mathbf{F}_{\geq 0.2}\, \neg a) \vee a\right) \;\wedge\; \mathbf{F}_{=1}\, \mathbf{G}_{=1}\, a \;\wedge\; \neg a \tag{4}$$

The formula (4) has the following simple model with three states:

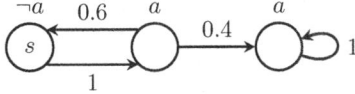

In [25], it is shown that *every* finite model of (4) has a non-bottom SCC with at least two states.

Intuitively, if a given PCTL formula φ enforces a model with a non-bottom SCC, then the top SCC must achieve some "progress" in satisfying φ, and successor SCCs are required to satisfy only "simpler" formulae. This idea is elaborated in [14] into a workable set of conditions defining *effectively progressive PCTL fragments* with computable upper bounds on the model's size. The resulting method applies to the fragments considered in [11,25] and also to other fragments not covered by these results (some of them include the formula (4)).

4 Negative Decidability Results

The first negative decidability result about PCTL satisfiability has been proven in [7]. It says that the finite and general PCTL satisfiability problem is undecidable and highly undecidable, respectively, if the class of eligible models is restricted to D_k for some $k \geq 2$ (see Sect. 2.3). The proof is based on reducing the strategy synthesis problem for Markov decision processes (MPDs) with PCTL objectives, which is undecidable for finite-memory strategies and highly undecidable for general strategies [6]. Roughly speaking, the reduction is based on modifying a given MDP so that every stochastic vertex has precisely k immediate successors, every non-deterministic vertex has precisely $k+1$ immediate successors, and every strategy satisfying a given PCTL objective must assign a positive probability to precisely k successors of every non-deterministic vertex. Then, the structure of the MDP and the PCTL objective are encoded into a single PCTL formula φ. Every model of φ from the class D_k corresponds to a Markov chain obtained from the original MDP by applying some strategy so that the original PCTL objective is satisfied. We refer to [8] for details.

Unfortunately, the above technique does not work for Markov chains with unrestricted branching degrees. The recent undecidability and high undecidability results for finite and general PCTL satisfiability [12,13] use different methods for encoding the counter values of Minsky machines and implementing the increment/decrement operations. Note that the undecidability of finite PCTL satisfiability also falsifies the (plausible) conjecture about the existence of an effectively computable upper bound on the size a finite model for a given finite satisfiable PCTL formula. In [12] a *fixed* PCTL formula ψ is constructed such that ψ can enforce an arbitrarily large finite model just by changing the numerical constants in two occurrences of the probabilistic operator in ψ. The details can be found in [12,13].

5 Conclusions

Although the decidability status of the general and finite PCTL satisfiability problems has been resolved, a detailed classification of the decidable/undecidable PCTL fragments is still missing. Another direction for future research is to determine the principal boundaries of effective probabilistic program synthesis from specifications written in probabilistic temporal logics.

References

1. Baier, C., Katoen, J.P.: Principles of Model Checking. The MIT Press, Cambridge (2008)
2. Baier, C., Kwiatkowska, M.: Model checking for a probabilistic branching time logic with fairness. Distrib. Comput. **11**(3), 125–155 (1998)
3. Bertrand, N., Fearnley, J., Schewe, S.: Bounded satisfiability for PCTL. In: Proceedings of CSL 2012. Leibniz International Proceedings in Informatics, vol. 16, pp. 92–106. Schloss Dagstuhl–Leibniz-Zentrum für Informatik (2012)
4. Bianco, A., de Alfaro, L.: Model checking of probabilistic and nondeterministic systems. In: Thiagarajan, P.S. (ed.) FSTTCS 1995. LNCS, vol. 1026, pp. 499–513. Springer, Heidelberg (1995). https://doi.org/10.1007/3-540-60692-0_70
5. Billingsley, P.: Probability and Measure. Wiley, Hoboken (1995)
6. Brázdil, T., Brožek, V., Forejt, V., Kučera, A.: Stochastic games with branching-time winning objectives. In: Proceedings of LICS 2006, pp. 349–358. IEEE Computer Society Press (2006)
7. Brázdil, T., Forejt, V., Křetínský, J., Kučera, A.: The satisfiability problem for probabilistic CTL. In: Proceedings of LICS 2008, pp. 391–402. IEEE Computer Society Press (2008)
8. Brázdil, T., Forejt, V., Křetínský, J., Kučera, A.: The satisfiability problem for probabilistic CTL. Technical report FIMU-RS-2008-03, Faculty of Informatics, Masaryk University (2008)
9. Brázdil, T., Forejt, V., Kučera, A.: Controller synthesis and verification for Markov decision processes with qualitative branching time objectives. In: Aceto, L., Damgård, I., Goldberg, L.A., Halldórsson, M.M., Ingólfsdóttir, A., Walukiewicz, I. (eds.) ICALP 2008. LNCS, vol. 5126, pp. 148–159. Springer, Heidelberg (2008). https://doi.org/10.1007/978-3-540-70583-3_13
10. Brázdil, T., Kučera, A., Stražovský, O.: On the decidability of temporal properties of probabilistic pushdown automata. In: Diekert, V., Durand, B. (eds.) STACS 2005. LNCS, vol. 3404, pp. 145–157. Springer, Heidelberg (2005). https://doi.org/10.1007/978-3-540-31856-9_12
11. Chakraborty, S., Katoen, J.: On the satisfiability of some simple probabilistic logics. In: Proceedings of LICS 2016, pp. 56–65 (2016)
12. Chodil, M., Kučera, A.: The finite satisfiability problem for PCTL is undecidable. In: Proceedings of LICS 2024, pp. 1–14. ACM Press (2024). Article No. 22
13. Chodil, M., Kučera, A.: The general and finite satisfiability problems for PCTL are undecidable. arXiv:2404.10648 (2024)
14. Chodil, M., Kučera, A.: The satisfiability problem for a quantitative fragment of PCTL. J. Comput. Syst. Sci. **139**, 103478 (2024)
15. Emerson, E.: Temporal and modal logic. In: Handbook of Theoretical Computer Science, pp. 995–1072 (1991)
16. Emerson, E., Halpern, J.: Decision procedures and expressiveness in the temporal logic of branching time. In: Proceedings of STOC 1982, pp. 169–180. ACM Press (1982)
17. Esparza, J., Kučera, A., Mayr, R.: Model-checking probabilistic pushdown automata. Log. Methods Comput. Sci. **2**(1:2), 1–31 (2006)
18. Etessami, K., Yannakakis, M.: Model checking of recursive probabilistic systems. ACM Trans. Comput. Log. **13**, 1–40 (2012)
19. Fischer, M., Ladner, R.: Propositional dynamic logic of regular programs. J. Comput. Syst. Sci. **18**, 194–211 (1979)

20. Hansson, H., Jonsson, B.: A logic for reasoning about time and reliability. Formal Aspects Comput. **6**, 512–535 (1994)
21. Hart, S., Sharir, M.: Probabilistic temporal logic for finite and bounded models. In: Proceedings of POPL 1984, pp. 1–13. ACM Press (1984)
22. Huth, M., Kwiatkowska, M.: Quantitative analysis and model checking. In: Proceedings of LICS 1997, pp. 111–122. IEEE Computer Society Press (1997)
23. Kozen, D.: A finite-model theorem for the propositional μ-calculus. Stud. Logica. **47**(3), 233–241 (1988)
24. Kraus, S., Lehmann, D.: Decision procedures for time and chance (extended abstract). In: Proceedings of FOCS 1983, pp. 202–209. IEEE Computer Society Press (1983)
25. Křetínský, J., Rotar, A.: The satisfiability problem for unbounded fragments of probabilistic CTL. In: Proceedings of CONCUR 2018. Leibniz International Proceedings in Informatics, vol. 118, pp. 32:1–32:16. Schloss Dagstuhl–Leibniz-Zentrum für Informatik (2018)
26. Lehmann, D., Shelah, S.: Reasoning with time and chance. Inf. Control **53**, 165–198 (1982)

Computability and Reachability

Computing Reachable Simulations on Transition Systems

Pierre Ganty[1], Nicolas Manini[1,2](✉), and Francesco Ranzato[3]

[1] IMDEA Software Institute, Pozuelo de Alarcón, Spain
{pierre.ganty,nicolas.manini}@imdea.org
[2] Universidad Politécnica de Madrid, Madrid, Spain
[3] University of Padova, Padova, Italy
francesco.ranzato@unipd.it

Abstract. We study the problem of computing the reachable principals of the simulation preorder and the reachable blocks of simulation equivalence. Following a theoretical investigation of this problem, which highlights a sharp contrast with the already settled case of bisimulation, we design algorithms to solve this problem by leveraging the idea of interleaving reachability and simulation computation while possibly avoiding the computation of all the reachable states or the whole simulation preorder. In particular, we put forward a symbolic algorithm processing state partitions and, in turn, relations between their blocks, which is suited for processing infinite-state systems.

Keywords: Simulation · Reachability · Reachable Simulation Problem · Simulation algorithm · Symbolic Data Structure

1 Introduction

Given a possibly infinite labeled transition system S, we study the problem of computing the reachable principals[1] of the greatest simulation preorder R_{sim}, and the reachable blocks of the induced simulation partition P_{sim}. By reachable, we mean the principals $R_{\text{sim}}(x)$ of the simulation preorder (resp., blocks $P_{\text{sim}}(x)$ of the simulation partition) that intersect the system's reachable states, therefore ignoring unreachable principals and blocks, which are typically of no/negligible interest. A naïve solution to this problem—that we call the reachable simulation problem—would be: first, compute the simulation preorder (partition), and then, filter out the principals (blocks) containing no reachable states. However, this requires the computation of the entire simulation preorder and, possibly, of all the reachable states. Here, we present a completely different solution relying on a convoluted interleaving of reachability and simulation computation, possibly avoiding the computation of all the reachable states and of the whole R_{sim}.

---

[1] The term principal comes from the well-known notion of *principal ideal* [13, Chapter I, sect. 3.4]. Detailed definitions are given in Sect. 3.

Contributions. We study the reachable simulation problem by showing, through an unsolvability proof (cf. Sect. 4), that there is a stark contrast w.r.t. the problem of computing the reachable blocks of the bisimulation partition, settled by Lee and Yannakakis [23] in STOC 1992. In Sect. 5, we put forward an algorithm computing the reachable part of the simulation preorder, and yielding an over-approximation for the partition. We prove correctness and termination on finite state systems, and extend correctness (under simple assumptions) to infinite state systems. Moreover, we provide examples showing termination on some infinite state systems[2]. Section 6 introduces the so-called 2PR (2 state Partitions and a Relation among them) triples which we use to design a symbolic algorithm for the reachable simulation problem. Besides inheriting the correctness guarantees of our first algorithm, we show that the 2PR-based algorithm terminates faster and more often for infinite state systems. In particular, we prove that the 2PR-based algorithm terminates on all systems having a finite bisimulation partition or when a local finiteness condition is satisfied. These termination results come with a runtime upper bound that is quadratic in the number of blocks of (a portion of) the bisimulation partition. As an auxiliary contribution, we define partitions induced by arbitrary relations (not limited to preorders or equivalences), generalizing previous definitions.

Applications. Computing reachable simulation principals and blocks has several practical applications. A noteworthy use case of the reachable principals of the simulation preorder is given by the determinization algorithms SUBSET(f) and TRANSSET(f) for nondeterministic finite automata designed by van Glabbeek and Ploeger [32]. Using the simulation preorder computationally enhances these procedures (f is picked to account for the simulation preorder). In particular, only the reachable principals are used since the automaton determinization proceeds forward starting from the initial states. It turns out that these simulation-based algorithms compute smaller deterministic automata compared to their plain versions [32]. In a different context, the reachable blocks of the simulation partition define the states of the reduced quotiented system. The question of computing the transitions between the blocks of the reduced system has been investigated in depth by Bustan and Grumberg [6], who explore the difference and trade-off of the $\exists\exists$ (i.e., $B \to^{\exists\exists} B'$ iff $\exists s \in B. \exists s' \in B'. s \to s'$) and $\forall\exists$ definitions (i.e., $B \to^{\forall\exists} B'$ iff $\forall s \in B. \exists s' \in B'. s \to s'$).

Furthermore, solutions to the reachable *simulation* problem have potential applications in program and hybrid systems verification, as past research [16,25, 26,33] has leveraged solutions to the reachable *bisimulation* problem.

Related Work. The closest work to ours is that by Lee and Yannakakis [23], who first designed an interleaving of reachability and bisimulation computation, here referred to as the LY algorithm. Their work is highly cited and has been applied and revisited several times (e.g. [1,10]), nevertheless, it remains an

---

[2] As shown in Sect. 4, an algorithm terminating on all infinite state systems cannot exist.

elaborate algorithm with hidden subtleties. Let us point out that we were not able to find, for some results claimed in the original paper [23], any proof argument (either searching online or contacting authors). To put the LY algorithm in its historical context, it was one of several algorithms to compute the reachable part of the bisimulation-based quotiented system [4,5,23]. These algorithms share the interleaving of the bisimulation computation—a partition refinement algorithm—with the computation determining block reachability. Interleaving reachability and bisimulation computation is remarkably interesting because the resulting algorithms terminate at least as often (possibly more often) as the naïve procedure consisting in first computing the bisimulation and next determining its reachable blocks. Later on, Alur and Henzinger revisited these algorithms for reachable bisimulation in their unpublished book on computer-aided verification [1, Chapter 4], while Fisler and Vardi conducted a theoretical and experimental evaluation of the LY algorithm [10]. We also mention that algorithms combining reachability and bisimulation computation inspired by the LY algorithm have been used in several different contexts ranging from program analysis [16,26] to hybrid systems verification, where [25] employs a LY-like approach for language preserving minimization for controller design.

We focus on simulation since it provides a better state space reduction than bisimulation, while retaining enough precision for checking all linear temporal logic formulas or branching temporal logic formulas without quantifier switches [2,6,8,14,15,24]. Moreover, infinite state systems like 2D rectangular automata may have infinite bisimilarity quotients, yet they always have finite similarity quotients (see [17,18]). There is a large body of work [3,7,9,11,12,17, 27–31] on efficiently computing the simulation preorder, through both explicit or symbolic algorithms. Kucera and Mayr [21,22] compared simulation and bisimulation equivalence using their computational complexity, and justified the claim that similarity is computationally harder than bisimilarity.

To the best of our knowledge, no previous work considered the problem of computing the reachable principals of the simulation preorder or the reachable blocks of the simulation partition. Due to lack of space, some material (e.g., proofs) is omitted.

2 Motivating Example

We introduce an example showing the challenges of the reachable simulation problem and some intuitive explanations on our solution. Consider the family of infinite transition systems parameterized over an integer $k \geq 0$ depicted below.

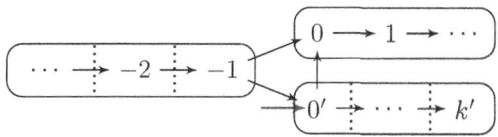

The set of states is $\mathbb{Z} \cup \{0', \ldots, k'\}$, and the transition relation is given by the arrows in the diagram. The initial state is $0'$, denoted by the incoming blue arrow. Consider the initial partition of states given by the three blocks $\{n \in \mathbb{Z} \mid n < 0\}$, $\{n \in \mathbb{Z} \mid n \geq 0\}$ and $\{n' \mid 0 \leq n \leq k\}$, depicted as solid boxes. Let us remark three observations about this family of transition systems: (1) there are infinitely many reachable states; (2) there are infinitely many simulation equivalence classes, depicted by dotted lines splitting the boxes; and (3) yet, there are finitely many simulation equivalence classes that are reachable, i.e., finitely many dotted blocks in the diagram include reachable states. In this paper, we tackle the challenge of effectively computing information such as in (3). Because of points (1) and (2), we must rule out naïve solutions that would include a computation of all the reachable states (a simple reachability computation would not terminate) or refining the blocks of the initial partition to the simulation partition (a simulation algorithm would not terminate). Yet, in this work, we define algorithms alternating bounded state space exploration and partition refinement. Such algorithms can indeed effectively compute information such as (3), while avoiding the pitfalls of computing all the reachable states or computing the full simulation partition. Section 5.1 shows a run of our algorithm on this example.

3 Background

Preorders and Partitions. Given a (possibly infinite) set Σ, we denote with $\wp(\Sigma)$ the powerset of Σ, and with $\mathrm{Rel}(\Sigma) \triangleq \wp(\Sigma \times \Sigma)$ the set of relations over Σ. If $R \in \mathrm{Rel}(\Sigma)$ then: for $S \subseteq \Sigma$, $R(S) \triangleq \{s' \in \Sigma \mid \exists s \in S.\, (s, s') \in R\}$; for $s \in \Sigma$, the set $R(s) \triangleq R(\{s\})$ is the *principal* of s; $\mathrm{Rel}(\Sigma) \ni R^{-1} \triangleq \{(y,x) \in \Sigma \times \Sigma \mid (x,y) \in R\}$ is the *converse* relation of R. Moreover, for a given set $S \subseteq \Sigma$, we denote by $R^S \triangleq \{R(x) \in \wp(\Sigma) \mid x \in \Sigma,\, R(x) \cap S \neq \varnothing\}$ the set of principals of R that intersect S. A relation $R \in \mathrm{Rel}(\Sigma)$ is a *preorder* if it is reflexive and transitive, and $\mathrm{PreO}(\Sigma) \triangleq \{R \in \mathrm{Rel}(\Sigma) \mid R \text{ is a preorder}\}$ denotes the set of preorders on Σ. Moreover, $R \in \mathrm{Rel}(\Sigma)$ is an *equivalence* on Σ if it is a symmetric preorder. A *partition* of Σ consists of pairwise disjoint nonempty subsets of Σ, called *blocks*, whose union is Σ, and $\mathrm{Part}(\Sigma)$ denotes the set of partitions of Σ. We consider finite partitions (i.e., consisting of finitely many blocks), unless otherwise specified. It is well known that a partition defines an equivalence relation, and vice versa, where blocks of the partition and equivalence classes coincide. Hence, given a partition $P \in \mathrm{Part}(\Sigma)$, $P(s)$, $P(S)$ and P^S (for $S \subseteq \Sigma, s \in \Sigma$) are well-defined thanks to the equivalence defined by P. In particular, $P(s)$ is the block including s, $P(S) = \cup\{P(s) \in P \mid s \in S\}$, and $P^S = \{P(s) \in P \mid s \in S\} \in \mathrm{Part}(P(S))$. Given two partitions $P, Q \in \mathrm{Part}(\Sigma)$, P is *coarser* than Q, denoted by $Q \preceq P$, if the equivalence relation underlying Q is a subset of the underlying equivalence of P. More in general, any relation $R \in \mathrm{Rel}(\Sigma)$ (not necessarily an equivalence or a preorder) *induces* a partition of Σ defined as $\{y \in \Sigma \mid R(y) = R(x)\}_{x \in \Sigma}$. Two elements belong to the same block of the induced partition if their image by R coincide. This general definition of partition induced by a relation has the following desirable properties. When R is an equivalence relation then the blocks of its induced partition coincide with the equivalence classes of R. Moreover, if R is a preorder then the blocks of its induced partition coincide with the equivalence classes of the equivalence relation given by $R \cap R^{-1}$. These induced partitions will be a key ingredient of our approach.

Simulation and Bisimulation. Let $G = (\Sigma, I, L, \rightarrow)$ be a (labeled) transition system (TS), where Σ is a (possibly infinite yet countable) set of states, $I \subseteq \Sigma$ are the *initial states*, L is a finite set of action labels, and $\rightarrow \, \subseteq \Sigma \times L \times \Sigma$ is the *labeled transition relation*, where we denote $(x, a, y) \in \, \rightarrow$ as $x \xrightarrow{a} y$. When L is a singleton set or when the label is unimportant we simply write $x \rightarrow y$. This comes handy in our examples where we assume L is a singleton. Given $a \in L$, $\mathsf{post}_a \colon \wp(\Sigma) \rightarrow \wp(\Sigma)$ denotes the usual successor transformer $\mathsf{post}_a(X) \triangleq \{y \in \Sigma \mid \exists x \in X. \, x \xrightarrow{a} y\}$, and, dually, $\mathsf{pre}_a \colon \wp(\Sigma) \rightarrow \wp(\Sigma)$ is the predecessor $\mathsf{pre}_a(Y) \triangleq \{x \in \Sigma \mid \exists y \in Y. \, x \xrightarrow{a} y\}$. Moreover, we define post: $\wp(\Sigma) \rightarrow \wp(\Sigma)$ as $\mathsf{post}(X) \triangleq \cup_{a \in L} \mathsf{post}_a(X)$ and, symmetrically, pre: $\wp(\Sigma) \rightarrow \wp(\Sigma)$ as $\mathsf{pre}(X) \triangleq \cup_{a \in L} \mathsf{pre}_a(X)$. Thus, $\mathsf{post}^*(I) \triangleq \cup_{n \in \mathbb{N}} \mathsf{post}^n(I)$ is the set of *reachable states*.

Given an (initial) preorder $R_i \in \mathrm{PreO}(\Sigma)$, a relation $R \in \mathrm{Rel}(\Sigma)$ is a *simulation* on G w.r.t. R_i if: (1) $R \subseteq R_i$; (2) $(s,t) \in R$ and $s \xrightarrow{a} s'$ imply $\exists t'. \, t \xrightarrow{a} t'$ and $(s', t') \in R$. Given two principals $R(s)$, $R(s')$ such that $s \xrightarrow{a} s'$, $R(s)$ is *a-stable* (or simply *stable*) w.r.t. $R(s')$ when $R(s) \subseteq \mathsf{pre}_a(R(s'))$, otherwise $R(s)$ is called *a-unstable* (or simply *unstable*) w.r.t. $R(s')$, and, in this case, $R(s')$ can *refine* $R(s)$ to $R(s) \cap \mathsf{pre}_a(R(s'))$. As a consequence, point (2) in the above simulation definition is equivalent to: (2′) for every transition $s \xrightarrow{a} s'$ in G, $R(s)$ is a-stable w.r.t. $R(s')$. The greatest (w.r.t. $\subseteq$) simulation relation on G exists and turns out to be a preorder called the *simulation preorder* of G w.r.t. R_i, denoted by $R_{\mathrm{sim}} \in \mathrm{PreO}(\Sigma)$. We denote by $P_{\mathrm{sim}} \in \mathrm{Part}(\Sigma)$ the partition induced by R_{sim} and call it the *simulation partition*[3] (or similarity). A relation $R \in \mathrm{Rel}(\Sigma)$ is a *bisimulation* on G w.r.t. an (initial) partition $P_i \in \mathrm{Part}(\Sigma)$ if both R and R^{-1} are simulations on G w.r.t. P_i. The greatest (w.r.t. $\subseteq$) bisimulation relation on G w.r.t. P_i exists, and turns out to be an equivalence called *bisimulation equivalence* (or bisimilarity), denoted by R_{bis}. The partition $P_{\mathrm{bis}} \in \mathrm{Part}(\Sigma)$ induced by R_{bis} is called the *bisimulation partition*.

Remark 3.1 (On the Initial Preorder). We point out that the role of the initial preorder R_i is that of having some a priori simulation information (e.g., accepting states in a finite state automaton simulate non-accepting ones but not the other way around [32]). For the bisimulation case, such a priori information is conveyed by an equivalence—e.g. specified by a labelling over the state space like in Kripke structures—and the natural generalization of the initial equivalence to the simulation case is an initial preorder. Nevertheless, as mentioned above, an equivalence relation can be used as R_i too, since it is a particular case of a preorder.

---

[3] Observe that since R_{sim} is a preorder, we have that $P_{\mathrm{sim}} \in \mathrm{Part}(\Sigma)$ coincides with the equivalence classes of the similarity equivalence $R_{\mathrm{sim}} \cap (R_{\mathrm{sim}})^{-1}$.

4 The Reachable Simulation Problem

We formally define the problem investigated in this work, which extends in a natural way the reachable bisimulation problem tackled by Lee and Yannakakis in [23].

*Problem 4.1 (**The Reachable Simulation Problem**).*
GIVEN: A labeled transition system $G = (\Sigma, I, L, \to)$ and $R_i \in \text{PreO}(\Sigma)$.
COMPUTE: The reachable principals of R_{sim} and the reachable blocks of P_{sim}.

A first challenge we face is related to the notion of reachability. The notion of reachable blocks (rb) for any partition $P \in \text{Part}(\Sigma)$, such as P_{sim}, is (trivially) defined as the set of blocks containing at least one reachable state, that is $P^{\text{post}^*(I)}$. Thus, the following equality holds:

$$P^{\text{post}^*(I)} = \{B \in P \mid B \cap \text{post}^*(I) \neq \varnothing\} = \{P(s) \in P \mid s \in \text{post}^*(I)\}. \quad \text{(rb)}$$

Moving from bisimulation to simulation, (rb) can be generalized to any of the two following definitions of *reachable principal* (rp) of a reflexive relation $R \in \text{Rel}(\Sigma)$, which can both be deemed adequate:

$$R^{\text{post}^*(I)} = \{R(s) \in \wp(\Sigma) \mid s \in \Sigma, \, R(s) \cap \text{post}^*(I) \neq \varnothing\}, \quad (\text{rp}_1)$$

$$R_{\text{alt}}^{\text{post}^*(I)} \triangleq \{R(s) \in \wp(\Sigma) \mid s \in \text{post}^*(I)\}. \quad (\text{rp}_2)$$

Clearly, when R is an equivalence relation, (rp_1) and (rp_2) coincide and boil down to (rb). In general, only $R_{\text{alt}}^{\text{post}^*(I)} \subseteq R^{\text{post}^*(I)}$ holds, and, moreover, the inclusion $(R_{\text{sim}})_{\text{alt}}^{\text{post}^*(I)} \subsetneq (R_{\text{sim}})^{\text{post}^*(I)}$ can hold strictly[4], as shown next.

Example 4.2. Consider the system depicted below, its initial preorder $R_i = \{0,1\} \times \{0,1\}$, and the block induced by R_i, represented as a box.

The simulation preorder w.r.t. R_i is $R_{\text{sim}}(0) = \{0,1\}$, $R_{\text{sim}}(1) = \{1\}$, and a dotted line delimits the blocks of P_{sim}. Thus, since $\text{post}^*(I) = \{1\}$, we get that the reachable principals as per (rp_2) are the singleton $\{R_{\text{sim}}(1)\}$, while the reachable principals as per (rp_1) are $\{R_{\text{sim}}(0), R_{\text{sim}}(1)\}$. ◇

Unsolvability. Problem 4.1 is, in general, unsolvable (i.e., no algorithm exists for computing the reachable principals and blocks of, resp., R_{sim} and P_{sim}), even under the assumption that P_{sim} is a finite partition. In particular, we observe that the subtask of Problem 4.1 involving the reachable blocks of P_{sim} is unsolvable. This result shows a difference with the problem of computing reachable blocks of P_{bis}, which Lee and Yannakakis [23] proved solvable for finite bisimulations.

---

[4] For some systems, (rp_1) could even be infinite, and (rp_2) be a finite set.

Theorem 4.3 (Unsolvability of the Reachable Simulation Problem).
Problem 4.1 is unsolvable, even under the assumption that P_{sim} is a finite partition.

Theorem 4.3 holds since solvability of Problem 4.1 would imply decidability of the well-known undecidable halting problem for 2-counter machines. In fact, with a few termination-preserving transforms, we can characterize the halting states for a 2-CM as a block of P_{sim}. It is worth noting that the halting problem for 2-counter machines is commonly used for proving related (yet different) undecidability results about simulation [20].

Remarks for Finite State Systems. We observe a further difference between Problem 4.1 and the corresponding problem for bisimulation, over *finite* transition systems. The reachability problem for blocks of both P_{bis} and P_{sim} is trivially decidable for finite systems, since we can simply compute independently $\text{post}^*(I)$ and P_{bis} or P_{sim}, and, check whether $\text{post}^*(I) \cap B = \varnothing$ holds for every block B in P_{bis} or P_{sim}. However, for the case of bisimulation, deciding reachability of a block $B \in P_{\text{bis}}$ can be done in $O(|P_{\text{bis}}^{\text{post}^*(I)}|)$ time by leveraging the definition of bisimulation: if x and y are bisimilar states, then x can reach B in one transition iff y can, hence picking any single state per block of P_{bis} suffices. For the simulation case, deciding the reachability of $B \in P_{\text{sim}}$ is more involved: as the following example hints, to decide whether a block $B \in P_{\text{sim}}$ is reachable we possibly have to check whether there is a path of arbitrary length (that is, in $O(|\Sigma|)$) in the transition system reaching B, so that no bound on the number of blocks of P_{sim} applies.

Example 4.4. Let G_1, G_2 be the transition systems depicted below.

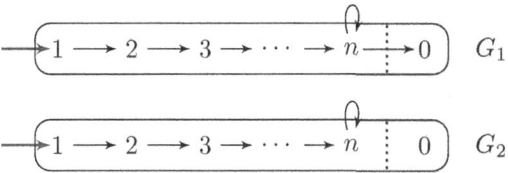

Blocks of states sharing the same principal in $R_i = \mathbb{N} \times \mathbb{N}$ are represented as boxes, while dotted lines are used to delimit the blocks of P_{sim} w.r.t. R_i. In fact, we have that $R_{\text{sim}}(0) = [0, n]$, and, for all $k \in [1, n]$, $R_{\text{sim}}(k) = [1, n]$, so we have that $P_{\text{sim}} = \{[0,0], [1,n]\}$ for G_1 and G_2. Observe that in G_1 the block $[0, 0] \in P_{\text{sim}}$ is reachable while in G_2 the block $[0, 0] \in P_{\text{sim}}$ is unreachable. Hence, to decide whether $[0, 0]$ is reachable in these two systems, we have to detect that 0 is reachable in G_1 and not in G_2. However, this is not the case for bisimulation, since it turns out that $P_{\text{bis}} = \{[k, k]\}_{k=0,\ldots,n}$ for G_1, while $P_{\text{bis}} = P_{\text{sim}}$ for G_2. ◇

5 A Reachable Simulation Algorithm

We define Algorithm 1 which, given a system G, a preorder R_i, and an initial (possibly empty) set of reachable states σ_i, computes the reachable principals of R_{sim} according to (rp_1), and over-approximates the reachable blocks of P_{sim}.

Algorithm 1: Relation-based Algorithm

Input: TS $G = (\Sigma, I, L, \rightarrow)$, $R_i \in \text{PreO}(\Sigma)$, an initial finite set $\sigma_i \subseteq \text{post}^*(I)$.
1 $\text{Rel}(\Sigma) \ni R := R_i$; $\wp(\Sigma) \ni \sigma := \sigma_i$;
2 **while true do**
    // INV_1: $\forall x \in \Sigma. \, R_{\text{sim}}(x) \subseteq R(x) \subseteq R_i(x)$
    // INV_2: $\sigma_i \subseteq \sigma \subseteq \text{post}^*(I)$ INV_3: $\forall x \in \Sigma. \, x \in R(x)$
3      $U := \{R(x) \mid R(x) \cap \sigma = \varnothing, \, R(x) \cap (I \cup \text{post}(\sigma)) \neq \varnothing\}$;
4      $V := \{\langle a, x, x'\rangle \in L \times \Sigma^2 \mid R(x) \cap \sigma \neq \varnothing, \, x \xrightarrow{a} x', \, R(x) \not\subseteq \text{pre}_a(R(x'))\}$;
5   **nif**
6          $(U \neq \varnothing) \longrightarrow$ *Search* :
7   **choose** $R(x) \in U, \, s \in R(x) \cap (I \cup \text{post}(\sigma))$;
8              $\sigma := \sigma \cup \{s\}$;
9          $(V \neq \varnothing) \longrightarrow$ *Refine* :
10   **choose** $\langle a, x, x'\rangle \in V$;
11            $R(x) := R(x) \cap \text{pre}_a(R(x'))$;
12         $(U = \varnothing \wedge V = \varnothing) \longrightarrow$ **return** $\langle R, \sigma \rangle$;

This algorithm maintains a relation $R \in \text{Rel}(\Sigma)$ specified through its principals $R(x) \in \wp(\Sigma)$, and a set $\sigma \subseteq \Sigma$ of reachable states[5], so that $R^\sigma = \{R(x) \mid R(x) \cap \sigma \neq \varnothing\}$ are the provably reachable principals of R. The algorithm computes the set U of principals which can be added to R^σ and the set V of unstable principal pairs. A principal $R(x)$ is in U if it contains an initial state or a successor of a provably reachable state. A triple $\langle a, x, x'\rangle$ is in V if the principal $R(x)$ is provably reachable and it can be refined by a principal $R(x')$, i.e., $x \xrightarrow{a} x'$ and $R(x) \not\subseteq \text{pre}_a(R(x'))$. Algorithm 1 is presented in *logical form*, meaning that in this pseudocode we do not require or provide a specific representation for the transition system G or for the sets maintained by the algorithm, namely the relation R, the provably reachable states of σ, and the sets U and V. Details on the specific representations are given in Sect. 6.

Algorithm 1 either updates, in the *Search* block, the reachability information, or stabilizes, in the *Refine* block, a pair of principals from V. The pseudocode of Algorithm 1 uses a nondeterministic choice between guarded commands (**nif**). We have three guarded commands: either the *Search* (lines 6–8) or the *Refine* blocks (lines 9–11) are executed, or, at line 12, when the other guards are false, the return statement is taken. Thus, every execution consists of an interleaving of *Search* and *Refine*, possibly followed by a return. Observe that the guards are such that when the algorithm terminates neither *Search* nor *Refine* are enabled.

A principal $R(x)$ is refined at line 11 provided it is provably reachable. Upon termination, the principals of R^σ and the principals of $R_{\text{sim}}^{\text{post}^*(I)}$ coincide (cf. (1.a) of Theorem 5.3 below)[6]. However, R may well contain unstable principals, so

---
[5] We distinguish the states of $\text{post}^*(I)$ from those in its subset σ by referring to the states in σ as *provably reachable*. We extend this notion to principals.
[6] Note that line 11 might break transitivity of R. In fact, R is not guaranteed to be a preorder during execution, and not even at termination.

that, in general, R and R_{sim} do not coincide. Turning to the simulation partition we face a more complex situation. To start with, we provide an example showing that the partition P induced by R is such that P^σ does not coincide with $P_{\text{sim}}^{\text{post}^*(I)}$. In fact, P^σ might lack some of the blocks of $P_{\text{sim}}^{\text{post}^*(I)}$.

Example 5.1. Consider the transition system depicted below, where $R_i = \{(1,1),(2,1),(2,2)\} = R_{\text{sim}}$, and $\sigma_i = \varnothing$.

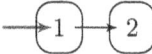

We have that $\text{post}^*(I) = \{1,2\}$, and the blocks induced by R_i coincide with $P_{\text{sim}} = \{\{1\},\{2\}\}$ and are depicted as boxes. Algorithm 1 on input R_i and σ_i returns $\langle R_{\text{sim}}, \{1\}\rangle$. Therefore, it turns out that $P_{\text{sim}}^{\text{post}^*(I)} = P_{\text{sim}}$ and $P^\sigma = \{\{1\}\}$, so that $P_{\text{sim}}^{\text{post}^*(I)} \not\subseteq P^\sigma$ holds. ◇

Algorithm 1 aims at populating σ with just enough states to correctly characterize the reachable principals (i.e., achieving (1.a)), but such states are, in general, not enough to intersect all the reachable blocks of P_{sim}. However, σ suffices to capture such blocks through a relaxation of the reachability notion which, in turn, induces a degree of over-approximation. This relaxed definition is given by $\{B \in P \mid R(B) \cap \sigma \neq \varnothing\}$ as defined in (1.b). Observe that reachable blocks are computed precisely: each block of $P_{\text{sim}}^{\text{post}^*(I)}$ is in P, but, in general, not all blocks in P belong to P_{sim}. Moreover, the following example shows that the converse inclusion of (1.b) does not always hold.

Example 5.2. Consider the system depicted below with $\text{post}^*(I) = \{1\}$, $R_i = \{1,2\} \times \{1,2\}$, $R_{\text{sim}} = \{(1,1),(2,1),(2,2)\}$, and $P_{\text{sim}} = \{\{1\},\{2\}\}$, where boxes depict blocks induced by R_i and dotted lines delimit the blocks of P_{sim}.

Algorithm 1 on input R_i and $\sigma_i = \varnothing$ outputs $R = R_{\text{sim}}$ and $\sigma = \{1\}$. Thus, the inclusion of (1.b) is strict since $\{\{1\}\} \subsetneq \{\{1\},\{2\}\}$. ◇

On the other hand, this inclusion is not arbitrarily loose since a block $B \in P$ such that $R(B) \cap \sigma \neq \varnothing$ (cf. (1.b)) is guaranteed to be simulated by some reachable state. Therefore, R_{sim} limits the magnitude of this over-approximation.

It turns out that Algorithm 1 is correct and terminates on finite state systems.

Theorem 5.3 (Correctness of Algorithm 1 for Finite Systems). *Let $\langle R, \sigma \rangle \in \text{Rel}(\Sigma) \times \wp(\Sigma)$ be the output of Algorithm 1 on input G with $|\Sigma| \in \mathbb{N}$, $R_i \in \text{PreO}(\Sigma)$, and $\sigma_i \subseteq \text{post}^*(I)$. Moreover, let $P \in \text{Part}(\Sigma)$ be the partition induced by R. Then:*

$$R_{\text{sim}}^{\text{post}^*(I)} = R^\sigma, \quad (1.a) \qquad P_{\text{sim}}^{\text{post}^*(I)} \subseteq \{B \in P \mid R(B) \cap \sigma \neq \varnothing\}. \quad (1.b)$$

Theorem 5.4 (Termination of Algorithm 1). *Let $G = (\Sigma, I, L, \rightarrow)$ with $|\Sigma| \in \mathbb{N}$, $R_i \in \text{PreO}(\Sigma)$, and $\sigma_i \subseteq \text{post}^*(I)$. Then, Algorithm 1 terminates on input G, R_i, and σ_i.*

Correctness and Termination for Infinite State Systems. Theorem 5.3 is introduced on finite systems for clarity reasons when formulating the proof. Nevertheless, the proof argument of Theorem 5.3 extends to infinite state systems—i.e., the condition $|\Sigma| \in \mathbb{N}$ can be removed from the hypotheses of the theorem—when the following assumption holds.

Assumption 5.5 (ω-Convergence for Simulation Approximants). Given a transition system $G = (\Sigma, I, L, \rightarrow)$, and an initial preorder $R_i \in \mathrm{PreO}(\Sigma)$, let $\preceq_0 \triangleq R_i$ and $\preceq_n$, for $n > 0$, be the strong simulation approximants as defined in [19, Points (2) and (3) of Definition 30]. Then, the ω-Convergence assumption holds iff $\preceq_\omega = R_{\mathrm{sim}}$, where ω is the first limit ordinal.

Note that Assumption 5.5 holds at least on all finitely branching systems, as stated in [19, Paragraph following Definition 30]. We can now formally state the correctness result for Algorithm 1 (on infinite systems) as follows.

Theorem 5.6 (Correctness of Algorithm 1). *Let $\langle R, \sigma \rangle \in \mathrm{Rel}(\Sigma) \times \wp(\Sigma)$ be the output of Algorithm 1 on input G, $R_i \in \mathrm{PreO}(\Sigma)$ and $\sigma_i \subseteq \mathrm{post}^*(I)$. Moreover, let $P \in \mathrm{Part}(\Sigma)$ be the partition induced by R. If Assumption 5.5 holds, then conditions (1.a) and (1.b) are satisfied.*

Concerning termination, we point out that Algorithm 1 can terminate on some infinite systems, such as the one of Sect. 2, and the following example.

Example 5.7. Consider the infinite transition system depicted below, where $R_i(0) = \{0\}$, $R_i(1) = \{0, 1\}$, and for all $n < 0$, $R_i(n) =]-\infty, -1]$. As usual, boxes denote the blocks induced by R_i.

The simulation preorder is therefore: $R_{\mathrm{sim}}(0) = \{0\}$, $R_{\mathrm{sim}}(1) = \{0, 1\}$, and $R_{\mathrm{sim}}(n) =]-\infty, n]$ for each $n < 0$. Notice that R_{sim} has infinitely many principals, and the corresponding blocks of P_{sim} are delimited by dotted lines in the above diagram. After two *Search* iterations of Algorithm 1 (on input R_i, $\sigma_i = \varnothing$), we get $\sigma = \{0, 1\}$. At this point, $V = \varnothing$ and $U = \varnothing$ holds, and the algorithm returns the correct result.

Consider instead the process of refining each principal $R_i(x)$ such that $R_i(x) \neq R_{\mathrm{sim}}(x)$. This process, which converges to R_{sim}, cannot terminate after finitely many steps since infinitely many principals would need to be refined. ◇

5.1 Execution on the Motivating Example

Let us run Algorithm 1 on the motivating example of Sect. 2. By fixing $k = 2$ we obtain the infinite state transition system depicted below.

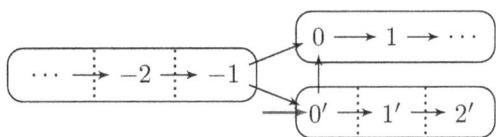

The initial relation is given by $R_i(x) = \mathbb{Z} \setminus \mathbb{N}$ for all $x < 0$, $R_i(x) = \mathbb{N}$ for all $x \geq 0$, and $R_i(0') = R_i(1') = R_i(2') = \{0', 1', 2'\}$. As usual, states sharing the same principal in R_i are depicted in boxes, while dotted lines delimit the blocks of P_{sim}. The remaining input is set to $\sigma_i = \varnothing$.

1st iteration: At the beginning, $U = \{R_i(0')\} = \{\{0', 1', 2'\}\}$, since the principals intersecting $I = \{0'\}$ are those corresponding to states $0'$, $1'$ and $2'$. Moreover, since $\sigma = \varnothing$, we have that $V = \varnothing$ holds. Hence, the *Search* block is executed and σ is updated: at line 8 we get $\sigma = \{0'\}$.

2nd iteration: Since $\sigma = \{0'\}$, we get $U = \{R_i(0)\} = \{\mathbb{N}\}$ because $\mathbb{N} \cap \text{post}(\sigma) \neq \varnothing$. Moreover, $V = \{\langle a, 0', 1'\rangle, \langle a, 1', 2'\rangle\}$ since $\text{pre}_a(R_i(1')) = \text{pre}_a(R_i(2')) = \{0', 1'\}$, and $R_i(0') = R_i(1') = \{0', 1', 2'\} \not\subseteq \{0', 1'\}$. Assume that Algorithm 1 nondeterministically executes a *Search* iteration, then it will update σ by using the principal in U: at line 8 we get $\sigma = \{0', 0\}$.

3rd iteration: We get $U = \varnothing$ since all the states reachable from σ are in principals intersecting σ. Moreover, V is as in the previous iteration. Note that $R_i(0)$, which now intersects σ, does not induce new unstable triples in V. Assume that Algorithm 1 picks $\langle a, 1', 2'\rangle$ from V (picking the other element leads to the same output), and executes the *Refine* block: at line 11 the relation is updated, so that $R(1') = R_i(1') \cap \text{pre}_a(R_i(2')) = \{0', 1'\}$.

4th iteration: Still, $U = \varnothing$, and $V = \{\langle a, 0', 1'\rangle\}$, since $R(0') = R_i(0') = \{0', 1', 2'\} \not\subseteq \text{pre}_a(R(1')) = \{0'\}$. Therefore, the algorithm executes a *Refine* step, and at line 11 we get $R(0') = \{0', 1', 2'\} \cap \{0'\} = \{0'\}$.

5th iteration: Again $U = \emptyset$ since $\sigma = \{0', 0\}$, and for every $x \in \mathbb{N} \cup \{0', 1', 2'\}$, it holds $R(x) \cap \sigma \neq \varnothing$. Moreover, we have that $V = \emptyset$, since all the transitions outgoing $0'$, $1'$, $2'$ and all the states in $\mathbb{N}$ are stable. Therefore, Algorithm 1 returns $\sigma = \{0', 0\}$, and R is as follows: $R(0') = \{0'\}$, $R(1') = \{0', 1'\}$, and $R(x) = R_i(x)$ for every other state. Observe that $|\sigma| = 2$, independently of the fixed parameter k. In fact, Algorithm 1 explores two reachable states (out of infinitely many) which suffice to characterize all the reachable principals.

6 2PR Triples for Designing a Symbolic Algorithm

Symbolic approaches for simulation algorithms based on state partitions are beneficial for algorithms manipulating infinite state systems, as shown by Henzinger et al.'s [17] symbolic simulation algorithm for infinite graphs, and, in particular, hybrid automata. Symbolic approaches are also advantageous in terms of space and time efficiency for finite state systems [7,9,27]. Accordingly, we introduce 2-Partitions-Relation triples (2PR), generalizing the partition-relation pairs used in the most efficient symbolic simulation algorithms [7,11,30] as symbolic representation of a relation between states. We exploit here 2PRs to design a symbolic version of Algorithm 1. The rationale behind the need for 2PRs rather than partition-relation pairs, i.e. 1PR, has more to do with enhancing the presentation and ease of understanding and less to do with limitations of 1PRs.

Definition 6.1 (2PR Triple). *Given an (infinite) set Σ, a triple $\langle P, \tau, Q\rangle$ with $P, Q \in \text{Part}(\Sigma)$ and $\tau \colon P \to \wp(Q)$, is a 2-Partitions-Relation (2PR) triple.* ◇

A relation $R \in \text{Rel}(\Sigma)$ induces a 2PR triple $\langle P_R, \tau_R, Q_R \rangle$ where P_R and Q_R are the partitions induced by R and R^{-1}, respectively, and the function τ_R is defined by $\tau_R(B) \triangleq \{C \in Q_R \mid C \subseteq R(B)\}$. Conversely, a 2PR triple $\langle P, \tau, Q \rangle$ defines a relation $R_{\langle P,\tau,Q \rangle} \in \text{Rel}(\Sigma)$ defined as $R_{\langle P,\tau,Q \rangle}(x) \triangleq \cup \tau(P(x))$ (namely, the union of the blocks in $\tau(P(x))$). In the following, $R_{\langle P,\tau,Q \rangle}$ is called the relation underlying the 2PR triple $\langle P, \tau, Q \rangle$ when no ambiguity arises. It is routine to check that $P \preceq P_{\langle P,\tau,Q \rangle}$ where $P_{\langle P,\tau,Q \rangle} \in \text{Part}(\Sigma)$ is induced by $R_{\langle P,\tau,Q \rangle}$.

We put forward Algorithm 2, designed as a refinement of Algorithm 1 representing R as a 2PR triple. Algorithm 2 is in symbolic logical form, meaning that it symbolically represents and processes state relations as 2PR triples $\langle P, \tau, Q \rangle$. The refinement process of this algorithm preserves reflexivity of the underlying relation, as in Algorithm 1, and during execution, for each state x, the set $R_{\langle P,\tau,Q \rangle}(x)$ includes the states which are candidate to simulate x, and the states in $P_{\langle P,\tau,Q \rangle}(x)$ are candidates to be simulation equivalent to x.

Algorithm 2: 2PR-based Algorithm

Input: TS $G = (\Sigma, I, L, \rightarrow)$, $R_i \in \text{PreO}(\Sigma)$, an initial finite set $\sigma_i \subseteq \text{post}^*(I)$

1 $\text{Part}(\Sigma) \ni P, Q :\!- \{y \in \Sigma \mid R_i(x) = R_i(y)\}_{x \in \Sigma}$;
2 **forall the** $B \in P$ **do** $\wp(Q) \ni \tau(B) :\!- \{C \in Q \mid C \subseteq R_i(B)\}$;
3 $\wp(\Sigma) \ni \sigma :\!- \sigma_i$;
4 **while** true **do**
  // $\text{INV}_1 : \forall x \in \Sigma. R_{\text{sim}}(x) \subseteq R_{\langle P,\tau,Q \rangle}(x) \subseteq R_i(x)$
  // $\text{INV}_2 : \sigma_i \subseteq \sigma \subseteq \text{post}^*(I)$ $\text{INV}_3 : \forall B \in P. B \subseteq \cup \tau(B)$
5    $U :\!- \{B \in P \mid \cup \tau(B) \cap \sigma = \varnothing, \cup \tau(B) \cap (I \cup \text{post}(\sigma)) \neq \varnothing\}$;
6    $V :\!- \{\langle a, B, C \rangle \in L \times P^2 \mid \cup \tau(B) \cap \sigma \neq \varnothing, B \cap \text{pre}_a(C) \neq \varnothing, \cup \tau(B) \not\subseteq \text{pre}_a(\cup \tau(C))\}$;
7   **nif**
8      $(U \neq \varnothing) \longrightarrow \textit{Search}$:
9   **choose** $B \in U$, $s \in (\cup \tau(B) \cap (I \cup \text{post}(\sigma)))$;
10       $\sigma :\!- \sigma \cup \{s\}$;
11     $(V \neq \varnothing) \longrightarrow \textit{Refine}$:
12   **choose** $\langle a, B, C \rangle \in V$; $S :\!- \text{pre}_a(\cup \tau(C))$;
13       $B' :\!- B \cap \text{pre}_a(C)$; $B'' :\!- B \smallsetminus \text{pre}_a(C)$;
14       $P.\text{replace}(B, \{B', B''\})$;
15       $\tau(B') :\!- \tau(B)$; $\tau(B'') :\!- \tau(B)$;
16   **forall the** $X \in \{E \in \tau(B') \mid E \cap S \neq \varnothing, E \not\subseteq S\}$ **do**
17         $Q.\text{replace}(X, \{X \cap S, X \smallsetminus S\})$;
18   **foreach** $A \in P$ **do** $\tau(A).\text{replace}(X, \{X \cap S, X \smallsetminus S\})$;
19       $\tau(B') :\!- \{E \in \tau(B') \mid E \subseteq S\}$;
20     $(U = \varnothing \wedge V = \varnothing) \longrightarrow$ **return** $\langle P, \tau, Q, \sigma \rangle$;

Following the idea of Algorithm 1, this symbolic procedure computes a set of principals $\{\cup \tau(B) \mid \cup \tau(B) \cap \sigma \neq \varnothing\}_{B \in P}$ each of which is provably reachable. A block B is in U at line 8 if $\cup \tau(B)$ contains no provably reachable state, while it contains either an initial state or a successor of a provably reachable state. Also, the set V at line 11 contains unstable triples, i.e., $\langle a, B, C\rangle$ is in V iff $\cup \tau(B)$ is provably reachable and there exist $b \in B, c \in C$ such that $R_{\langle P,\tau,Q\rangle}(b)$ is a-unstable w.r.t. $R_{\langle P,\tau,Q\rangle}(c)$. Algorithm 2 either updates reachability for the principal of some block in U by executing a *Search* iteration or stabilizes the pair of blocks associated to some triple in V by executing a *Refine* iteration. *Refine* iterations a-stabilize a pair of blocks (B, C) by possibly splitting B into $B \cap \mathsf{pre}_a(C)$ and $B \smallsetminus \mathsf{pre}_a(C)$ (lines 13–15). Then, it refines the principal $\cup\tau(B \cap \mathsf{pre}_a(C))$ at lines 16–18 by first splitting blocks of Q if they contain states occurring in different sets of principals for the current relation $R_{\langle P,\tau,Q\rangle}$, and, successively, by removing at line 19 all the blocks not contained in $\mathsf{pre}_a(\cup\tau(C))$.

At termination, we have that $R^\sigma_{\langle P,\tau,Q\rangle}$ coincides with the set of reachable principals of R_{sim} (cf. (2.a) below), while we obtain an over-approximation of $P^{\mathsf{post}^*(I)}_{\mathsf{sim}}$ (cf. (2.b) below). Similarly to Algorithm 1, the term "over-approximation" is used w.r.t. the set $P^{\mathsf{post}^*(I)}_{\mathsf{sim}}$ itself, and not to the contained blocks, whose elements are computed in an exact way. The reader might find surprising the need of an $\exists$ quantifier in (2.b). We have that (2.b) provides a statement equivalent to (2.a), but since $P_{\langle P,\tau,Q\rangle}$ is, in general, coarser than P (as stated previously), then B is not guaranteed to be in the domain of τ, hence the existential quantifier.

Theorem 6.2 (Correctness of Algorithm 2 for Finite Systems). *Let $\langle P, \tau, Q, \sigma\rangle$ be the output of Algorithm 2 on input G with $|\Sigma| \in \mathbb{N}$, $R_i \in \mathrm{PreO}(\Sigma)$, and $\sigma_i \subseteq \mathsf{post}^*(I)$. Moreover, let $P_{\langle P,\tau,Q\rangle} \in \mathrm{Part}(\Sigma)$ be the partition induced by $R_{\langle P,\tau,Q\rangle}$. Then:*

$$R^{\mathsf{post}^*(I)}_{\mathsf{sim}} = R^\sigma_{\langle P,\tau,Q\rangle} \; , \tag{2.a}$$

$$P^{\mathsf{post}^*(I)}_{\mathsf{sim}} \subseteq \{B \in P_{\langle P,\tau,Q\rangle} \mid \exists E \in P.\ E \subseteq B \wedge (\cup \tau(E)) \cap \sigma \neq \varnothing\} \; . \tag{2.b}$$

As done for Algorithm 1 and Theorem 5.3, we extend Theorem 6.2 to infinite transition systems as follows.

Theorem 6.3 (Correctness of Algorithm 2). *Let $\langle P, \tau, Q, \sigma\rangle$ be the output of Algorithm 2 on input G, $R_i \in \mathrm{PreO}(\Sigma)$, and $\sigma_i \subseteq \mathsf{post}^*(I)$. Moreover, let $P_{\langle P,\tau,Q\rangle} \in \mathrm{Part}(\Sigma)$ be the partition induced by $R_{\langle P,\tau,Q\rangle}$. If Assumption 5.5 holds, then conditions (2.a) and (2.b) are satisfied.*

Termination and Complexity of Algorithm 2. We provide two conditional termination results for Algorithm 2 together with complexity bounds on the total number of its iterations. These results rely on progression guarantees and on the fact that the partitions P and Q are coarser than P_{bis} throughout execution. Our first conditional termination result is akin to that of Lee and Yannakakis [23, Th. 3.1 and the following paragraph therein], and applies when the bisimulation partition P_{bis} is finite. Moreover, it turns out that Algorithm 2 carries out a total number of iterations which is at most quadratic in the size of P_{bis}.

Theorem 6.4 (Termination of Algorithm 2). *Let G, $R_i \in \mathsf{PreO}(\Sigma)$ and $\sigma_i \subseteq \mathsf{post}^*(I)$ be the input of Algorithm 2. Moreover, let $P_{\mathsf{bis}} \in \mathsf{Part}(\Sigma)$ be the bisimulation partition w.r.t. P_i, the partition induced by R_i. If P_{bis} consists of finitely many blocks then the number of iterations of Algorithm 2 is in $O(|P_{\mathsf{bis}}|^2)$.*

A second termination result holds under a local finiteness condition: Algorithm 2 terminates if the number of blocks of P_{bis} contained in $R_i(\mathsf{post}^*(I))$ is finite. The result comes with a quadratic bound on the number of iterations.

Theorem 6.5 (Alternative Termination of Algorithm 2). *Let G, $R_i \in \mathsf{Part}(\Sigma)$ and $\sigma_i \subseteq \mathsf{post}^*(I)$ be the input of Algorithm 2 (note that R_i is a partition). Let $P_{\mathsf{bis}} \in \mathsf{Part}(\Sigma)$ be the bisimulation partition w.r.t. R_i, and let $T \triangleq \{Y \in P_{\mathsf{bis}} \mid R_i(Y) \cap \mathsf{post}^*(I) \neq \varnothing\}$ ($= \{Y \in P_{\mathsf{bis}} \mid Y \subseteq R_i(\mathsf{post}^*(I))\}$). If T is finite then the number of iterations of Algorithm 2 is in $O(|T|^2)$.*

It follows from Theorem 6.4 that Algorithm 2 terminates on all finite state systems (as does Algorithm 1). It turns out that these two termination results for Algorithm 2 are indeed stronger and extend to many infinite state systems as well. In fact, the use of 2PR triples as a representation structure in Algorithm 2 brings significant benefits in terms of termination on infinite systems, since there exist many inputs on which Algorithm 1 does not terminate, while Algorithm 2 does. The intuition here is that, through the use of 2PR triples, Algorithm 2 can refine infinitely many principals of the relation encoded by $\langle P, \tau, Q \rangle$, in a single *Refine* iteration. Below we show two examples which compare executions of Algorithms 1 and 2 on one input, illustrating this behaviour.

Example 6.6. Consider the following infinite transition system, where $\sigma_i = I = \{1\}$, and $R_i = \mathbb{N} \times \mathbb{N}$.

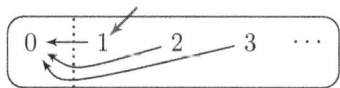

The simulation preorder R_{sim} is such that $R_{\mathsf{sim}}(0) = \mathbb{N}$, and $R_{\mathsf{sim}}(n) = \mathbb{N} \smallsetminus \{0\}$ for $n \geq 1$, meaning that $P_{\mathsf{sim}} = \{\{0\}, \mathbb{N} \smallsetminus \{0\}\}$. As usual, the block induced by R_i is represented as a box, and the dotted line delimits the blocks of P_{sim}. During execution of Algorithm 1, the set U is empty, since the state 1 is in every principal. On the other hand, the set V contains a triple $\langle a, i, 0 \rangle$ for every $i \in \mathbb{N} \smallsetminus \{0\}$. Executing a *Refine* iteration refines exactly one principal: if $\langle a, i, 0 \rangle \in V$ is selected, then $R(i)$ is updated to $\mathbb{N} \smallsetminus \{0\}$. It is easily seen that Algorithm 1 never terminates because V has infinitely many elements, and each iteration removes exactly one element from it, thus V is never empty. ◇

On the other hand, Algorithm 2 is able to refine infinitely many principals of the underlying relation in a single *Refine* step. To illustrate this, we consider the input from Example 6.6 and show that Algorithm 2 converges in a few iterations.

Example 6.7. Consider the input of Example 6.6. The initial 2PR is given by $P = Q = \{\mathbb{N}\}$ and $\tau(\mathbb{N}) = \{\mathbb{N}\}$, $U = \emptyset$, and V is $\{\langle a, \mathbb{N}, \mathbb{N}\rangle\}$. Executing one iteration refines both P, Q and τ, so that $P = Q = \{\{0\}, \mathbb{N} \smallsetminus \{0\}\}$, $\tau(\{0\}) = Q$ and $\tau(\mathbb{N} \smallsetminus \{0\}) = \{\mathbb{N} \smallsetminus \{0\}\}$. At this point, the underlying relation is such that $R_{\langle P, \tau, Q\rangle} = R_{\text{sim}}$, and Algorithm 2 terminates. ◇

7 Conclusion and Future Work

We introduced and proved the correctness and termination of algorithms solving the reachable simulation problem. We showed fundamental differences w.r.t. to the analogous problem for bisimulation studied by Lee and Yannakakis [23] in 1992. To the best of our knowledge, this is the first investigation of Lee and Yannakakis' problem recast to the simulation preorder and partition. Algorithm 2 is the most relevant one for practical purposes, since this procedure converges on all finite state systems and is well-suited to handle infinite state systems through its symbolic representation, being able to converge on some—but not all, due to undecidability—such infinite systems. In particular, we have shown that Algorithm 2 offers the same termination guarantee as the LY algorithm [23]. On top of that, our algorithm also terminates under a local finiteness hypothesis, while LY [23] has no counterpart to such a termination guarantee.

Future work will explore possible domains in which the algorithm can be applied. Choosing a specific class of implicitly[7] represented systems gives rise to a multitude of questions which are domain dependent. Solving these are crucial to obtain an efficient implementation. Orthogonally, further generalizations of our algorithm—for instance, to other behavioral relations such as branching bisimilarity for labeled transition systems or stuttering equivalence for Kripke structures—constitute future research paths.

Acknowledgements. Francesco Ranzato was partially funded by: the *Italian MUR*, under the PRIN 2022 PNRR project no. P2022HXNSC; *Meta* (formerly *Facebook*) *Research*, under a "Probability and Programming Research Award" and under a *WhatsApp Research Award* on "Privacy-aware Program Analysis"; by an *Amazon Research Award* for "AWS Automated Reasoning". Nicolas Manini is supported by the grant PIPF-2022/COM-24370, funded by the Madrid Regional Government. This publication is part of the grant PID2022-138072OB-I00, funded by MCIN/AEI/10.13039/501100011033/FEDER, UE and part of the PRODIGY Project (TED2021-132464B-I00) funded by MCIN/AEI/10.13039/501100011033/and the European Union NextGenerationEU/PRTR.

---

[7] Reachability analysis is mostly superfluous on explicitly represented systems as they usually do not encode unreachable states.

References

1. Alur, R., Henzinger, T.A.: Computer-Aided Verification (1999). chapter 4: Graph minimization (Unpublished manuscript)
2. Bensalem, S., Bouajjani, A., Loiseaux, C., Sifakis, J.: Property preserving simulations. In: von Bochmann, G., Probst, D.K. (eds.) CAV 1992. LNCS, vol. 663, pp. 260–273. Springer, Heidelberg (1993). https://doi.org/10.1007/3-540-56496-9_21
3. Bloom, B., Paige, R.: Transformational design and implementation of a new efficient solution to the ready simulation problem. Sci. Comput. Program. **24**(3), 189–220 (1995). https://doi.org/10.1016/0167-6423(95)00003-B
4. Bouajjani, A., Fernandez, J.-C., Halbwachs, N.: Minimal model generation. In: Clarke, E.M., Kurshan, R.P. (eds.) CAV 1990. LNCS, vol. 531, pp. 197–203. Springer, Heidelberg (1991). https://doi.org/10.1007/BFb0023733
5. Bouajjani, A., Fernandez, J.C., Halbwachs, N., Raymond, P., Ratel, C.: Minimal state graph generation. Sci. Comput. Program. **18**(3), 247–269 (1992). https://doi.org/10.1016/0167-6423(92)90018-7
6. Bustan, D., Grumberg, O.: Simulation-based minimization. ACM Trans. Comput. Log. **4**(2), 181–206 (2003). https://doi.org/10.1145/635499.635502
7. Cécé, G.: Foundation for a series of efficient simulation algorithms. In: Proceedings of the 32nd Annual ACM/IEEE Symposium on Logic in Computer Science, LICS 2017, pp. 1–12. IEEE Computer Society (2017). https://doi.org/10.1109/LICS.2017.8005069
8. Clarke, E.M., Henzinger, T.A., Veith, H., Bloem, R.: Handbook of Model Checking, 1st edn. Springer, Cham (2018)
9. Crafa, S., Ranzato, F., Tapparo, F.: Saving space in a time efficient simulation algorithm. Fundam. Informaticae **108**(1–2), 23–42 (2011). https://doi.org/10.3233/FI-2011-412
10. Fisler, K., Vardi, M.Y.: Bisimulation minimization and symbolic model checking. Formal Methods Syst. Des. **21**(1), 39–78 (2002). https://doi.org/10.1023/A:1016091902809
11. Gentilini, R., Piazza, C., Policriti, A.: From bisimulation to simulation: coarsest partition problems. J. Autom. Reason. **31**(1), 73–103 (2003). https://doi.org/10.1023/A:1027328830731
12. van Glabbeek, R., Ploeger, B.: Correcting a space-efficient simulation algorithm. In: Gupta, A., Malik, S. (eds.) CAV 2008. LNCS, vol. 5123, pp. 517–529. Springer, Heidelberg (2008). https://doi.org/10.1007/978-3-540-70545-1_49
13. Gratzer, G.A.: Lattice Theory: Foundation. Springer, Basel (2011)
14. Grumberg, O., Long, D.E.: Model checking and modular verification. In: Baeten, J.C.M., Groote, J.F. (eds.) CONCUR 1991. LNCS, vol. 527, pp. 250–265. Springer, Heidelberg (1991). https://doi.org/10.1007/3-540-54430-5_93
15. Grumberg, O., Long, D.E.: Model checking and modular verification. ACM Trans. Program. Lang. Syst. (TOPLAS) **16**(3), 843–871 (1994). https://doi.org/10.1145/177492.177725
16. Gulavani, B.S., Henzinger, T.A., Kannan, Y., Nori, A.V., Rajamani, S.K.: SYNERGY: a new algorithm for property checking. In: Proceedings of the 14th ACM SIGSOFT International Symposium on Foundations of Software Engineering, FSE 2006, pp. 117–127. ACM (2006). https://doi.org/10.1145/1181775.1181790
17. Henzinger, M.R., Henzinger, T.A., Kopke, P.W.: Computing simulations on finite and infinite graphs. In: Proceedings of IEEE 36th Annual Foundations of Computer Science, FOCS 1995, pp. 453–462 (1995). https://doi.org/10.1109/SFCS.1995.492576

18. Henzinger, T.A., Kopke, P.W.: Hybrid automata with finite mutual simulations. Technical report, TR-95-1497, Computer Science Department (1995)
19. Hofman, P., Lasota, S., Mayr, R., Totzke, P.: Simulation problems over one-counter nets. Log. Methods Comput. Sci. **12**, 1–46 (2016). https://doi.org/10.2168/LMCS-12(1:6)2016
20. Kučera, A., Jančar, P.: Equivalence-checking on infinite-state systems: techniques and results. Theory Pract. Log. Program. **6**(3), 227–264 (2006). https://doi.org/10.1017/S1471068406002651
21. Kucera, A., Mayr, R.: Simulation preorder over simple process algebras. Inf. Comput. **173**(2), 184–198 (2002). https://doi.org/10.1006/inco.2001.3122
22. Kučera, A., Mayr, R.: Why is simulation harder than bisimulation? In: Brim, L., Křetínský, M., Kučera, A., Jančar, P. (eds.) CONCUR 2002. LNCS, vol. 2421, pp. 594–609. Springer, Heidelberg (2002). https://doi.org/10.1007/3-540-45694-5_39
23. Lee, D., Yannakakis, M.: Online minimization of transition systems. In: Proceedings of the 24th Annual ACM Symposium on Theory of Computing, STOC 1992, pp. 264–274. ACM (1992). https://doi.org/10.1145/129712.129738
24. Loiseaux, C., Graf, S., Sifakis, J., Bouajjani, A., Bensalem, S.: Property preserving abstractions for the verification of concurrent systems. Formal Methods Syst. Des. **6**(1), 11–44 (1995). https://doi.org/10.1007/BF01384313
25. Majumdar, R., Ozay, N., Schmuck, A.K.: On abstraction-based controller design with output feedback. In: Proceedings of the 23rd International Conference on Hybrid Systems: Computation and Control, HSCC 2020, pp. 1–11. ACM (2020). https://doi.org/10.1145/3365365.3382219
26. Păsăreanu, C.S., Pelánek, R., Visser, W.: Concrete model checking with abstract matching and refinement. In: Etessami, K., Rajamani, S.K. (eds.) CAV 2005. LNCS, vol. 3576, pp. 52–66. Springer, Heidelberg (2005). https://doi.org/10.1007/11513988_7
27. Ranzato, F.: A more efficient simulation algorithm on Kripke structures. In: Chatterjee, K., Sgall, J. (eds.) MFCS 2013. LNCS, vol. 8087, pp. 753–764. Springer, Heidelberg (2013). https://doi.org/10.1007/978-3-642-40313-2_66
28. Ranzato, F.: An efficient simulation algorithm on Kripke structures. Acta Informatica **51**(2), 107–125 (2014). https://doi.org/10.1007/s00236-014-0195-9
29. Ranzato, F., Tapparo, F.: A new efficient simulation equivalence algorithm. In: Proceedings of the 22nd IEEE Symposium on Logic in Computer Science, LICS 2007, pp. 171–180. IEEE Computer Society (2007). https://doi.org/10.1109/LICS.2007.8
30. Ranzato, F., Tapparo, F.: An efficient simulation algorithm based on abstract interpretation. Inf. Comput. **208**(1), 1–22 (2010). https://doi.org/10.1016/j.ic.2009.06.002
31. Tan, L., Cleaveland, R.: Simulation revisited. In: Margaria, T., Yi, W. (eds.) TACAS 2001. LNCS, vol. 2031, pp. 480–495. Springer, Heidelberg (2001). https://doi.org/10.1007/3-540-45319-9_33
32. van Glabbeek, R., Ploeger, B.: Five determinisation algorithms. In: Ibarra, O.H., Ravikumar, B. (eds.) CIAA 2008. LNCS, vol. 5148, pp. 161–170. Springer, Heidelberg (2008). https://doi.org/10.1007/978-3-540-70844-5_17
33. Yannakakis, M., Lee, D.: An efficient algorithm for minimizing real-time transition systems: extended abstract. Formal Methods Syst. Des. **11**(2), 113–136 (1997). https://doi.org/10.1023/A:1008621829508

Computing All Minimal Ways to Reach a Context-Free Language

Florian Bruse[(⊠)] and Martin Lange

Theoretical Computer Science/Formal Methods, University of Kassel,
Kassel, Germany
{florian.bruse,martin.lange}@uni-kassel.de

Abstract. We present a theory of rewriting a word into a given target language. We show that the natural notion of equivalence between corrections as sequences of edit operations can be captured syntactically by means of a rather simple rewrite system. Completeness relies on a normal form for corrections that is then also used to develop a notion of minimality for corrections. This is not based on edit distance between words and languages but on a subsequence order on corrections, capturing the intuitive notion of doing a minimal number of rewriting steps. We show that the number of minimal corrections is always finite, and that they are computable for context-free languages.

Keywords: formal languages · edit distance · term rewriting

1 Introduction

The term *error-correcting parsing* describes the following problem. Given a formal language L over some alphabet Σ and a word $w \in \Sigma^*$, generally assuming $w \notin L$, compute some $v \in L$ and a transformation ρ of w into v using basic editing operations. Since every word can somehow be turned into any other word using sufficiently many such operations, we are typically interested in some minimal v and/or ρ, and sensible notions of minimality are for instance based on minimal edit distance [11]. Virtually all known parsing algorithms have been extended to the error-correcting framework [2,9,12,13].

In this paper, we consider a further generalisation of this problem: given a language $L \subseteq \Sigma^*$ and a word $w \in \Sigma^*$, compute *all* transformations (which we call *corrections*) ρ that turn w into some $v \in L$. Clearly, this is not well-defined in general because there are infinitely many such ρ whenever L is infinite. Hence, we also need to restrict attention to a subset, for example by employing minimality as above. However, we propose an entirely different notion of minimality than the one based on edit distance: a correction ρ is minimal (for L, w) if there is no way to apply a subset of the atomic operations in ρ to reach the target language earlier. This means that all paths that lead from w to L and that represent ρ (modulo natural rearrangements and simplifications) in the graph Σ^*, with edges between words given by elementary edit operations, never cross L in between.

Intuitively, a correction is therefore considered to be minimal not necessarily when it creates a word in the target language at minimal distance, but when the set of operations contained in it is minimal in the sense that leaving anything out will lead to not reaching the target language.

The motivation for this more complex notion of minimality is seen in the following small example using English as the target. Let $w =$ "*I am happy I am rich.*" Consider the corrections ρ_1 that inserts "*if*" and "*then*", and ρ_2 that inserts "*because*" appropriately. Amongst these, ρ_2 is minimal w.r.t. edit distance because it consists of a single operation only while ρ_1 consists of two. Since both introduce different causal connections between the two properties expressed in w, it may be impertinent to discard ρ_1 in favour of ρ_2, and both are in fact minimal w.r.t. the notion introduced formally later on. Now consider ρ_3 that turns w into "*I am happy because I am very rich*" by inserting "*because*" and "*very.*" This is not minimal anymore, because after inserting "*because*" we have already hit L. Intuitively, it is *less* likely that the author intended to write the word obtained through ρ_3 because it extends what is being achieved with ρ_2. This could be fooled, though, by *first* inserting "*very*", not hitting L yet, and *then* inserting "*because.*" This is why we develop a notion of minimality that is invariant under such rearrangements. Unlike minimality based on edit distance, it is not obvious that the number of minimal solutions to the problem of computing all corrections is necessarily finite. We will prove that it is indeed the case.

The paper is organised as follows. Section 2 recalls necessary preliminaries from the theory of formal languages and Sect. 3 introduces corrections formally. The main result there is a syntactic characterisation of the natural notion of equivalence between corrections by means of a simple word rewriting system. This algebraic theory allows us to formalise a notion of "subcorrection" which is invariant under syntactic representations and leads to a formalisation of the notion of minimality in Sect. 4. We show how to compute all minimal corrections for a context-free language L and give a sufficient condition on L for the problem to become undecidable. Proof details are omitted for reasons of space limitations. Section 5 reports on an implementation and presents some empirical results. Section 6 concludes with remarks on further work.

2 Preliminaries

We first recall some preliminaries from the theory of words and formal languages, then formally introduce corrections.

We use Σ to denote an alphabet which is always finite. As usual $\Sigma^*, \Sigma^+, \Sigma^\ell$, $\Sigma^{\leq \ell}$ for some $\ell \geq 0$ denote the set of all, resp. all non-empty words, resp. those of length ℓ, resp. of length at most ℓ. The empty word is denoted ε. We write $w(i)$ for the i-th letter of w, beginning with index 0. A language is a subset of Σ^*.

Since $|\Sigma| < \infty$, we can always assume, w.l.o.g., that its elements are totally ordered by some order $<$. This will help us determinise one particular operation, as follows. Given two words $w, w' \in \Sigma^*$ and a letter $a \in \Sigma$, a *split* of w and w' is a quintuple (x, y, a, x', y') for $x, y, x', y' \in \Sigma^*$ and $a \in \Sigma$ such that $w = xay$

and $w' = x'ay'$. It is a *rightmost split* if, additionally, $y, y' \in (\Sigma \setminus \{a\})^*$ and y, y' are letter-disjoint. A *minimal split* is a rightmost split (x, y, a, x', y') for a letter a that is minimal w.r.t. the order $<$ on the underlying alphabet.

Example 1. Let $\Sigma = \{\mathsf{A}, \mathsf{C}, \mathsf{G}, \mathsf{T}\}$. The following depicts the set of all splits of two words u and v, depending on whether $\mathsf{C} < \mathsf{G}$ (left) or $\mathsf{G} < \mathsf{C}$ (right).

$$u = \mathsf{C\ G\ C\ A\ A} \qquad\qquad u = \mathsf{C\ G\ C\ A\ A}$$
$$v = \mathsf{T\ G\ C\ C\ G\ T} \qquad\qquad v = \mathsf{T\ G\ C\ C\ G\ T}$$

A line from a position in u to a position in v determines a split consisting of each of the two halves left and right of the line's endpoint, as well as the letter on the endpoints. Dotted lines represent non-rightmost splits, and a double line represents the (necessarily unique) minimal split.

Lemma 1. *Let w, w' be words over a totally ordered alphabet. There is no split for w and w' if w and w' are letter-disjoint. Otherwise, there is a unique minimal split for w and w'.*

Because of Lemma 1, which is rather easy to see, we write $msp(w, w')$ to denote the unique minimal split of w and w' if it exists.

Instead of the atomic operations of insertion, deletion and replacement of a letter in a word that underly common notions of transforming a word into another, we work – for the sake of simplicity – with a single replacement operation $u/_i v$ where $i \in \mathbb{N}$ and $u, v \in \Sigma^*$. We call i the *index* of $u/_i v$. The effect of $u/_i v$ on a word $w = a_0 \ldots a_{n-1}$ is explained formally as follows.

$$u/_i v(w) := \begin{cases} a_0 \ldots a_{i-1} v a_{i+|u|} \ldots a_{n-1}, & \text{if } i + |u| \le n, u = a_i \ldots a_{i+|u|-1}, \\ \bot, & \text{otherwise.} \end{cases}$$

Hence, $u/_i v$ replaces the subword u starting at position i with the word v, and fails if u cannot be found at this position.

It should be clear that this encompasses more atomic operations: an insertion of letter a at position i is given as $\varepsilon/_i a$, a corresponding deletion is $a/_i \varepsilon$, and a replacement of a single letter a by b is simply given as $a/_i b$. Likewise, an operation $u/_i v$ can be mimicked by a (not necessarily unique) sequence of letter replacements, together with either deletions or insertions. Hence, corrections based on generalised replacements instead of single insertions, deletions and replacements have the same expressive power. In addition, they are technically simpler to handle because having a single type of operation instead of three types avoids many case distinctions in the following.

A *correction* is a (possibly empty) sequence $\rho = \langle \alpha_1, \ldots, \alpha_m \rangle$ of replacements. The empty correction is denoted as $\langle \rangle$. The effect of the application of a correction to a word, or simply *correcting* the word, is explained by a homomorphic extension of the effect of replacements: $\langle \rangle(w) := w$ and $\langle \rho, \alpha \rangle(w) := \rho(\alpha(w))$

(P) $u_2/_j v_2, u_1/_i v_1 \to u_1/_i v_1, u_2/_{j-|v_1|+|u_1|} v_2$ if $i + |v_1| \leq j$
(O) $yz/_{i-|y|} v, u/_i zx \to yu/_{i-|y|} vx$
($\overline{\text{O}}$) $zy/_{i+|x|} v, u/_i xz \to uy/_i xv$
(D) $xzy/_{i-|x|} v, u/_i z \to xuy/_{i-|x|} v$
($\overline{\text{D}}$) $z/_{i+|x|} v, u/_i xzy \to u/_i xvy$
(S) $xau/_i yav \to x/_i y, a/_{i+|x|} a, u/_{i+|x|+1} v$
(ε) $\varepsilon/_i \varepsilon, \alpha \to \alpha$ if $\mathsf{idx}(\alpha) \geq i$

Fig. 1. Rewriting corrections, where $a \in \Sigma$ and all other variables range over Σ^*.

with the provision that $\rho(\bot) = \bot$ for any correction ρ. Note that this means that corrections are to be read from right to left. The *domain* of a correction ρ is $dom(\rho) := \{w \mid \rho(w) \neq \bot\}$. Two corrections ρ, ρ' are *equivalent*, written $\rho \equiv \rho'$, if $\rho(w) = \rho'(w)$ for all $w \in \Sigma^*$. It is not hard to see that $\equiv$ is indeed a congruence for composition of corrections.

Example 2. Let $\rho := \langle a/_3 \varepsilon, \varepsilon/_3 a \rangle$ and $\rho' := \langle \varepsilon/_3 a, a/_3 \varepsilon \rangle$. We have $\rho \not\equiv \langle \rangle$, i.e. inserting a at position 3 and then deleting it is not equivalent to not doing anything at all because $dom(\langle \rangle) = \Sigma^*$ but $dom(\rho) = \Sigma^{\geq 3}$. Note that inserting a letter at position i requires at least position $i - 1$ to be available. We also have $\rho' \not\equiv \langle \rangle$. So first deleting a letter and then reinserting it is also not the same as not doing anything because this not only implicitly checks for the position to be available, but it also checks whether that position is occupied by some letter (or subword in general). So $dom(\rho') = \Sigma^3 a \Sigma^*$.

Now let $\rho_1 := \langle b/_2 \varepsilon, a/_1 a, \varepsilon/_0 b \rangle$ and $\rho_2 := \langle ab/_0 ba \rangle$. We have $\rho_1 \equiv \rho_2$ since both realise the same replacement at the beginning of a word: for both $i \in \{1, 2\}$ we have $\rho_i(w) = bau$ if $w = abu$ for some $u \in \Sigma^*$, and $\rho_i(w) = \bot$ otherwise.

The latter shows that operations of the form $u/_i u$ for any $u \in \Sigma^+$ should not be considered to be uneffective. They merely are tests disguised as replacements.

3 Rewriting Corrections

The key to subsequent developments, in particular that of a notion of minimality and the associated computation problem, is a syntactic characterisation of $\equiv$. We briefly recall some basic notions from term rewriting and refer to the literature, cf. [3], for a more detailed introduction.

A rewriting relation $\to$ is given by a finite number of rules of the form $\sigma \to \rho$. It induces a relation on terms (here: sequences) by congruence. We have $\langle \tau, \sigma, \tau' \rangle \to \langle \tau, \rho, \tau' \rangle$ if there is a rule $\sigma \to \tau$. No substitutions are needed here. We use $\leftarrow := \to^{-1}$, $\leftrightarrow := \to \cup \leftarrow$ and $\to^+, \to^*$ for the transitive, resp. reflexive-transitive closure of $\to$, and likewise for $\leftrightarrow^+, \leftrightarrow^*$.

We introduce seven rewriting rules, shown in Fig. 1, operating on one or two successive operations in a correction. There are five possibilities regarding the position of a subword to be replaced, relative to the previous subword to be replaced, or its replacement. This leads to the five rules (P) – (O) that are used to disentangle replacements and put them in a particular order. Rules (S) and (ε) are mainly useful in providing a unique form of corrections. We will briefly explain the functionality and depict the effect of each of these rules.

Rule (P) is used to swap the order of two operations in which the first *precedes* the second in a word, resulting in the first one to be applied to be executed further towards the end of a word than the second. Since the index in a replacement refers to the position at the moment that this replacement gets applied, swapping the order incurs index shifts in general. This can be visualised as follows.

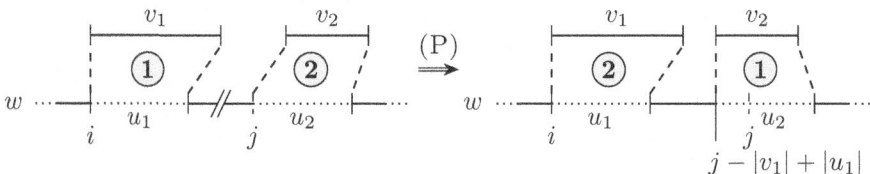

The left side shows the execution of $u_1/_j u_2$ first, then followed by the execution of $u_2/_i v_2$. The point to note here is that, by the time the first operation has been executed, all positions from $j + |u_1|$ on get shifted by $|v_1| - |u_1|$ to the right. In particular, what is shown as the subword u_2 gets shifted to position i and can then be replaced by v_2. Hence, what is *then* position i was originally position $i - |v_1| + |u_1|$. Hence, it is possible to execute these two in opposite order, but then the now first one gets applied at that position directly, as shown on the right side.

The following example shows why such a swap may be desirable.

Example 3. Let $w = baca$ and $\rho = \langle a/_3 \varepsilon, b/_0 bbb \rangle$. We have $\langle a/_3 \varepsilon, b/_0 bbb \rangle(baca) = \langle a/_3 \varepsilon \rangle(bbbaca) = bbbca$ and $\rho \to \langle b/_0 bbb, a/_1 \varepsilon \rangle$ with rule (P). Note how the indices of the operations in the rewritten correction correspond to the actual positions at which they are applied, which was not the case in ρ since it did not list its operations in order of application from right to left in a word. Rule (P) can thus be used to adjust this.

Rules (O) and ($\overline{\text{O}}$) are used to combine the effect of two replacements in which the first *overlaps* the second (to the left resp. right). This can be achieved in a single replacement, eliminating the common factor that is inserted by the first and then removed by the second replacement, seen as follows for rule (O)

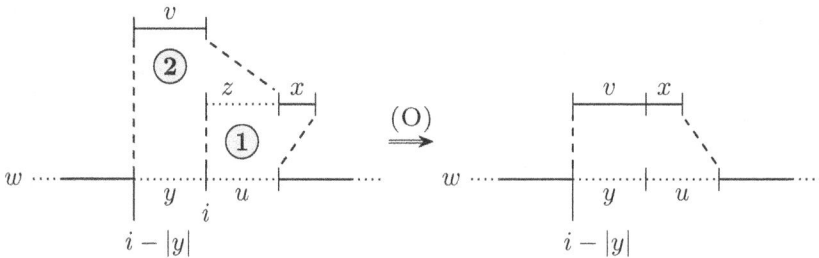

and for rule $(\overline{O})$:

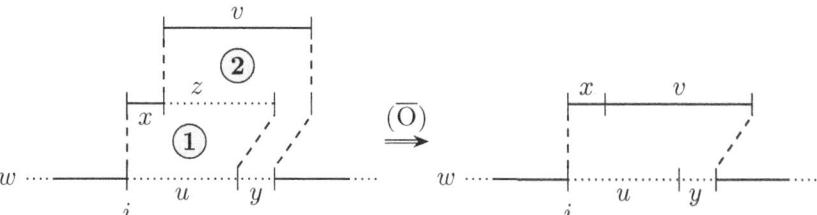

Also note the special case of $z = \varepsilon$ in these rules that makes them act as mergers. For instance, we have $\langle ab/_3c, b/_5ac\rangle \stackrel{(O)}{\rightarrow} \langle abb/_3cac\rangle$. Similarly, rules (D) and $(\overline{D})$ cover situations in which one replacement operates on a subword that is contained within that of the other, or lies *during* the other, seen as follows for rule (D)

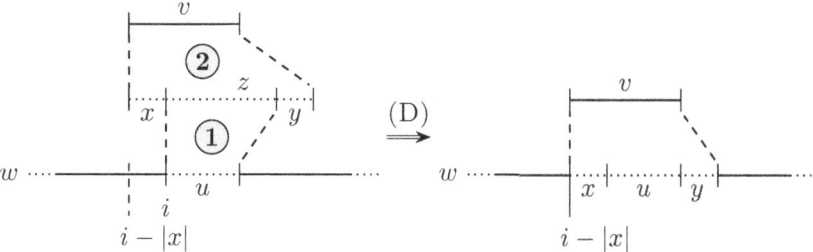

and for rule $(\overline{D})$:

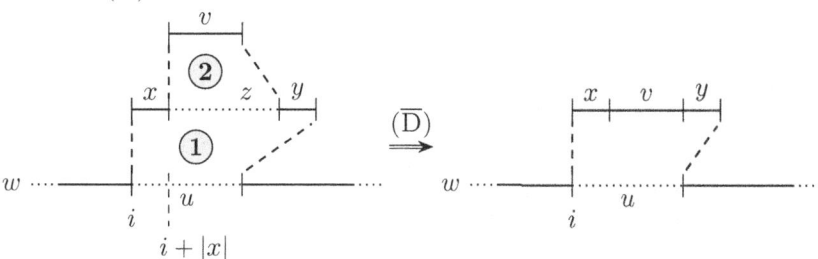

Unlike the previous four rules that can be used to combine multiple replacements into one, rule (S) can be used to *split* the replacement of a subword xau by a subword yav – sharing at least one common letter – into three parts. While this increases the number of replacement operations, the result can be seen as simpler in the sense that it reduces the number of letters that are actually being replaced. Its effect is visualised as follows.

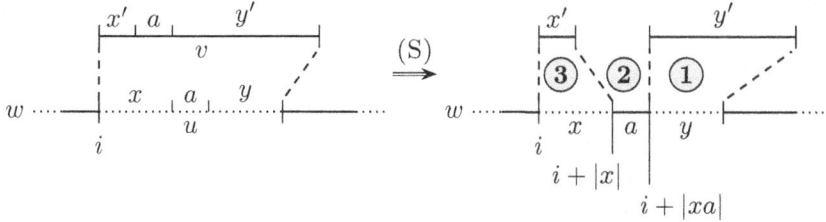

Note that there may not be a unique way to apply rule (S). For example, we have
$$\langle \varepsilon/_0 b, a/_0 a, b/_1 \varepsilon \rangle \xleftarrow{(S)} \langle ab/_0 ba \rangle \xrightarrow{(S)} \langle a/_0 \varepsilon, b/_1 b, \varepsilon/_2 a \rangle.$$

At last, an operation $\varepsilon/_i \varepsilon$ simply tests for the existence of position i. In the context of another operation that supersedes this by implicitly testing for a position that is at least i, the test for position i can be dropped using rule (ε). Rule (S) may produce such operations.

These seven rules are not chosen arbitrarily, but they can be used to ensure that corrections are given in a specific normalised form.

Definition 1. *A correction $\rho = \langle \alpha_1, \ldots, \alpha_m \rangle$ with $\alpha_j = u_j/_{i_j} v_j$ for all $j = 1, \ldots, m$ is* weakly normalised, *if for all $j = 1, \ldots, m-1$ we have $i_{j+1} \geq i_j + |u_j|$. We say that there is* no gap *between α_j and α_{j+1}, written $\alpha_j \mid \alpha_{j+1}$ if $i_{j+1} = i_j + |u_j|$. Likewise, there is* a gap, *written $\alpha_j \parallel \alpha_{j+1}$, if $i_{j+1} > i_j + |u_j|$.*

Weak normalisation guarantees that the result of applying correction ρ to a word w can immediately be read off ρ and w.

Lemma 2. *Let $\rho = \langle \alpha_1, \ldots, \alpha_m \rangle$ with $\alpha_j = u_j/_{i_j} v_j$ be weakly normalised, and $w = a_0 \ldots a_{n-1} \in \Sigma^*$. If $\rho(w) \neq \bot$ then $w = x_0 u_1 x_1 u_2 x_2 \ldots x_{m-1} u_m x_m$ for some $x_0, \ldots, x_m$ such that*

- *for all $j = 1, \ldots, m$: $i_j = |x_0 u_1 \ldots u_{j-1} x_{j-1}|$,*
- *for all $j = 1, \ldots, m-1$: $x_j = \varepsilon$ iff $\alpha_j \mid \alpha_{j+1}$,*
- *$\rho(w) = x_0 v_1 x_1 v_2 x_2 \ldots x_{m-1} v_m x_m$.*

Note that non-weakly-normalised corrections do not have this property. Consider $\rho = \langle b/_0 \varepsilon, a/_0 \varepsilon \rangle$ and $w = ab$. Then $\rho(w) = \varepsilon \neq \bot$ but $w(0) \neq b$.

Theorem 1 (Soundness). *For all σ, ρ: if $\sigma \leftrightarrow^* \rho$ then $\sigma \equiv \rho$.*

Proof (Sketch). This follows from the fact that $\equiv$ is an equivalence relation and that each rule application preserves equivalence in the sense that if $\sigma \to \rho$ then $\sigma \equiv \rho$ which is easily proved by close inspection of each rule. □

Lemma 3 (Weak Normalisation). *For every correction ρ with nonempty domain there is a weakly normalised ρ' such that $\rho \to^* \rho'$.*

Weak normalisation does not guarantee unique form of corrections, though.

Example 4. Consider the transformation of the word ab into ba. It is achieved by any of the following corrections, all of them being weakly normalised: $\langle ab/_0 ba \rangle$, $\langle a/_0 b, b/_1 a \rangle$, $\langle \varepsilon/_0 b, b/_1 \varepsilon \rangle$, $\langle a/_0 \varepsilon, \varepsilon/_2 a \rangle$, $\langle ab/_0 \varepsilon, \varepsilon/_2 ba \rangle$, $\langle a/_0 \varepsilon, b/_1 ba \rangle$, $\langle a/_0 ba, b/_1 \varepsilon \rangle$. There also are non-weakly-normalised corrections like $\langle \varepsilon/_1 a, a/_0 \varepsilon \rangle$.

Take an arbitrary replacement operation like $abbac/_4 dbaa$. It can be seen as doing a mixture of replacements and preservations, in this case keeping a b and an a, deleting another a, another b and a c and inserting a d and an

a, or keeping two a's, deleting two b's and a c and inserting a d and a b. We are particularly interested in replacement operations at two ends of the scales, i.e. with no mixtures in this respect. A *tester* is an operation $u/_i v$ where $u = v$, i.e. it replaces nothing and only tests for the occurrence of u at position i. A *sweeper* is a $u/_i v$ such that u and v are letter-disjoint. Hence, it keeps nothing, deletes all of u and inserts all of v. Note that the operation $\varepsilon/_i \varepsilon$ is both a tester and a sweeper, and it is the only one that falls under both categories.

Definition 2. *Let $\rho = \langle \alpha_1, \ldots, \alpha_m \rangle$ be a weakly normalised correction with $\alpha_j = u_j /_{i_j} v_j$ for all $j = 1, \ldots, m$. It is* strongly normalised, *if the following are true.*

1. *For all $j = 1, \ldots, m$: α_j is either a tester or a sweeper, and $u_j v_j = \varepsilon$ only if $j = m$ and, if $m > 1$, $\alpha_{m-1} \, \| \, \alpha_m$.*
2. *For all $j = 1, \ldots, m-1$: if α_j and α_{j+1} are both testers or both sweepers then $\alpha_j \, \| \, \alpha_{j+1}$.*
3. *For all $j = 2, \ldots, m-1$ with α_j being a tester, and $u_j = xa = v_j$ for some $x \in \Sigma^*$, $a \in \Sigma$, one of the following three cases holds:*
   - *$\alpha_{j-1} \, | \, \alpha_j \, | \, \alpha_{j+1}$ and $(u_{j-1} x, v_{j-1} x, a, u_{j+1}, v_{j+1}) = \mathrm{msp}(u_{j-1} u_j u_{j+1}, v_{j-1} v_j v_{j+1})$,*
   - *$\alpha_{j-1} \, \| \, \alpha_j \, | \, \alpha_{j+1}$ and $(x, x, a, u_{j+1}, v_{j+1}) = \mathrm{msp}(u_j u_{j+1}, v_j v_{j+1})$,*
   - *$\alpha_{j-1} \, | \, \alpha_j \, \| \, \alpha_{j+1}$ and $(u_{j-1} x, v_{j-1} x, a, \varepsilon, \varepsilon) = \mathrm{msp}(u_{j-1} u_j, v_{j-1} v_j)$.*

Strong normalisation thus requires every consecutive part of replacements that are not divided by gaps to be alternating between testers and sweepers. Moreover, it replaces subwords along minimal splits from right to left.

We need an explicit definition of normal form here because the term rewriting system $\to$ is not terminating. Take for instance $\rho = \langle ab/_0 ba \rangle$. We have

$$\rho \stackrel{(S)}{\to} \langle \varepsilon/_0 b, a/_0 a, b/_1 \varepsilon \rangle \stackrel{(O)}{\to} \langle \varepsilon/_0 b, ab/_0 a \rangle \stackrel{(\overline{O})}{\to} \rho.$$

It is not hard to see, though, that such inverse effects of the use of rules (S) and (O) or ($\overline{O}$) are the only way to cause nonterminating rewriting paths. Definition 2 then singles out one particular form of those that can be obtained and at the same time determinises the way that the splits with rule (S) need to be done.

Theorem 2 (Strong Normalisation). *For every ρ with nonempty domain there is a strongly normalised ρ' such that $\rho \to^* \rho'$.*

Proof (Sketch). According to Lemma 3, ρ can be assumed to be weakly normalised. It only remains to see how it can then be strongly normalised using rule applications. The condition on empty testers is easily satisfied using rule (ε). The trick is then to divide ρ into maximal subsequences of operations with no gaps between them, called blocks, and to rewrite these blocks successively from right to left.

Rule (O) can be used to transform each block into a single replacement operation $\beta = u/_i v$. If β is a tester or a sweeper we are finished and can proceed with the next block. Otherwise u and v are neither equal nor letter-disjoint, whence there is a minimal split $(u, u', av, v') = msp(u, v)$. In particular, u' and v' are letter-disjoint. Split β accordingly using rule (S). This splits off a (possibly empty) sweeper and leaves a smaller block to continue with. □

Our next goal is to show that the rewriting rules in Fig. 1 can be used to prove equivalence by rewriting.

Lemma 4. *Let ρ, ρ' be strongly normalised. If $\rho \neq \rho'$ then $\rho \not\equiv \rho'$.*

We write $\rho^\downarrow$ for the normalisation of ρ which, according to this, is necessarily unique. Completeness is then a rather immediate consequence of the fact that each correction can be strongly normalised. So given two equivalent ones, their normalisations need to be equal according to Lemma 4, and this provides a correction into which both can be rewritten.

Theorem 3 (Completeness). *For all ρ, σ: if $\sigma \equiv \rho$ then $\sigma \leftrightarrow^* \rho$.*

Proof. Let σ, ρ be given s.t. $\sigma \equiv \rho$. According to Theorem 2, both have a strong normalisation σ', resp. ρ', i.e. we have $\sigma \rightarrow^* \sigma'$ and $\rho \rightarrow^* \rho'$.

Now assume that $\rho' \neq \sigma'$. According to Lemma 4 we would have $\sigma' \not\equiv \rho'$, and therefore, with Theorem 1, $\rho \equiv \rho' \not\equiv \sigma' \equiv \sigma$ contradicting the assumption that $\sigma \equiv \rho$. Hence, we must have $\sigma' = \rho'$, but then we get $\sigma \rightarrow^* \sigma' = \rho' \leftarrow^* \rho$ and therefore $\sigma \leftrightarrow^* \rho$. □

4 Minimal Corrections

We now turn to the following problem: given $w \in \Sigma^*$ and $L \subseteq \Sigma^*$, find *all* ρ such that $\rho(w) \in L$. Clearly the number of such corrections is always infinite unless $L = \emptyset$, so restrictions are needed in general in order to make the problem well-defined. We propose a notion of minimality in corrections that is not just the edit distance between source and target word. Instead, we intuitively consider a correction minimal (for w and L) if, by executing its operations in an arbitrary order, the target language is only reached once the entire correction is completely applied. The study of corrections done in the previous section, in particular the role of normalisation, is useful in making this notion formal.

Note that here we are only interested in corrections operating on a given w. Consequently, corrections containing testers should not be considered to be minimal since leaving them out achieves the same effect. Also note that the format of replacements $u/_i v$ with words u, v instead of letters makes it more difficult to formally capture the above intuition of executing only parts of the operations. For instance, $a/_2 b$ should be considered as doing only parts of what $a/_2 bc$ does. We need to break up sweepers for the purpose of such comparisons.

Definition 3. *Replacements of the form $\varepsilon/_i a$ resp. $a/_i \varepsilon$ for $a \in \Sigma$ are called* insertions *resp.* deletions *and are abbreviated as $\nearrow_i a$ resp. $a\searrow_i$. A correction is* primitive *if it is a sequence of insertions and deletions. A primitive correction ρ represents the operation $\alpha = u/_i v$ if $\rho(wuw') = \alpha(wuw')$ for all $w, w' \in \Sigma^*$ with $|w| = i$, and ρ contains exactly $|u|$ many deletions and $|v|$ many insertions.*

The notion of representation generalises in the obvious way to corrections containing several replacements.

Lemma 5. *Every normalised correction is represented by a primitive correction.*

Primitive corrections can then be used to define minimality of those corrections that they represent. For $u, v \in \Sigma^*$ we write $u \preceq v$ if u is a subsequence of v, and $u\!\downarrow_v$ for the projection of u onto letters occurring in v.

Definition 4. *Let $\alpha = u/_i v$ and $\beta = u/_i w$ such that α is a sweeper. We write $\beta \trianglelefteq \alpha$, if $w\!\downarrow_u \preceq u$ and $w\!\downarrow_v \preceq v$.*

Intuitively, $\beta \trianglelefteq \alpha$ if β deletes at most as much, for instance by keeping some letters ($w\!\downarrow_u \preceq u$), or it inserts as most as much ($w\!\downarrow_v \preceq v$) as α.

Example 5. Let $\alpha = abab/_0 cdcd$, $\beta_1 = abab/_0 abadcd$, $\beta_2 = abab/_0 dcabab$, $\beta_3 = abab/_0 ab$ and $\beta_4 = abab/_0 c$. We have $\beta_i \trianglelefteq \alpha$ for $i \in \{1, \ldots, 4\}$. Consider

$$\alpha' = \langle \mathord{\uparrow}_0 c, \mathord{\uparrow}_0 d, \mathord{\uparrow}_0 c, \mathord{\uparrow}_0 d, a\!\downarrow_0, b\!\downarrow_1, a\!\downarrow_2, b\!\downarrow_3 \rangle,$$
$$\alpha'' = \langle a\!\downarrow_0, b\!\downarrow_1, a\!\downarrow_2, b\!\downarrow_3, \mathord{\uparrow}_4 c, \mathord{\uparrow}_4 d, \mathord{\uparrow}_4 c, \mathord{\uparrow}_4 d \rangle.$$

Then α is represented by both α' and α'', and it is not hard to find subsequences of α' representing β_2, β_3 and β_4, or of α'' representing $\beta_1, \beta_3, \beta_4$.

We now lift $\trianglelefteq$ to corrections. Let $\alpha = u/_i v$, $\beta = u'/_{i'} v'$ such that α is a sweeper. Then β *fits into* α if $i \leq i'$ and $i' + |u'| \leq i + |u|$ and $u(i' + j) = u'(i' + j)$ for all $0 \leq j < |u'|$.

Definition 5. *Let $u = x_0 u_0 x_1 u_1 \cdots x_k u_k x_{k+1}$, $\alpha = u/_i v$ be a sweeper, and $\rho = \langle \beta_0, \ldots, \beta_k \rangle$ consist of sweepers with $\beta_j = u_j/_{i_j} v_j$ such that $i_0 \geq i$, $i_k + |u_k| \leq i + |u|$ and, for all $j < k$, we have $i_j + |u_j| < i_{j+1}$. The (u, i)-padding of ρ is the replacement $u/_i x_0 v_0 x_1 v_1 \ldots v_k x_{k+1}$ (which is not a sweeper in general).*

Definition 6. *Let $\rho = \langle \beta_0, \ldots, \beta_k \rangle$ and $\sigma = \langle \alpha_0, \ldots, \alpha_m \rangle$ be (strongly) normalised corrections. We have $\rho \trianglelefteq \sigma$ if the following hold:*

1. *for all $0 \leq j \leq k$, β_j fits into some $\alpha_{j'}$,*
2. *if $\beta_j, \ldots, \beta_{j+k'}$ is the set of sweepers that fit into $\alpha_{j'} = u/_i v$, and if β is the (u, i)-padding of $\beta_j, \ldots, \beta_{j+k'}$, then $\beta \trianglelefteq \alpha$.*

Example 6. Consider the strongly normalised corrections $\sigma = \langle aba/_0 cc, ab/_4 d \rangle$ and $\rho = \langle a/_0 c, a/_2 \varepsilon, ab/_4 d \rangle$. Then $\rho \trianglelefteq \sigma$ because $a/_0 c, a/_2 \varepsilon$ fits into $aba/_0 cc$, and $ab/_4 d$ fits into $ab/_4 d$. Moreover, $aba/_0 cb$ is the $(aba, 0)$-padding of $\langle a/_0 c, a/_2 \varepsilon \rangle$, and we have $aba/_0 cb \trianglelefteq aba/_0 cc$ according to Definition 5. The need for padding is apparent due to the mismatch in domains of the replacements. On the other hand, σ is represented by the primitive correction $\langle \mathord{\uparrow}_0 c, a\!\downarrow_0, b\!\downarrow_1, a\!\downarrow_2, \mathord{\uparrow}_3 c, \mathord{\uparrow}_4 d, a\!\downarrow_4, b\!\downarrow_5 \rangle$ while ρ is represented, among others, by $\langle \mathord{\uparrow}_0 c, a\!\downarrow_0, a\!\downarrow_2, \mathord{\uparrow}_4 d, a\!\downarrow_4, b\!\downarrow_5 \rangle$ which is clearly a subsequence.

Lemma 6. *Deciding whether $\sigma \trianglelefteq \rho$ for two strongly normalised corrections σ, ρ can be done in time $\mathcal{O}(|\rho| + |\sigma|)$.*

Definition 7. *Let $L \subseteq \Sigma^*, w \in \Sigma$. A testerless, strongly normalised correction ρ for w, L is* minimal *if there is no strongly normalised correction $\sigma \neq \rho$ with $\sigma \trianglelefteq \rho$ and $\sigma(w) \in L$.*

We write $MinCorr(w, L)$ for the set of minimal corrections for w and L. The problem ALLCORRECT then is: given w, L, compute $MinCorr(w, L)$.

Theorem 4. *Let $w \in \Sigma^*$, $L \subseteq \Sigma^*$. Then $|MinCorr(w, L)| < \infty$.*

Proof (Sketch). This follows from an application of Higman's Lemma [7]. Normalisation, in particular the application of operations from right to left in a word, guarantees that corrections themselves are words over a finite alphabet (of elementary insertion, deletion and replacement operations), and minimality as corrections essentially translated into minimality with respect to the subsequence order. Then Higman's Lemma states that amongst infinitely many corrections there are only finitely many minimal ones. □

We include another example which reveals some perhaps unexpected phenomenon occurring with this notion of minimality.

Example 7. Let $L = \{aba\}$ and $w = \{aa\}$. One may assume that there is only one minimal correction for w and L, namely $\rho_0 = \langle \nearrow_1 b \rangle$. There are others, though, and the reason is the double occurrence of a in w. Note that ρ_0 leaves both a's untouched, so it essentially matches the first a in w with the first in $\rho_0(w)$ and likewise for the second.

Consider $\rho_1 = \langle a/_0 \varepsilon, \varepsilon/_2 ba \rangle$. It is also minimal, even though it also contains the insertion of b. However, this happens at a different position: while ρ_0 inserts b *between* the two existing a's, ρ_1 inserts it *behind* them, and this is why we have $\rho_0 \not\trianglelefteq \rho_1$. Intuitively, ρ_1 matches the second a in w with the first one in aba, and then the insertion of b at position 2 (in the context of ba) appears to be different to the insertion of b at position 1 in ρ_0.

We now turn to the problem of solving ALLCORRECT algorithmically. Decidability of ALLCORRECT depends on the underlying language class. We first give a lower bound.

Theorem 5. *ALLCORRECT is computationally at least as hard for a class $\mathcal{C} \subseteq 2^{\Sigma^*}$ as the emptiness problem for $\mathcal{C}$.*

Proof. Suppose algorithm $A(w, L)$ solves ALLCORRECT over $\mathcal{C}$. Pick an arbitrary word $w_0 \in \Sigma^*$ and consider algorithm $B(L) := A(w_0, L)$. Since there is no correction for w_0 and L if $L = \emptyset$, and there is some correction for w and L if $L \neq \emptyset$, $B(L)$ produces some correction iff $L \neq \emptyset$. □

Moreover, it is possible to characterise those language classes for which ALLCORRECT is indeed decidable.

Theorem 6. *Let $\mathcal{L}$ be a class of languages. Then* ALLCORRECT *is computable for $\mathcal{L}$ if the following hold.*

- *the emptiness problem for $\mathcal{L}$ is decidable, and*
- *$\mathcal{L}$ is closed under intersections with regular languages.*

Proof (Sketch). Let $w \in \Sigma^*$ and $L \in \mathcal{L}$ be given. First note that there are only finitely many subsequences $v_0, \ldots, v_k$ of w, and each can be seen as being obtained by deleting letters from w. Note that a minimal correction can be regarded as doing some deletions and then some insertions between the letters that are left over. We can then check, for each such $v_i = a_1 a_2 \ldots a_{m_i}$, whether it is possible to form a word in L from v_i by insertions only which boils down to checking whether $L \cap \Sigma^* a_1 \Sigma^* a_2 \Sigma^* \ldots \Sigma^* a_{m_i} \Sigma^* \neq \emptyset$. By doing these checks for enumerated words from Σ^* one can even guarantee to obtain shortest witnesses for nonemptiness. This then yields a finite number of corrections which can mutually be compared in order to filter out non-minimal ones. □

Since emptiness is decidable for the class of CFL context-free languages and already undecidable for its intersection closure, CFL is a reasonable candidate for devising a concrete algorithmic solution for ALLCORRECT. The one given in the following is an extension of the well-known CYK parser [6,10,14].

To briefly recall, a context-free grammar (CFG) $G = (N, \Sigma, \to, S)$ can be assumed to be in Chomsky Normal Form (CNF) [5] meaning that all rules are of the form $A \to BC$ or $A \to a$ for nonterminals A, B, C, and $a \in \Sigma$, cf. [8]. We exclude the case of an additional rule $S \to \varepsilon$ here to keep the presentation short. We write $\alpha \Rightarrow^* \beta$ if the sentential form β is derivable from α, as usual. Given a nonterminal A, we write $L(A)$ for the language $\{w \in \Sigma^* \mid A \Rightarrow^* w\}$.

Algorithm 1 takes a word $w = a_0 \ldots a_{n-1}$ and a CFG $G = (N, \Sigma, \to, S)$ in CNF. It maintains a two-dimensional table $\mathcal{T}$ whose entries at position (i, j) with $0 \leq i \leq j < n$ are sets of pairs (A, ρ) s.t. $\rho \in \mathit{MinCorr}(a_i \ldots a_{j-1}, L(A))$, here built in primitive form (for better readability). It extends CYK in the sense that, when $w \in L(G)$, then all computed corrections are $\langle \rangle$.

A key ingredient in Algorithm 1 and conceptual difference to an ordinary CYK-based parser is procedure CLOSE which takes indices i, j and computes a closure of table entry $\mathcal{T}(i, j)$. It assumes that table entries $\mathcal{T}(i, i)$ and $\mathcal{T}(j, j)$ contain corrections (A, ρ) s.t. $A \Rightarrow^* \rho(\varepsilon)$. Note that ρ necessarily needs to be represented by insertions only, for otherwise it would not be normalised. Moreover, if $A \Rightarrow BC$, $B \Rightarrow \sigma(u)$ and $C \Rightarrow^* \rho(v)$ then $A \Rightarrow^* \langle \sigma, \rho \rangle^\downarrow(uv)$ This is also true if $v = \varepsilon$ or $u = \varepsilon$. This is the reason why binary productions of the form $A \to BC$ do not necessarily only derive longer subwords, as it is the case in ordinary CYK-parsing on grammars in CNF. At last, procedure CLOSE is a saturation scheme that maintains a set of corrections (w.r.t. each nonterminal) that are minimal within this set. Any newly obtained correction is compared to already existing ones w.r.t. $\trianglelefteq$. Only the minimal ones survive, i.e. a new one can replace an old one, and it can later get replaced by yet another one.

Theorem 7. *Algorithm 1 computes $\mathit{MinCorr}(w, L(G))$ for a CFG G in CNF, and a word w.*

Algorithm 1. Computing all minimal corrections for a word and a context-free language.

```
 1: procedure MinCorrs(w = a₀ ... aₙ₋₁, G = (N, Σ, →, S))
 2:     T(0,0) ← {(A, ⟨/₀a⟩) | A → a}
 3:     Close(0,0)
 4:     for i = 1, ..., n do
 5:         T(i,i) ← {(A, shftᵢ(ρ)) | (A, ρ) ∈ T(0,0)}
 6:     for i = 0, ..., n − 1 do
 7:         T(i,i+1) ← {(A, ⟨⟩) | A → aᵢ} ∪ {(A, ⟨rplᵢ(aᵢ,b)⟩) | A → b}
 8:         Close(i, i+1)
 9:     for ℓ = 2, ..., n do
10:         for i = 0, ..., n − ℓ do
11:             j ← i + ℓ
12:             T(i,j) ← {(A, ⟨ρ, aⱼ₋₁/ⱼ₋₁⟩↓) | (A, ρ) ∈ T(i, j−1)}
13:                 ∪ {(A, ⟨aᵢ/ᵢ, ρ⟩↓) | (A, ρ) ∈ T(i+1, j)}
14:                 ∪ {(A, ⟨σ, ρ⟩↓) | A → BC, i ≤ h < j, (B, σ) ∈ T(i,h), (C, ρ) ∈ T(h,j)}
15:             Close(i, j)
16:     return {ρ | (S, ρ) ∈ T(0,n)}
17: procedure Close(i, j)
18:     repeat
19:         N ← {(A, ⟨σ, ρ⟩↓) | A → BC, (B, σ) ∈ T(i,j), (C, ρ) ∈ T(j,j)}
20:             ∪ {(A, ⟨σ, ρ⟩↓) | A → BC, (B, σ) ∈ T(i,i), (C, ρ) ∈ T(i,j)}
21:         T(i,j) ← {(A, ρ) ∈ T(i,j) ∪ N | ∄σ : σ ≺ ρ, (A, σ) ∈ T(i,j) ∪ N}
22:     until T(i,j) stabilises
```

Proof (Sketch). One shows, by induction on the length of subwords, that Algorithm 1 computes all minimal corrections for the languages of each nonterminal and this subword, and that minimal corrections for longer words can in general be obtained from minimal corrections for its subwords. □

We remark that the cubic dependence on the word length is not anymore the dominating factor in the algorithm's runtime. Let G_m be the standard CFG for $L_m := \{a^k \mid k \geq 2^m\}$ of size $\mathcal{O}(m)$. Consider $w_n = a^n$ for some $n < 2^m$. Then $\mathit{MinCorr}(w_n, L_m) = \{\langle \varepsilon /_i a^{2^m - n} \rangle \mid 0 \leq i \leq n\}$. Each minimal correction is of size $2^{\mathcal{O}(m)}$, and their explicit construction will take such time.

5 Empirical Evaluation

We provide an implementation of Algorithm 1 written in OCaml and publicly available via a Gitlab server.[1] It features a number of example grammars from arithmetical expressions to fragments of natural language. It also has some extensions that are not presented here in detail due to space limitations, for instance for filtering corrections more aggressively than just w.r.t. minimality.

---

[1] https://cumbernauld.tifm.cs.uni-kassel.de/mlange/minimal-corrections-cyk.

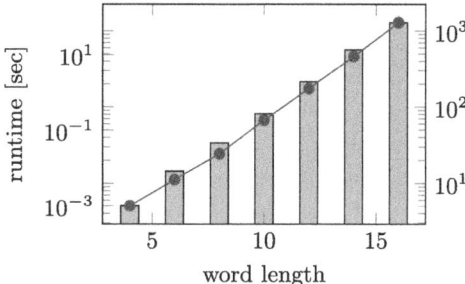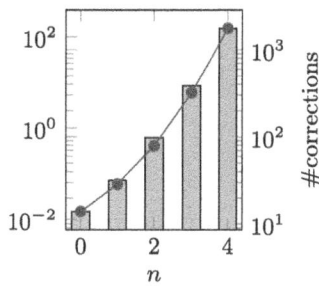

Fig. 2. Running times (line plot with bullet points, left y-axis) and numbers of minimal corrections (vertical bar plot, right y-axis) for $L_1 = \{a^n b^n \mid n \geq 1\}$ and random words of length $2n$ (left), and L_2 and words w_n as described below (right).

We present empirical data obtained on two benchmark CFLs. The first one is the standard example $L_1 = \{a^n b^n \mid n \geq 1\}$. Tests are created as follows. Given $n \geq 1$, let $w_n = a^n b^n$ and w'_n result from it by successively applying an arbitrary permutation of a letter in the first half with one in the second half. The number of such permutations is determined randomly by a geometric probability distribution with expected value 5. Each test is run 20 times for each n. Figure 2 (left) shows the average running times to compute $MinCorr(w'_n, L_1)$ as well as its average size, plotted against the word length $2n$. The tests were run on a MacBook Air with 16 GB RAM and CPU speed up to 3.5 GHz. The plots, drawn on a logarithmic scale each, clearly shows that the runtime is exponential. They also indicate that the number of minimal corrections, which is growing exponentially in the word length here, is the factor dominating the runtime.

The second benchmark is used to examine the effect of obtaining several minimal corrections when the source word contains repeated parts, as it is the case with the linguistic example mentioned in the introduction. Let

$$L_2 := \{ \textit{if I am very}^n \textit{ happy then I am very}^m \textit{ rich} \mid n \leq m \}$$
$$\cup \{ \textit{I am very}^n \textit{ happy because I am very}^m \textit{ rich} \mid n \geq m \}.$$

and $w_n = $ "*I am very^n happy I am very^n rich*" for $n \geq 0$. The number of corrections computed for L_2 and w_n as well as the runtime needed for this is shown in Fig. 2 on the right. The perhaps unexpectedly high numbers of minimal corrections for this seemingly simple example can be explained as follows. $MinCorr(w_n, L_2)$ can be divided into two equal-sized halves – inserting "*because*", resp. "*if*" and "*then*". Each half contains one minimal correction that does exactly that and intuitively maintains both substrings "*I am very^n*". The others match parts of the first occurrence of this substring to parts of the other, for instance forming "*I am*" from the first "*I*" and the second "*am*", etc.

6 Conclusion

We extended the problem of computing *some* (minimal) correction for a word to hit a target language, also known as error-correcting parsing, to the problem of computing *all* such corrections. We also presented a suggestion of a notion of minimality that captures the intuitive notion of executing a shortest path from the input to the target language that is invariant under reorderings of the atomic operations in the correction. The number of minimal corrections in this sense is always guaranteed to be finite. We also showed how it can be computed when the target language is context-free.

The motivation for the development presented here comes from an attempt to digitalise interactive learning environments. One can regard the target language L as (word encodings of) a concept to be learned, and the source word w as an attempt to create an example of that concept. The set $MinCorr(w,l)$ then represents recoveries for possible errors on a syntactical level that the learner has made in trying to create examples of that concept. A learning tool can then use that information to create individualised feedback.

On the other hand, one can also regard this work from a purely theoretical point of view as a decidability result about a certain reachability problem in automatic structures [4], something that is undecidable in general. Note that the graph of nodes from Σ^* and edges formed by edit operations is infinite but automatic, and ALLCORRECT simply asks for the computation of all the most direct ways to get from one node to a context-free target area. It remains to be seen whether this has applications in infinite-state model checking, perhaps for branching-time logics.

There are three directions in which this work should be extended. From a formal-language point of view, it remains to be seen whether ALLCORRECT's decidability can be extended to richer classes of formal languages. One can prove that it is decidable for a class $\mathcal{C}$ iff $\mathcal{C}$'s emptiness problem is decidable and $\mathcal{C}$ is effectively closed under intersections with piecewise testable languages, so indexed languages are a natural candidate for such an extension. This then also leads to the second strand of extension in the educational domain: indexed languages [1], as opposed to context-free ones, allow certain semantical concepts to be syntactified. For instance, while the set of Boolean expressions is context-free, the set of satisfiable Boolean expressions is not but it is an indexed language. Such an extension could then be used in a learning tool to help understand such semantic concepts.

At last, but related to the motivation of its use in an educational context, ALLCORRECT does not provide any deeper structural information about the presence of mistakes w.r.t. a word and a target language that could be used to classify mistakes. However, the output could be used to train classifiers that can then exhibit certain patterns of mistakes. Again, it remains to be seen if such a continuation of this work in a machine-learning direction will bring out further benefits of being able to automatically compute all minimal ways to correct a word w.r.t. target language.

References

1. Aho, A.V.: Indexed grammars - an extension of context-free grammars. J. ACM **15**(4), 647–671 (1968)
2. Aho, A.V., Peterson, T.G.: A minimum distance error correcting parser for context-free languages. SIAM J. Comput. **1**(4), 305–312 (1972). https://doi.org/10.1137/0201022
3. Baader, F., Nipkow, T.: Term Rewriting and All That. Cambridge University Press, Cambridge (1998). https://doi.org/10.1017/CBO9781139172752
4. Blumensath, A., Grädel, E.: Automatic structures. In: Proceedings 15th Symposium on Logic in Computer Science, LICS 2000, pp. 51–62. IEEE (2000). https://doi.org/10.1109/LICS.2000.855755
5. Chomsky, N.: On certain formal properties of grammars. Inf. Control **2**(2), 137–167 (1959). https://doi.org/10.1016/S0019-9958(59)90362-6
6. Cocke, J., Schwartz, J.T.: Programming Languages and Their Compilers. Courant Instituite of Mathematical Sciences, New York (1970)
7. Higman, G.: Ordering by divisibility in abstract algebras. Proc. Lond. Math. Soc. **s3-2**(1), 326–336 (1952). https://doi.org/10.1112/plms/s3-2.1.326
8. Hopcroft, J.E., Motwani, R., Ullman, J.D.: Introduction to Automata Theory, Languages, and Computation, 3rd edn. Pearson (2013)
9. Irons, E.T.: An error-correcting parse algorithm. Commun. ACM **6**(11), 669–673 (1963). https://doi.org/10.1145/368310.368385
10. Kasami, T.: An efficient recognition and syntax analysis algorithm for context-free languages. Technical report, AFCRL-65-758, Air Force Cambridge Research Lab. (1965)
11. Levenshtein, V.I.: Binary codes capable of correcting deletions, insertions and reversals. Sov. Phys. Dokl. **10**, 707–710 (1966)
12. Myers, G.: Approximately matching context-free languages. Inf. Process. Lett. **54**(2), 85–92 (1995). https://doi.org/10.1016/0020-0190(95)00007-Y
13. Rajasekaran, S., Nicolae, M.: An error correcting parser for context free grammars that takes less than cubic time. In: Dediu, A.-H., Janoušek, J., Martín-Vide, C., Truthe, B. (eds.) LATA 2016. LNCS, vol. 9618, pp. 533–546. Springer, Cham (2016). https://doi.org/10.1007/978-3-319-30000-9_41
14. Younger, D.H.: Recognition and parsing of context-free languages in time n^3. Inf. Control **10**(2), 372–375 (1967). https://doi.org/10.1016/s0019-9958(67)80007-x

On Solving All-Path Reachability Problems for Starvation Freedom of Concurrent Rewrite Systems Under Process Fairness

Misaki Kojima[✉][iD] and Naoki Nishida[iD]

Graduate School of Informatics, Nagoya University,
Furo-cho, Chikusa-ku, Nagoya 4648601, Japan
k-misaki@nagoya-u.jp, nishida@i.nagoya-u.ac.jp

Abstract. Logically constrained term rewrite systems are useful models of not only sequential but also concurrent programs. We have proposed a framework to soundly reduce starvation freedom of a concurrent program to an all-path reachability problem of the corresponding logically constrained term rewrite system, where process fairness is not considered. In this paper, we show a disproof criterion for starvation freedom and then extend the framework to starvation freedom under process fairness.

Keywords: Constrained rewriting · Runtime-error verification · Cyclic proof

1 Introduction

Recently, approaches to program verification by means of *logically constrained term rewrite systems* (LCTRSs, for short) [17] are well investigated [5,6,8–10,20,27]. LCTRSs are useful as computation models of not only functional but also imperative programs. For instance, equivalence checking by means of LCTRSs is useful to ensure the correctness of terminating functions (cf. [8]). The method of transforming sequential programs into LCTRSs has been extended to concurrent programs with semaphore-based exclusive control [16]. In applying verification techniques for LCTRSs to the verification of practical programs such as automotive embedded systems, as for equivalence verification, verification of the non-occurrence of a specified runtime error in a given system is an important task for, e.g., developing concurrent systems.

An *all-path reachability problem* (APR problem, for short) [6,25,26] of a transition system is a pair $P \Rightarrow Q$ of state sets P, Q and is *demonically valid*

M. Kojima—JSPS Research Fellow.
This work was partially supported by JSPS KAKENHI Grant Numbers 18K11160 and 24K02900, Grant-in-Aid for JSPS Fellows Grant Number 24KJ1240, and DENSO Corporation.

if every finite execution path—a transition sequence starting with a state in P and ending with a terminating state—includes a state in Q. The problem is called *constant-directed* if Q is a singleton set of an irreducible constant [13,14]. A framework to reduce the non-occurrence of a specified runtime error in a concurrent program to a constant-directed APR problem of the corresponding LCTRS has been proposed [12,14]. For a concurrent program with exclusive control, the framework can reduce, e.g., the race freedom of mutual exclusion to a constant-directed APR problem of the corresponding LCTRS. The proof system c-DCC for constant-directed APR problems, which is a restricted variant of the proof system DCC [6], has been proposed [14]. Proofs and disproofs for constant-directed APR problems have been formulated in the *cyclic proof* style [4] as *APR proofs* and *APR disproofs*, respectively [15].

Starvation for a process w.r.t. its critical section, which is a typical *liveness* property, is a situation where the process wants to enter but never enters the critical section. Note that such starvation is called *individual starvation* [3]. To apply the framework to the verification of *starvation freedom*, a modification of the LCTRS obtained from a concurrent program by the aforementioned transformation [16] has been proposed in [13]: A counter for the waiting time of the process is introduced as a new argument into the function symbol representing configurations; a positive integer $c_{\max}$ used as an upper limit of the waiting time is given by a user, and both states where the counter exceeds the given positive integer and terminating states where the process is waiting are considered error states; the framework reduces the starvation freedom for the process to a constant-directed APR problem of the modified LCTRS w.r.t. the error states.

The reduction is sound but not refutationally complete because error states depend on the *tentative* upper limit $c_{\max}$ for the waiting time and the upper limit may be too small to ensure starvation freedom, while the use of the upper limit is practical in some cases such as real-time systems. For some APR problems that are not demonically valid, we may obtain APR disproofs. However, due to refutational incompleteness, we cannot know the cause of the disproof, e.g., $c_{\max}$ is too small or starvation indeed happens. For this reason, we have a research question (RQ1) how to find a witness of starvation from the obtained disproof.

On the other hand, it is usual to assume *process fairness*—every process that is enabled infinitely often gets its turn infinitely often [3]. However, the aforementioned framework [13] does not assume process fairness.

Example 1.1. Consider an *asynchronous integer transition system* (AITS, for short) [3] consisting of two process P_1, P_2 represented by the program graph in Fig. 1 [3, Example 2.24], where P_1 and P_2 share a binary semaphore y which is a lock to access the critical sections crit_1 and crit_2 (see [14] for details). The LCTRS $\mathcal{R}_1$ for the AITS consisting of PG_1 and PG_2 is illustrated in Fig. 2, where $\mathcal{R}_1$ is over $\Sigma_1 = \{\text{state} : loc \times loc \times int \to state, \text{noncrit}_1, \text{wait}_1, \text{crit}_1, \text{noncrit}_2, \text{wait}_2, \text{crit}_2 : loc\}$ [14]. The initial state is the term $\text{state}(\text{noncrit}_1, \text{noncrit}_2, 1)$. To reduce starvation freedom for process P_1 to a constant-directed APR problem, a counter c for the waiting time of P_1 is introduced and, given a positive integer $c_{\max}$ as a tentative upper limit of the waiting time (i.e., the counter), the LCTRS $\mathcal{R}_1^\forall$ over $\Sigma_1' = \{\text{state} : loc \times loc \times int \times int \to state,$

```
      ──→ noncrit_i ⇄ wait_i ──y>0:y:=y−1──→ crit_i
                      y:=y+1
```

Fig. 1. Program graph PG_i of process P_i for mutual exclusion with semaphore y.

$$\mathcal{R}_1 = \left\{ \begin{array}{l} \mathsf{state}(\mathsf{noncrit}_1, p_2, y) \to \mathsf{state}(\mathsf{wait}_1, p_2, y) \\ \mathsf{state}(\mathsf{wait}_1, p_2, y) \to \mathsf{state}(\mathsf{crit}_1, p_2, y') \quad [\, y > 0 \wedge y' = y - 1 \,] \\ \mathsf{state}(\mathsf{crit}_1, p_2, y) \to \mathsf{state}(\mathsf{noncrit}_1, p_2, y') \, [\, y' = y + 1 \,] \\ \mathsf{state}(p_1, \mathsf{noncrit}_2, y) \to \mathsf{state}(p_1, \mathsf{wait}_2, y) \\ \mathsf{state}(p_1, \mathsf{wait}_2, y) \to \mathsf{state}(p_1, \mathsf{crit}_2, y') \quad [\, y > 0 \wedge y' = y - 1 \,] \\ \mathsf{state}(p_1, \mathsf{crit}_2, y) \to \mathsf{state}(p_1, \mathsf{noncrit}_2, y') \, [\, y' = y + 1 \,] \end{array} \right\}$$

Fig. 2. LCTRS $\mathcal{R}_1$ for the AITS consisting of PG_1 and PG_2 in Fig. 1.

$\mathsf{noncrit}_1, \mathsf{wait}_1, \mathsf{crit}_1, \mathsf{noncrit}_2, \mathsf{wait}_2, \mathsf{crit}_2 : loc\}$ in Fig. 3 is generated, where the error states are those with $c \geq \mathsf{c}_{\max}$ [13]. If the constant-directed APR problem $\langle \mathsf{state}(\mathsf{noncrit}_1, \mathsf{noncrit}_2, 1, 0) \mid \mathsf{true} \rangle \Rightarrow \mathsf{success}$ is demonically valid, then starvation freedom for process P_1 is ensured. Unfortunately, the APR problem is not demonically valid. Indeed, the AITS causes starvation for process P_1, and it is not demonically valid for any positive integer $\mathsf{c}_{\max}$, while the AITS is starvation free for P_1 under process fairness.

Now, we have a research question (RQ2) how to drop process-unfair paths from APR (dis)proofs in order to prove starvation freedom under process fairness.

In this paper, we aim at developing an APR-based framework for the verification of starvation freedom under process fairness. To this end, for *concurrent LCTRSs* which are LCTRSs obtained from concurrent programs, we first show a disproof criterion for starvation freedom focusing on a path that reaches an error state (Sect. 4). Then, we extend the APR framework in [13] to starvation freedom under process fairness (Sect. 5). In the extended framework, in constructing an APR (dis)proof, if we find a path satisfying the disproof criterion and causing a process-unfair execution path, then we ignore the end node of the path. In recalling the APR framework for starvation freedom in [13], we improve the modification of a given concurrent LCTRS in order to avoid redundant paths starting from nodes for error states (Sect. 3).

The framework is applicable to LCTRSs obtained from, e.g., C-like concurrent programs with semaphore-based exclusive control [16], which are written in a concurrent extension of SIMP [7,9]. As in [13], to simplify discussions and to concentrate on the essence of the research questions, this paper deals with LCTRSs obtained from concurrent programs such as AITSs for mutual exclusion, which are usually transformed into simpler LCTRSs than those obtained from concurrent SIMP programs. In addition, we assume that LCTRSs are deadlock free w.r.t. terms representing the initial state of the corresponding programs, i.e., there is no ground normal form that represents a state reachable from the initial state. The treatment of LCTRSs that may have deadlock can be seen in [13].

$$\mathcal{R}_1^\forall = \left\{\begin{array}{ll} \mathsf{state}(\mathsf{noncrit}_1, p_2, y, c) \to \mathsf{state}(\mathsf{wait}_1, p_2, y, 1) & \\ \mathsf{state}(\mathsf{wait}_1, p_2, y, c) \to \mathsf{state}(\mathsf{crit}_1, p_2, y', 0) & [\,y{>}0 \wedge y'{=}y{-}1\,] \\ \mathsf{state}(\mathsf{crit}_1, p_2, y, c) \to \mathsf{state}(\mathsf{noncrit}_1, p_2, y', c) & [\,y'{=}y{+}1\,] \\ \mathsf{state}(p_1, \mathsf{noncrit}_2, y, c) \to \mathsf{state}(p_1, \mathsf{wait}_2, y, c') & [\,c'{=}\mathsf{inc}_{\geq 1}(c)\,] \\ \mathsf{state}(p_1, \mathsf{wait}_2, y, c) \to \mathsf{state}(p_1, \mathsf{crit}_2, y', c') & [\,y{>}0 \wedge y'{=}y{-}1 \wedge c'{=}\mathsf{inc}_{\geq 1}(c)\,] \\ \mathsf{state}(p_1, \mathsf{crit}_2, y, c) \to \mathsf{state}(p_1, \mathsf{noncrit}_2, y', c') & [\,y'{=}y{+}1 \wedge c'{=}\mathsf{inc}_{\geq 1}(c)\,] \\ \mathsf{state}(p_1, p_2, y, c) \to \mathsf{success} & \\ \mathsf{state}(p_1, p_2, y, c) \to \mathsf{error} & [\,c{>}c_{\max}\,] \end{array}\right\}$$

Fig. 3. LCTRS $\mathcal{R}_1^\forall$ obtained from $\mathcal{R}_1$ by the method in [13] (see Definition 3.1).

The main contributions of this paper are to improve the sound reduction of starvation freedom to APR problems, to show a disproof criterion for starvation by means of APR disproofs, and to propose a method to take process fairness into account in the framework of APR problems for starvation freedom.

Related Work. The $\mathbb{K}$ framework [24] is a more general setting of rewriting than LCTRSs, and the proof system for all-path reachability has been implemented in $\mathbb{K}$ [25,26]. The race freedom of Peterson's mutual exclusion algorithm [22] has been proved in [26] by means of the APR approach, while, to the best of our knowledge, neither starvation freedom nor process fairness has been considered. Unlike our framework, for the algorithm in [26], any infinite execution path does not have to be taken into account. A comparison of all-path reachability logic with CTL* can be seen in [26]. *Model checking* is the most well-known and well-investigated method for verification of safety and liveness properties such as race freedom and starvation freedom. Some model checkers have a function to consider fairness, NuSMV[1] [1] can declare `FAIRNESS`. Our proof method for APR problems is based on coinduction on rewriting steps of constrained terms. This enables us to deal with infinite-state systems to be verified. On the other hand, the implementation based on the APR framework is inefficient compared with model checkers because the case analysis on symbolic execution with SMT solving is very expensive. For this reason, our method is not scalable enough, while our approach can deal with more concrete models of programs.

2 Preliminaries

In this section, we briefly recall LCTRSs [8,17] and all-path reachability [6,25,26]. Familiarity with basic notions on term rewriting [2,21] is assumed.

2.1 Logically Constrained Term Rewrite Systems

To define an LCTRS [8,17] over an $\mathcal{S}$-sorted signature Σ, we consider the following sorts, signatures, mappings, and constants: *Theory sorts* in $\mathcal{S}_{theory}$ and *term*

---
[1] https://nusmv.fbk.eu.

sorts in $\mathcal{S}_{term}$ such that $\mathcal{S} = \mathcal{S}_{theory} \uplus \mathcal{S}_{term}$; a *theory signature* Σ_{theory} and a *term signature* Σ_{terms} such that $\Sigma = \Sigma_{theory} \cup \Sigma_{terms}$ and $\iota_1, \ldots, \iota_n, \iota \in \mathcal{S}_{theory}$ for any symbol $f : \iota_1 \times \cdots \times \iota_n \to \iota \in \Sigma_{theory}$; a mapping $\mathcal{I}$ that assigns to each theory sort ι a (non-empty) set $\mathcal{A}_\iota$, so-called the universe of ι (i.e., $\mathcal{I}(\iota) = \mathcal{A}_\iota$); a mapping $\mathcal{J}$, so-called an interpretation for Σ_{theory}, that assigns to each function symbol $f : \iota_1 \times \cdots \times \iota_n \Rightarrow \iota \in \Sigma_{theory}$ a function $f^\mathcal{J}$ in $\mathcal{I}(\iota_1) \times \cdots \times \mathcal{I}(\iota_n) \Rightarrow \mathcal{I}(\iota)$ (i.e., $\mathcal{J}(f) = f^\mathcal{J}$); a set $\mathcal{V}al_\iota \subseteq \Sigma_{theory}$ of *value-constants* $a : \iota$ for each theory sort ι such that $\mathcal{J}$ gives a bijection from $\mathcal{V}al_\iota$ to $\mathcal{I}(\iota)$. We denote $\bigcup_{\iota \in \mathcal{S}_{theory}} \mathcal{V}al_\iota$ by $\mathcal{V}al$. Note that $\mathcal{V}al \subseteq \Sigma_{theory}$. For readability, we may not distinguish $\mathcal{V}al_\iota$ and $\mathcal{I}(\iota)$, i.e., for each $v \in \mathcal{V}al_\iota$, v and $\mathcal{J}(v)$ may be identified. We require that $\Sigma_{terms} \cap \Sigma_{theory} \subseteq \mathcal{V}al$. Symbols in $\Sigma_{theory} \setminus \mathcal{V}al$ are *calculation symbols*, for which we may use infix notation. A term in $T(\Sigma_{theory}, \mathcal{V})$ is called a *theory term*, where $\mathcal{V}$ is a (countably infinite) set of variables. We define the *interpretation* $[\![\cdot]\!]_\mathcal{J}$ of ground theory terms as $[\![f(s_1, \ldots, s_n)]\!]_\mathcal{J} = \mathcal{J}(f)([\![s_1]\!]_\mathcal{J}, \ldots, [\![s_n]\!]_\mathcal{J})$. Note that for every ground theory term s, there is a unique value-constant c such that $[\![s]\!]_\mathcal{J} = [\![c]\!]_\mathcal{J}$.

We typically choose a theory signature such that $\mathcal{S}_{theory} \supseteq \mathcal{S}_{theory}^{core} = \{bool\}$, $\Sigma_{theory} \supseteq \Sigma_{theory}^{core} = \mathcal{V}al_{bool} \cup \{\wedge, \vee, \Longrightarrow, \Longleftrightarrow : bool \times bool \Rightarrow bool, \neg : bool \Rightarrow bool\} \cup \{=_\iota, \neq_\iota : \iota \times \iota \Rightarrow bool \mid \iota \in \mathcal{S}_{theory}\}$, $\mathcal{I}(bool) = \{\top, \bot\}$, and $\mathcal{J}$ interprets these symbols as expected: $\mathcal{J}(\mathsf{true}) = \top$ and $\mathcal{J}(\mathsf{false}) = \bot$. We omit the sort subscripts ι from $=_\iota$ and $\neq_\iota$ when they are clear from context. A theory term with sort $bool$ is called a *constraint*.

A *constrained rewrite rule* is a triple $\ell \to r \, [\varphi]$ such that ℓ and r are terms of the same sort, φ is a constraint, and ℓ is not a theory term (i.e., not a variable). If $\varphi = \mathsf{true}$, then we may write $\ell \to r$. We define $\mathcal{L}\mathcal{V}ar(\ell \to r \, [\varphi])$ as $\mathcal{V}ar(\varphi) \cup (\mathcal{V}ar(r) \setminus \mathcal{V}ar(\ell))$, the set of *logical variables* in $\ell \to r \, [\varphi]$ which are variables instantiated with values in rewriting terms. We say that a substitution γ *respects* $\ell \to r \, [\varphi]$ if $\gamma(x) \in \mathcal{V}al$ for all $x \in \mathcal{L}\mathcal{V}ar(\ell \to r \, [\varphi])$ and $[\![\varphi\gamma]\!]_\mathcal{J} = \top$. Given a set $\mathcal{R}$ of constrained rewrite rules, we denote the set $\{f(x_1, \ldots, x_n) \to y \, [y = f(x_1, \ldots, x_n)] \mid f \in \Sigma_{theory} \setminus \mathcal{V}al, x_1, \ldots, x_n, y \text{ are pairwise distinct variables}\}$ by $\mathcal{R}_{calc}$. The elements of $\mathcal{R}_{calc}$ are also called constrained rewrite rules (or *calculation rules*) even though their left-hand sides are theory terms. The *rewrite relation* $\to_\mathcal{R}$ is a binary relation over terms, defined as follows: For a term s, $s[\ell\gamma]_p \to_\mathcal{R} s[r\gamma]_p$ if and only if $\ell \to r \, [\varphi] \in \mathcal{R} \cup \mathcal{R}_{calc}$ and γ respects $\ell \to r \, [\varphi]$. We may say that the reduction *occurs at position* p and may write $\to_{p,\mathcal{R}}$ or $\to_{p,\ell \to r \, [\varphi]}$ instead of $\to_\mathcal{R}$. A reduction step with $\mathcal{R}_{calc}$ is called a *calculation*. Now we define a *logically constrained term rewrite system* (LCTRS, for short) as an abstract reduction system $(T(\Sigma, \mathcal{V}), \to_\mathcal{R})$, simply denoted by $\mathcal{R}$, where $\mathcal{R}$ is a set of constrained rewrite rules. An LCTRS is usually given by supplying Σ, $\mathcal{R}$, and an informal description of $\mathcal{I}$ and $\mathcal{J}$ if these are not clear from context. The set of *normal forms* of $\mathcal{R}$ is denoted by $NF_\mathcal{R}$.

A *constrained term* is a pair $\langle t \mid \phi \rangle$ of a term t and a constraint ϕ, which can be considered the set of all instances of t w.r.t. substitutions that respect ϕ: $\langle t \mid \phi \rangle = \{t\gamma \mid \gamma \text{ respects } \phi\}$. The set of *derivatives* of a constrained term w.r.t. an LCTRS $\mathcal{R}$ is defined as follows: $\Delta_\mathcal{R}(\langle s \mid \phi \rangle) = \bigcup_{\ell \to r \, [\varphi] \in \mathcal{R}} \Delta_{\ell, r, \varphi}(\langle s \mid \phi \rangle)$, where $\ell \to r \, [\varphi]$

has no shared variable with $\langle s \,|\, \phi\rangle$[2] and $\Delta_{\ell,r,\varphi}(\langle s\,|\,\phi\rangle) = \{\langle (s[r]_p)\gamma \,|\, (\phi \wedge \varphi)\gamma\rangle \,|\, p \in \mathcal{P}os(s),\ s|_p \notin \mathcal{V},\ s|_p$ and ℓ are unifiable, $\gamma = mgu(s|_p, \ell),\ \mathcal{R}an(\gamma|_{\mathcal{V}ar(\phi,\varphi)}) \subseteq \mathcal{V}al \cup \mathcal{V},\ (\phi \wedge \varphi)\gamma$ is satisfiable$\}$.

The *standard integer signature* Σ_{theory}^{int} is $\Sigma_{theory}^{core} \cup \{+, -, \times, \exp, \mathsf{div}, \mathsf{mod} : int \times int \Rightarrow int\} \cup \{\geq, > \,:\, int \times int \Rightarrow bool\} \cup \mathcal{V}al_{int}$ where $\mathcal{S}_{theory} \supseteq \{int, bool\}$, $\mathcal{V}al_{int} = \{\mathsf{n} \,|\, n \in \mathbb{Z}\}$, $\mathcal{I}(int) = \mathbb{Z}$, and $\mathcal{J}(\mathsf{n}) = n$ for any $n \in \mathbb{Z}$—we use n (in sans-serif font) as the value-constant for $n \in \mathbb{Z}$ (in *math* font). We define $\mathcal{J}$ in a natural way. An LCTRS over a signature $\Sigma \supseteq \Sigma_{theory}^{int}$ is called an *integer LCTRS*.

Concurrent LCTRSs. To deal with LCTRSs representing concurrent programs, this paper considers an integer LCTRS $\mathcal{R}$ such that

- $state, loc \in \mathcal{S}_{term}$, where $state$ and loc are sorts for states and processes, respectively,
- $\mathsf{state} : \iota_1 \times \cdots \times \iota_m \to state \in \Sigma_{terms}$ with $1 \leq n \leq m$, where $\iota_1 = \cdots = \iota_n = loc$, $\iota_{n+1}, \ldots, \iota_m \in \{bool, int\}$, and
- for each rule $\ell \to r\ [\varphi] \in \mathcal{R}$, ℓ is rooted by state and both ℓ and r are calculation free, i.e., no calculation symbol appears in either ℓ or r.[3]

Function symbol state represents states such that

- for any $i \in \{1, \ldots, n\}$, the i-th argument of state is a location of the i-th process, and
- for any $j \in \{n+1, \ldots, m\}$, the j-th argument of state is a value stored in a program variable.

We call such an LCTRS a *concurrent LCTRS*. A concurrent LCTRS is said to be *deadlock free* if there is no ground normal form rooted by state.

Example 2.1. The LCTRSs $\mathcal{R}_1, \mathcal{R}_1^\forall$ in Example 1.1 are deadlock-free concurrent LCTRSs having, e.g., the reduction sequence $\mathsf{state}(\mathsf{noncrit}_1, \mathsf{noncrit}_2, 1) \to_{\mathcal{R}_1} \mathsf{state}(\mathsf{wait}_1, \mathsf{noncrit}_2, 1) \to_{\mathcal{R}_1} \mathsf{state}(\mathsf{crit}_1, \mathsf{noncrit}_2, 0) \to_{\mathcal{R}_1} \cdots$.

In the rest of the paper, we use $p_1, \ldots, p_n$ and $x_{n+1}, \ldots, x_m$ as pairwise distinct variables for processes and theory terms (i.e., $p_i : loc$ and either $x_j : int$ or $x_j : bool$). We denote the sequences $p_1, \ldots, p_n$ and $x_{n+1}, \ldots, x_m$ by $\vec{p}$ and $\vec{x}$, respectively. We also denote sequences $t_1, \ldots, t_n$ and $u_{n+1}, \ldots, u_m$ of terms by $\vec{t}$ and $\vec{u}$, respectively.

2.2 All-Path Reachability

An *all-path reachability problem* (APR problem, for short) of an LCTRS $\mathcal{R}$ is a pair $\langle s\,|\,\phi\rangle \Rightarrow \langle t\,|\,\psi\rangle$ of constrained terms $\langle s\,|\,\phi\rangle, \langle t\,|\,\psi\rangle$ for state sets [6,25,26]. We call $\langle s\,|\,\phi\rangle \Rightarrow \langle t\,|\,\psi\rangle$ *constant-directed* if $\langle t\,|\,\psi\rangle = \{d\}$ for some constant normal form d [13]. We abbreviate it to $\langle s\,|\,\phi\rangle \Rightarrow d$.

---

[2] When there exists a shared variable, we rename the variables in $\ell \to r\ [\varphi]$.
[3] We assume calculation freedom to reduce the search space in proving APR problems.

Example 2.2. The APR problem $\langle \mathsf{state}(\mathsf{noncrit}_1, \mathsf{noncrit}_2, 1, 0) \mid \mathsf{true}\rangle \Rightarrow \mathsf{success}$ in Example 1.1 is constant-directed.

As in our previous work, in the rest of this paper, we only deal with constant-directed APR problems and use the terminology "APR problems" for constant-directed APR problems. A (possibly infinite) reduction sequence of an LCTRS $\mathcal{R}$ with some initial states is called an *execution path* if the reduction sequence either is infinite or ends with an irreducible term. An APR problem $\langle s \mid \phi \rangle \Rightarrow d$ is *demonically valid* w.r.t. $\mathcal{R}$ if every *finite* execution path includes d, i.e., every execution path ends with d.

The proof system c-DCC [13–15] for constant-directed APR problems is a weakened variant of the proof system DCC [6] for arbitrary APR problems.

Definition 2.3 (c-DCC [13–15]**).** *Given an LCTRS $\mathcal{R}$ and a finite set G of APR problems, the proof system c-DCC$(\mathcal{R}, G)$ consists of the following rules:*

$$\frac{}{\langle s \mid \phi \rangle \Rightarrow d}(\text{axiom/c-subs}) \text{ if } \phi \text{ is unsatisfiable or } s = d$$

$$\frac{\langle s_1 \mid \phi_1 \rangle \Rightarrow d \quad \ldots \quad \langle s_n \mid \phi_n \rangle \Rightarrow d}{\langle s \mid \phi \rangle \Rightarrow d}(\text{c-der}) \text{ if } \langle s_n \mid \phi_n \rangle \cap NF_{\mathcal{R}} = \emptyset$$

where $\Delta_{\mathcal{R}}(\langle s \mid \phi \rangle) = \{\langle s_i \mid \phi_i \rangle \mid 1 \leq i \leq n\}$ *for some* $n > 0$.

$$\frac{}{\langle s \mid \phi \rangle \Rightarrow d}(\text{weak circ}) \text{ if } \exists (\langle s' \mid \phi' \rangle \Rightarrow d) \in G. \ \langle s \mid \phi \rangle \subseteq \langle s' \mid \phi' \rangle$$

$$\frac{\bot}{\langle s \mid \phi \rangle \Rightarrow d}(\text{dis}) \text{ if } d \notin \langle s \mid \phi \rangle \text{ and} \langle s \mid \phi \rangle \cap NF_{\mathcal{R}} \neq \emptyset$$

For a single APR problem $\langle s \mid \phi \rangle \Rightarrow d$, a construction of a single proof tree in the style of *cyclic proofs* [4] has been proposed [15].

Definition 2.4 (APR (dis)proof) *Let $\mathcal{R}$ be an LCTRS. An* APR derivation tree *(w.r.t. $\mathcal{R}$) is a finite tree $\mathcal{T} = (N, a, r, p)$ consisting of a finite set N of nodes, a total mapping a from N to the set of APR problems of $\mathcal{R}$, a partial mapping r from N to the rule names $\{\mathsf{axiom/c\text{-}der}, \mathsf{rmc\text{-}der}, \mathsf{dis}\}$, and a partial mapping p from N to N^* (we write $p_j(v)$ for the j-th component of $p(v)$) such that*

- *for all nodes $v \in N$, $p_j(v)$ is defined just in case $r(v)$ is a rule with k premises $(1 \leq j \leq k)$, and $\frac{a(p_1(v)) \ldots a(p_k(v))}{a(v)}$ is an instance of rule $r(v)$, and*
- *if $r(v) = \mathsf{dis}$, then $p(v)$ is undefined.*

A leaf v of $\mathcal{T}$ is said to be open *if $r(v)$ and $p(v)$ are undefined. The set of open leaves of $\mathcal{T}$ is denoted by N_{open}.*

An APR pre-proof *of an APR problem $\langle s \mid \phi \rangle \Rightarrow d$ is a pair $(\mathcal{T}, \xi)$ of a derivation tree $\mathcal{T} = (N, a, r, p)$ (with v_0 the root node) and a partial mapping ξ from N_{open} to $N \setminus N_{open}$ such that $a(v_0) = (\langle s \mid \phi \rangle \Rightarrow d)$, and for any open leaf $v \in N_{open}$, if $\xi(v)$ is defined, then $\xi(v)$ is an inner node of $\mathcal{T}$ with $\langle s' \mid \phi' \rangle \subseteq$*

$\langle s'' | \phi'' \rangle$, where $a(v) = (\langle s' | \phi' \rangle \Rightarrow d)$ and $a(\xi(v)) = (\langle s'' | \phi'' \rangle \Rightarrow d)$. An open leaf $v \in N_{open}$ with $\xi(v)$ defined is called a bud node of $\mathcal{T}$, and the node $\xi(v)$ is called a companion of v.[4] We denote the set of bud nodes in N by N_{bud} ($\subseteq N$). An APR pre-proof $(\mathcal{T}, \xi)$ of $\langle s | \phi \rangle \Rightarrow d$ is called an APR proof if the domain of ξ is N_{open}, i.e., all open nodes are bud nodes. and an APR disproof if there exists an open leaf $v \in N_{open}$ such that $r(v) = \mathsf{dis}$.

To simplify the discussion, in constructing APR pre-proofs, we often reduce constraints to equivalent simpler ones without notice.

Theorem 2.5 (correctness of APR (dis)proofs [15]). *Let $\mathcal{R}$ be an LCTRS, and $\langle s | \phi \rangle \Rightarrow d$ an APR problem of $\mathcal{R}$. If there exists an APR (dis)proof of $\langle s | \phi \rangle \Rightarrow d$, then $\langle s | \phi \rangle \Rightarrow d$ is (not) demonically valid w.r.t. $\mathcal{R}$.*

Example 2.6. Consider $\mathcal{R}_1^\forall$ in Example 1.1 again. To make an APR pre-proof as small as possible, we consider the APR problem $\langle \mathsf{state}(\mathsf{noncrit}_1, \mathsf{noncrit}_2, y', c) | y'=1 \wedge c=0 \rangle \Rightarrow \mathsf{success}$. An APR pre-proof $\mathcal{P}_1 = ((\{(1), (2), \ldots\}, a, r, p), \xi)$ illustrated in Fig. 4 is an APR disproof for the APR problem of $\mathcal{R}_1^\forall$ with $c_{\max} = 3$, where nodes v with $a(v) = (\langle \mathsf{success} | \phi \rangle \Rightarrow \mathsf{success})$ are omitted, each bud node and its companion are identified, and † indicates bud nodes.

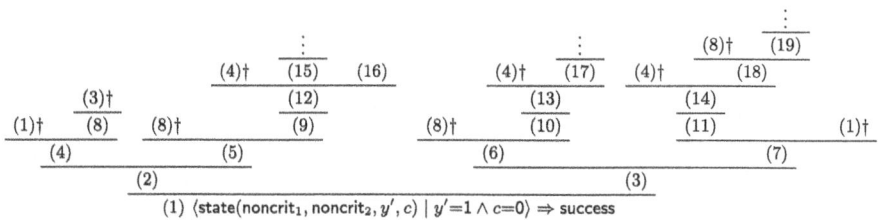

(1) $\langle \mathsf{state}(\mathsf{noncrit}_1, \mathsf{noncrit}_2, y', c) | y'=1 \wedge c=0 \rangle \Rightarrow \mathsf{success}$

where the left-hand sides of APR problems attached to the nodes are the following:

(2) $\langle \mathsf{state}(\mathsf{wait}_1, \mathsf{noncrit}_2, y', c) | y'=1 \wedge c=1 \rangle$
(3) $\langle \mathsf{state}(\mathsf{noncrit}_1, \mathsf{wait}_2, y', c) | y'=1 \wedge c=0 \rangle$
(4) $\langle \mathsf{state}(\mathsf{crit}_1, \mathsf{noncrit}_2, y', c) | y'=0 \wedge c=0 \rangle$
(5) $\langle \mathsf{state}(\mathsf{wait}_1, \mathsf{wait}_2, y', c) | y'=1 \wedge c=2 \rangle$
(6) $\langle \mathsf{state}(\mathsf{wait}_1, \mathsf{wait}_2, y', c) | y'=1 \wedge c=1 \rangle$
(7) $\langle \mathsf{state}(\mathsf{noncrit}_1, \mathsf{crit}_2, y', c) | y'=0 \wedge c=0 \rangle$
(8) $\langle \mathsf{state}(\mathsf{crit}_1, \mathsf{wait}_2, y', c) | y'=0 \wedge c=0 \rangle$
(9) $\langle \mathsf{state}(\mathsf{wait}_1, \mathsf{crit}_2, y', c) | y'=0 \wedge c=3 \rangle$
(10) $\langle \mathsf{state}(\mathsf{wait}_1, \mathsf{crit}_2, y', c) | y'=0 \wedge c=2 \rangle$
(11) $\langle \mathsf{state}(\mathsf{wait}_1, \mathsf{crit}_2, y', c) | y'=0 \wedge c=1 \rangle$
(12) $\langle \mathsf{state}(\mathsf{wait}_1, \mathsf{noncrit}_2, y', c) | y'=1 \wedge c=4 \rangle$
(13) $\langle \mathsf{state}(\mathsf{wait}_1, \mathsf{noncrit}_2, y', c) | y'=1 \wedge c=3 \rangle$
(14) $\langle \mathsf{state}(\mathsf{wait}_1, \mathsf{noncrit}_2, y', c) | y'=1 \wedge c=2 \rangle$
(15) $\langle \mathsf{state}(\mathsf{wait}_1, \mathsf{wait}_2, y', c) | y'=1 \wedge c=5 \rangle$
(16) $\langle \mathsf{error} | y'=1 \wedge c=4 \rangle$
(17) $\langle \mathsf{state}(\mathsf{wait}_1, \mathsf{wait}_2, y', c) | y'=1 \wedge c=4 \rangle$
(18) $\langle \mathsf{state}(\mathsf{wait}_1, \mathsf{wait}_2, y', c) | y'=1 \wedge c=3 \rangle$
(19) $\langle \mathsf{state}(\mathsf{wait}_1, \mathsf{crit}_2, y', c) | y'=0 \wedge c=4 \rangle$
$\ldots$

Fig. 4. A APR disproof for $\langle \mathsf{state}(\mathsf{noncrit}_1, \mathsf{noncrit}_2, y', c) | y'=1 \wedge c=0 \rangle \Rightarrow \mathsf{success}$ of $\mathcal{R}_1^\forall$ with $c_{\max} = 3$.

Paths of APR pre-proofs correspond to narrowing derivations.

---

[4] From the viewpoint of graphs, we identify a bud node and its companion.

Definition 2.7 ([15]). *Let $\mathcal{R}$ be an LCTRS. We say that a constrained term $\langle s \,|\, \phi \rangle$ narrows to a constrained term $\langle t \,|\, \psi \rangle$ at a non-variable position p of s with a substitution γ, written as $\langle s \,|\, \phi \rangle \rightsquigarrow_{\gamma|_{\mathcal{V}ar(s,\phi)},\mathcal{R}} \langle t \,|\, \psi \rangle$, if there exists a constrained rewrite rule $\ell \to r \,[\varphi] \in \mathcal{R} \cup \mathcal{R}_{calc}$ such that $s|_p$ and ℓ are unifiable, $\gamma = mgu(s|_p, \ell)$, $\mathcal{R}an(\gamma|_{\mathcal{V}ar(\phi,\varphi)}) \subseteq \mathcal{V}al \cup \mathcal{V}$, $t = (s[r]_p)\gamma$, and $\psi = (\phi \wedge \varphi)\gamma$, where $\ell \to r \,[\varphi]$ is assumed to have no shared variables with $\langle s \,|\, \phi \rangle$. To make the applied position p and rule $\ell \to r \,[\varphi]$ explicit, we may write $\langle s \,|\, \phi \rangle \rightsquigarrow_{\gamma|_{\mathcal{V}ar(s,\phi)}, \ell \to r \,[\varphi]} \langle t \,|\, \psi \rangle$ or $\langle s \,|\, \phi \rangle \rightsquigarrow_{\gamma|_{\mathcal{V}ar(s,\phi)}, p, \ell \to r \,[\varphi]} \langle t \,|\, \psi \rangle$. Note that we do not require ψ to be satisfiable in the definition because rewriting of constrained terms with unsatisfiable constraints is possible. We denote by $\langle s_0 \,|\, \phi_0 \rangle \rightsquigarrow^*_{\gamma,\mathcal{R}} \langle s_n \,|\, \phi_n \rangle$ or $\langle s_0 \,|\, \phi_0 \rangle \rightsquigarrow^n_{\gamma,\mathcal{R}} \langle s_n \,|\, \phi_n \rangle$ a sequence of narrowing steps $\langle s_0 \,|\, \phi_0 \rangle \rightsquigarrow_{\gamma_1,\mathcal{R}} \langle s_1 \,|\, \phi_1 \rangle \rightsquigarrow_{\gamma_2,\mathcal{R}} \cdots \rightsquigarrow_{\gamma_n,\mathcal{R}} \langle s_n \,|\, \phi_n \rangle$, where $\gamma = \gamma_n \circ \cdots \circ \gamma_2 \circ \gamma_1$. Note that if $n = 0$, then γ is the identity substitution (i.e., $\mathcal{D}om(\gamma) = \emptyset$). We may omit γ from $\langle s_0 \,|\, \phi_0 \rangle \rightsquigarrow^*_{\gamma,\mathcal{R}} \langle s_n \,|\, \phi_n \rangle$ and $\langle s_0 \,|\, \phi_0 \rangle \rightsquigarrow^n_{\gamma,\mathcal{R}} \langle s_n \,|\, \phi_n \rangle$.*

Lemma 2.8 ([15]). *Let $\mathcal{R}$ be an LCTRS, and $\langle s \,|\, \phi \rangle, \langle t \,|\, \psi \rangle$ constrained terms.*

(a) For any $n \in \mathbb{N}$, if $\langle s \,|\, \phi \rangle \rightsquigarrow^n_{\gamma,\mathcal{R}} \langle t \,|\, \psi \rangle$, then $\langle s\gamma \,|\, \psi \rangle \to^n_{\mathcal{R}} \langle t \,|\, \psi \rangle$.
(b) $\Delta_{\mathcal{R}}(\langle s \,|\, \phi \rangle) = \{\langle t \,|\, \psi \rangle \mid \langle s \,|\, \phi \rangle \rightsquigarrow_{\mathcal{R}} \langle t \,|\, \psi \rangle, \psi \text{ is satisfiable}\}$ up to variable renaming.
(c) Let $(\mathcal{T}, \xi)$ be an APR pre-proof w.r.t. $\mathcal{R}$, and $(v_0, v_1, \ldots, v_n)$ a path of $\mathcal{T}$ such that $a(v_i) = (\langle s_i \,|\, \phi_i \rangle \Rightarrow d)$ for $i \in \{0, \ldots, n\}$. Then, $\langle s_0 \,|\, \phi_0 \rangle \rightsquigarrow_{\gamma_1,\mathcal{R}} \langle s_1 \,|\, \phi_1 \rangle \rightsquigarrow_{\gamma_2,\mathcal{R}} \cdots \rightsquigarrow_{\gamma_n,\mathcal{R}} \langle s_n \,|\, \phi_n \rangle$ for some substitutions $\gamma_1, \ldots, \gamma_n$.

3 Reduction of Starvation Freedom to APR Problems

In this section, we recall and improve the framework in [13] to reduce starvation freedom to an APR problem. In the rest of the paper, w.l.o.g. we consider starvation freedom for the first process P_1 w.r.t. a critical section in a concurrent LCTRS $\mathcal{R}$ with initials states $\langle s_0 \,|\, \phi_0 \rangle$. Let crit be a location which is the entry point of the critical section, and $\mathsf{wait}_1, \ldots, \mathsf{wait}_k$ the locations with rewrite rules of the form $\mathsf{state}(\mathsf{wait}_j, p_2, \ldots, p_n, \vec{x}) \to \mathsf{state}(\mathsf{crit}, p_2, \ldots, p_m, \vec{t}) \,[\phi_j] \in \mathcal{R}$ ($1 \leq j \leq k$). Note that we can verify starvation freedom of the entire system by verifying individual starvation freedom for every process and its critical section.

We first show how to reduce starvation freedom to an APR problem. Starvation freedom guarantees that P_1 can enter the critical section in finite time, whenever P_1 wants to enter it. To approximately characterize starvation (freedom) in $\mathcal{R}$, we represent a waiting time of P_1 at some waiting location wait_j until entering the critical section (i.e., transitioning to crit). To this end, we introduce a counter for the waiting time as a new argument of $\mathsf{state} : \iota_1 \times \cdots \times \iota_m \to \mathsf{state}$, where the new sort declaration of state is $\iota_1 \times \cdots \times \iota_m \times \mathit{int} \to \mathsf{state}$. Note that the "$m+1$"-th argument of state is used for the counter. In the following, we use c, c' as variables for the counter. We modify rewrite rules in $\mathcal{R}$ so that

- the counting starts when P_1 enters a waiting state (some wait_j), and resets when the process enters the critical section (i.e., transitions to crit), and

– the value of the counter is incremented at any transition step of other processes whenever P_1 is waiting.

Constrained terms $\langle \mathsf{state}(\vec{p}, \vec{x}, c) \mid c > \mathsf{c}_{\max}\rangle$—states with the value of the counter exceeding some upper limit $\mathsf{c}_{\max}$ which specified by a user in advance—are considered error states. Introducing fresh constants success and error, we add the rewrite rules $\mathsf{state}(\vec{p}, \vec{x}) \to \mathsf{success}$ and $\mathsf{state}(\vec{p}, \vec{x}, c) \to \mathsf{error}\ [c > \mathsf{c}_{\max}]$ to the modified LCTRS. Starvation freedom for P_1 is reduced to the APR problem $\langle s_0 \mid \phi_0 \rangle \Rightarrow \mathsf{success}$ of the modified LCTRS $\mathcal{R}^\forall_{>\mathsf{c}_{\max}}$. Recall that $\langle s_0 \mid \phi_0 \rangle$ is the set of initial terms of $\mathcal{R}$. For a term $s = \mathsf{state}(\vec{t}, \vec{u})$ and a term $s' : int$, $s \blacktriangleleft s'$ denotes the term $\mathsf{state}(\vec{t}, \vec{u}, s')$.

The approach above considers the states in $\langle \mathsf{state}(\vec{p}, \vec{x}, c) \mid c > \mathsf{c}_{\max}\rangle$ to be error states. Though, some rules $(\ell \blacktriangleleft c) \to (r \blacktriangleleft c')\ [\varphi \wedge \varphi'_{c'}] \in \mathcal{R}^\forall_{>\mathsf{c}_{\max}}$ with $\ell \to r\ [\varphi] \in \mathcal{R}$ are applicable to the error states, where $\varphi'_{c'}$ is a constraint to update c to c'. For example, the term $\mathsf{state}(\mathsf{wait}_1, \mathsf{noncrit}_2, 1, 4)$ of $\mathcal{R}^\forall_1$ is an error state, to which both the second and fourth rules of $\mathcal{R}^\forall_1$ are applicable, and the constrained term (12) in Fig. 4 has the children (4) and (15), in addition to (16). Allowing such reduction is redundant and does not affect the construction of APR pre-proofs for demonically valid APR problems because no error state is reachable from initial terms. To avoid the redundant reduction of error states, in introducing a counter to a rewrite rule of a given LCTRS $\mathcal{R}$, we conjunct the formula $c \leq \mathsf{c}_{\max}$ with the constraint of the rule. Using this improvement, the reduction of starvation freedom to APR problems in [13] is reformulated as follows.

Definition 3.1 ($\mathcal{R}^\forall_{>\mathsf{c}_{\max}}$). *Let $\mathsf{c}_{\max}$ be a positive integer. Introducing fresh constants* success *and* error *with sort* state *into the signature Σ, we define the LCTRS $\mathcal{R}^\forall_{>\mathsf{c}_{\max}}$ as the following set:*

$\{\,(s[\ell]_1 \blacktriangleleft c) \to (s'[\mathsf{wait}_j]_1 \blacktriangleleft 1)\ [\phi \wedge c \leq \mathsf{c}_{\max}] \mid s[\ell]_1 \to s'[\mathsf{wait}_j]_1\ [\phi] \in \mathcal{R}\,\}$
$\cup \{\,(s[\mathsf{wait}_j]_1 \blacktriangleleft c) \to (s'[\mathsf{crit}]_1 \blacktriangleleft 0)\ [\phi_j \wedge c \leq \mathsf{c}_{\max}] \mid s[\mathsf{wait}_j]_1 \to s'[\mathsf{crit}]_1\ [\phi_j] \in \mathcal{R}\,\}$
$\cup \{\,(s[\ell]_1 \blacktriangleleft c) \to (s'[\ell']_1 \blacktriangleleft c)\ [\phi \wedge c \leq \mathsf{c}_{\max}]$
$\hspace{4em} \mid s[\ell]_1 \to s'[\ell']_1\ [\phi] \in \mathcal{R},\ \ell \notin \{\mathsf{wait}_1, \ldots, \mathsf{wait}_k\}\,\}$
$\cup \{\,(s[\ell]_j \blacktriangleleft c) \to (s'[\ell']_j \blacktriangleleft c')\ [\phi \wedge c' = \mathsf{inc}_{\geq 1}(c)] \mid j > 1,\ s[\ell]_j \to s'[\ell']_j\ [\phi] \in \mathcal{R}\,\}$
$\cup \{\,\mathsf{state}(\vec{p}, \vec{x}, c) \to \mathsf{success},\ \mathsf{state}(\vec{p}, \vec{x}, c) \to \mathsf{error}\ [c > \mathsf{c}_{\max}]\,\}$

where $\mathsf{inc}_{\geq 1}(c)$ *denotes* $\mathsf{ite}(c \geq 1, c+1, c)$. We omit "$>\mathsf{c}_{\max}$" from $\mathcal{R}^\forall_{>\mathsf{c}_{\max}}$, writing $\mathcal{R}^\forall$, because we can know $\mathsf{c}_{\max}$ from $\mathcal{R}^\forall_{>\mathsf{c}_{\max}} \setminus \mathcal{R}$.

Starvation freedom for P_1 is reduced to the APR problem $\langle s_0 \blacktriangleleft 0 \mid \phi_0 \rangle \Rightarrow \mathsf{success}$ of $\mathcal{R}^\forall$. Recall that $\langle s_0 \mid \phi_0 \rangle$ is the set of initial terms of $\mathcal{R}$.

The rule $\mathsf{state}(\vec{p}, \vec{x}) \to \mathsf{success}$ is not necessary for deadlock-free concurrent LCTRSs. The downside of the demonical validity of APR problems is to consider finite execution paths only. The rule makes all finite prefix of (possibly infinite) execution paths of $\mathcal{R}$ finite execution paths of $\mathcal{R}^\forall_{>\mathsf{c}_{\max}}$. On the other hand, for runtime-error verification, it suffices to make finite execution paths of $\mathcal{R}$ end with success; if a specified error occurs, then there exists a finite execution path

$$\tilde{\mathcal{R}}_1^\forall = \begin{cases} \mathsf{state}(\mathsf{noncrit}_1, p_2, y, c) \to \mathsf{state}(\mathsf{wait}_1, p_2, y, 1) & [\, c \leq \mathsf{c_{max}} \,] \\ \mathsf{state}(\mathsf{wait}_1, p_2, y, c) \to \mathsf{state}(\mathsf{crit}_1, p_2, y', 0) & [\, y{>}0 \wedge y'{=}y{-}1 \wedge c \leq \mathsf{c_{max}} \,] \\ \mathsf{state}(\mathsf{crit}_1, p_2, y, c) \to \mathsf{state}(\mathsf{noncrit}_1, p_2, y', c) & [\, y'{=}y{+}1 \wedge c \leq \mathsf{c_{max}} \,] \\ \mathsf{state}(p_1, \mathsf{noncrit}_2, y, c) \to \mathsf{state}(p_1, \mathsf{wait}_2, y, c') & [\, c'{=}\mathsf{inc}_{\geq 1}(c) \wedge c \leq \mathsf{c_{max}} \,] \\ \mathsf{state}(p_1, \mathsf{wait}_2, y, c) \to \mathsf{state}(p_1, \mathsf{crit}_2, y', c') & [\, y{>}0 \wedge y'{=}y{-}1 \wedge c'{=}\mathsf{inc}_{\geq 1}(c) \\ & \quad \wedge c \leq \mathsf{c_{max}} \,] \\ \mathsf{state}(p_1, \mathsf{crit}_2, y, c) \to \mathsf{state}(p_1, \mathsf{noncrit}_2, y', c') & [\, y'{=}y{+}1 \wedge c'{=}\mathsf{inc}_{\geq 1}(c) \\ & \quad \wedge c \leq \mathsf{c_{max}} \,] \\ \mathsf{state}(p_1, p_2, y, c) \to \mathsf{error} & [\, c > \mathsf{c_{max}} \,] \end{cases}$$

Fig. 5. LCTRS $\tilde{\mathcal{R}}_1^\forall$ obtained from $\mathcal{R}_1$ by Definition 3.1.

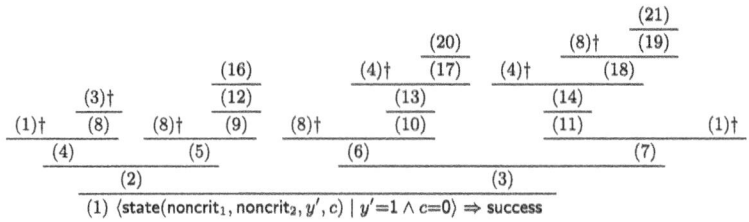

Fig. 6. A finite APR disproof for $\langle \mathsf{state}(\mathsf{noncrit}_1, \mathsf{noncrit}_2, y', c) \mid y'{=}1 \wedge c{=}0 \rangle \Rightarrow \mathsf{success}$ of $\tilde{\mathcal{R}}_1^\forall$ with $\mathsf{c_{max}} = 3$, where $a((20)), a((21))$ are of the form $(\langle \mathsf{error} \mid \ldots \wedge c{=}4 \rangle \Rightarrow \mathsf{success})$.

of $\mathcal{R}_{>\mathsf{c_{max}}}^\forall$ that ends with error, i.e., the APR problem is not demonically valid. Therefore, if $\mathcal{R}$ is deadlock free, then we drop $\mathsf{state}(\vec{p}, \vec{x}) \to \mathsf{success}$ from $\mathcal{R}_{>\mathsf{c_{max}}}^\forall$.

Example 3.2. Given a positive integer $\mathsf{c_{max}}$, the LCTRS $\mathcal{R}_1$ in Example 1.1 is modified to the LCTRS $\tilde{\mathcal{R}}_1^\forall$ in Fig. 5. Starvation freedom for P_1 w.r.t. $\mathcal{R}_1$ is reduced to the APR problem $\langle \mathsf{state}(\mathsf{noncrit}_1, \mathsf{noncrit}_2, y', c) \mid y' = 1 \wedge c = 0 \rangle \Rightarrow \mathsf{success}$ of $\tilde{\mathcal{R}}_1^\forall$. The APR problem is not demonically valid w.r.t. $\tilde{\mathcal{R}}_1^\forall$ with $\mathsf{c_{max}} = 2$. In such a case, we increase the value of $\mathsf{c_{max}}$ and try to prove again. In the case where $\mathsf{c_{max}} = 3$, we obtain a finite APR disproof in Fig. 6, which is the APR pre-proof obtained from the one in Fig. 4 by removing some nodes. On the other hand, starvation for P_1 occurs due to unfair execution paths of $\mathcal{R}_1$, e.g., $\mathsf{state}(\mathsf{noncrit}_1, \mathsf{noncrit}_2, 1) \to_{\mathcal{R}_1}^* \mathsf{state}(\mathsf{wait}_1, \mathsf{wait}_2, 1) \to_{\mathcal{R}_1} \mathsf{state}(\mathsf{wait}_1, \mathsf{crit}_2, 0) \to_{\mathcal{R}_1} \mathsf{state}(\mathsf{wait}_1, \mathsf{noncrit}_2, 1) \to_{\mathcal{R}_1} \mathsf{state}(\mathsf{wait}_1, \mathsf{wait}_2, 1) \to_{\mathcal{R}_1} \cdots$. For this reason, for any $\mathsf{c_{max}} \in \mathbb{N}$, the APR problem is not demonically valid w.r.t. $\tilde{\mathcal{R}}_1^\forall$. However, we cannot find any witness of the demonical invalidity, and we cannot disprove starvation freedom, repeating the attempt by increasing the upper limit $\mathsf{c_{max}}$.

For an APR problem with an APR proof, error states are not reachable from initial terms. For this reason, the soundness theorem in [13] of the modification of LCTRSs still holds for $\mathcal{R}_{>\mathsf{c_{max}}}^\forall$ in Definition 3.1.

Theorem 3.3. *Let $\mathsf{c_{max}}$ be a positive integer. If $\langle s_0 \triangleleft 0 \mid \phi_0 \rangle \Rightarrow \mathsf{success}$ is demonically valid w.r.t. $\mathcal{R}_{>\mathsf{c_{max}}}^\forall$, then for any ground term $t_0 \in \langle s_0 \mid \phi_0 \rangle$ and any ground term t such that $t_0 \to_{\mathcal{R}}^* t$ and $t|_1 = \mathsf{wait}_j$, every (possibly infinite)* execution path starting from t includes a ground term u such that $u|_1 = \text{crit}$ (i.e., $\mathcal{R}$ is starvation free for P_1).

Proof. We proceed by contradiction. Assume that $\langle s_0 \triangleleft 0 \mid \phi_0 \rangle \Rightarrow \text{success}$ is demonically valid w.r.t. $\mathcal{R}^\forall_{>c_{\max}}$, and there exist a ground term $t_0 \in \langle s_0 \mid \phi_0 \rangle$ and ground terms $t, t_1, t_2, \ldots$ such that $t_0 \to^*_\mathcal{R} t$, $t|_1 = \text{wait}_j$, $t \to_\mathcal{R} t_1 \to_\mathcal{R} t_2 \to_\mathcal{R} \cdots$, and $t_i|_1 \neq \text{crit}$ for all $i \geq 1$. It follows from the construction of $\mathcal{R}^\forall_{>c_{\max}}$ that $(t_0 \triangleleft 0) \to_{\mathcal{R}^\forall_{>c_{\max}}} (t \triangleleft n) \to_{\mathcal{R}^\forall_{>c_{\max}}} (t_1 \triangleleft c_1) \to_{\mathcal{R}^\forall_{>c_{\max}}} (t_2 \triangleleft c_2) \to_{\mathcal{R}^\forall_{>c_{\max}}} \cdots$ for some $n \leq c_{\max}$, where $c_1 = n$ and $c_{i+1} = c_i + 1$ for all $i > 1$. This contradicts the assumption that $\langle s_0 \triangleleft 0 \mid \phi_0 \rangle \Rightarrow \text{success}$ is demonically valid w.r.t. $\mathcal{R}^\forall_{>c_{\max}}$. □

Note that in the above proof, the rule $\text{state}(\vec{p}, \vec{x}) \to \text{success}$ in $\mathcal{R}^\forall_{>c_{\max}}$ is not used, which is necessary to obtain an APR proof; if $\mathcal{R}$ is not deadlock free, there exists a reachable normal form t, and $\mathcal{R}$ does not include $\text{state}(\vec{p}, \vec{x}) \to \text{success}$, then we cannot apply any inference rule to an APR problem $\langle s \mid \phi \rangle \Rightarrow c$ with $t \in \langle s \mid \phi \rangle$ and thus, we cannot obtain any APR proof.

Regarding an APR proof of $\mathcal{R}^\forall_{>c_{\max}}$, by definition, the parent node of a node with $\langle \text{error} \mid \phi \rangle \Rightarrow \text{success}$ has exactly one child.

Proposition 3.4. *Let $\mathcal{P}$ be an APR pre-proof $((N, a, r, p), \xi)$ w.r.t. $\mathcal{R}^\forall_{>c_{\max}}$, and v, v' nodes in N such that v' is the parent of v and $a(v) = (\langle \text{error} \mid \phi \rangle \Rightarrow \text{success})$. Then, v' does not have any other child, i.e., $p(v') = v$.*

4 A Disproof Criterion for Starvation Freedom

Using the method shown in Sect. 3, we try to prove APR problems for starvation freedom. As described in Sect. 1, the method may fail because of several reasons, even when we obtain an APR disproof which may not disprove starvation freedom. In this section, we show a disproof criterion for starvation freedom focusing on a path of APR pre-proofs that reaches a node for error states.

Consider an execution path $t_0 \to_{\mathcal{R}^\forall} t_1 \to_{\mathcal{R}^\forall} \cdots \to_{\mathcal{R}^\forall} t_n \to_{\mathcal{R}^\forall} \text{error}$ of $\mathcal{R}^\forall$ such that $t_0 \in \langle s_0 \triangleleft 0 \mid \phi_0 \rangle$. Let $(t'_j \triangleleft c_j) = t_j$ for all $0 \leq j \leq n$. If there exists some $i \in \{0, \ldots, n-1\}$ such that t_i and t_n agree except for their counter values (i.e., $t'_i = t'_n$), then there exists a loop $t'_i \to_\mathcal{R} t'_{i+1} \to_\mathcal{R} \cdots \to_\mathcal{R} t'_n (= t'_i)$ of $\mathcal{R}$. If $c_n > 0$ and P_1 is waiting in the loop, then P_1 wants to enter the critical section crit but does not enter crit, i.e., starvation for P_1 occurs in the infinite execution path $t'_i \to_\mathcal{R} t'_{i+1} \to_\mathcal{R} \cdots \to_\mathcal{R} t'_n (= t'_i) \to_\mathcal{R} t'_{i+1} \to_\mathcal{R} \cdots$. The premise "$c_n > 0$ and P_1 is waiting in the loop" can be reduced to whether $c_n - c_i = n - i$ holds: Since $0 \leq i < n$, at the state t_n, P_1 is waiting for entering crit; if $c_n - c_i = n - i$ holds, then the counter is incremented at every step of the loop, and thus, P_1 does not enter crit in the loop, waiting for entering crit.

As shown in Example 3.2, the APR problem $\langle \text{state}(\text{noncrit}_1, \text{noncrit}_2, y', c) \mid y'=1 \land c=0 \rangle \Rightarrow \text{success}$ is not demonically valid w.r.t. $\tilde{\mathcal{R}}^\forall_1$ with $c_{\max} = 3$ due to, e.g., the following execution path

$$\text{state}(\text{noncrit}_1, \text{noncrit}_2, 1, 0) \to_{\tilde{\mathcal{R}}^\forall_1} \text{state}(\text{wait}_1, \text{noncrit}_2, 1, 1)$$
$$\to_{\tilde{\mathcal{R}}^\forall_1} \text{state}(\text{wait}_1, \text{wait}_2, 1, 2) \to_{\tilde{\mathcal{R}}^\forall_1} \text{state}(\text{wait}_1, \text{crit}_2, 0, 3)$$
$$\to_{\tilde{\mathcal{R}}^\forall_1} \text{state}(\text{wait}_1, \text{noncrit}_2, 1, 4) \to_{\tilde{\mathcal{R}}^\forall_1} \text{error}$$

The second and fifth terms $\mathsf{state}(\mathsf{wait}_1, \mathsf{noncrit}_2, 1, 1), \mathsf{state}(\mathsf{wait}_1, \mathsf{noncrit}_2, 1, 4)$ in the above sequence agree except for their counter values and the difference between the counter values, $4-1$, coincides with the number of steps, i.e., 3. This means that P_2 can keep transitioning while P_1 is waiting. Hence, starvation for P_1 occurs in $\mathcal{R}_1$. Note that the above execution path corresponds to the path (1),(2),(5),(9),(12),(16) of the APR disproof in Fig. 6.

Theorem 4.1. *Let $\mathcal{P} = (\mathcal{T}, \xi)$ with $\mathcal{T} = (N, a, r, p)$ be an APR pre-proof of the APR problem $\langle s_0 \triangleleft 0 \mid \phi_0 \rangle \Rightarrow \mathsf{success}$ of $\mathcal{R}^\forall$, and $t_0 = (s_0 \triangleleft 0)$. Suppose that there exists a path $v_0, v_1, \ldots, v_n, v$ $(n > 1)$ of $\mathcal{T}$ such that $a(v_j) = (\langle t_j \mid \phi_j \rangle \Rightarrow \mathsf{success})$ for all $0 \leq j \leq n$, and $a(v) = (\langle \mathsf{error} \mid \phi' \rangle \Rightarrow \mathsf{success})$. Let $(t_j' \triangleleft c_j) = t_j$ for all $0 \leq j \leq n$. If $\langle t_i' \mid \phi_i \rangle \supseteq \langle t_n' \mid \phi_n \rangle$ and $c_n - c_i = n - i$ for some $i \in \{0, \ldots, n-2\}$, then there exists an infinite execution path $t_0' \gamma_0 \to_\mathcal{R} t_1' \gamma_1 \to_\mathcal{R} \cdots \to_\mathcal{R} t_i' \gamma_i \to_\mathcal{R} t_{i+1}' \gamma_{i+1} \to_\mathcal{R} \cdots \to_\mathcal{R} t_n' \gamma_n (= t_i' \gamma_i') \to_\mathcal{R} t_{i+1}' \gamma_{i+1}' \to_\mathcal{R} \cdots$ for some substitutions $\gamma_0, \gamma_1, \ldots, \gamma_n, \gamma_i', \gamma_{i+1}', \ldots$ such that $t_0' \gamma_0 \in \langle s_0 \mid \phi_0 \rangle$ (i.e., starvation for P_1 occurs in $\mathcal{R}$).*

Proof. It follows from Lemma 2.8 that there exists some reduction sequence $t_0 \gamma_0 \to_{\mathcal{R}^\forall_{>c_{\max}}} t_1 \gamma_1 \to_{\mathcal{R}^\forall_{>c_{\max}}} \cdots \to_{\mathcal{R}^\forall_{>c_{\max}}} t_n \gamma_n$ such that $t_j \gamma_j \in \langle t_j \mid \phi_j \rangle$ for all $0 \leq j \leq n$. In constructing $\mathcal{R}^\forall_{>c_{\max}}$ from $\mathcal{R}$, for any rule $\ell \to r \ [\varphi]$ in $\mathcal{R}$, a constraint to update the counter is conjunct to φ. Thus, the conjunct constraint does not affect the applicability of the generate rule in $\mathcal{R}^\forall_{>c_{\max}}$ and hence we have the reduction sequence $t_0' \gamma_0 \to_\mathcal{R} t_1' \gamma_1 \to_\mathcal{R} \cdots \to_\mathcal{R} t_i' \gamma_i \to_\mathcal{R} t_{i+1}' \gamma_{i+1} \to_\mathcal{R} \cdots \to_\mathcal{R} t_n' \gamma_n$. It follows from $c_n - c_i = n - i$ that the reduction sequence of $\mathcal{R}$ includes no reduction step by a rewrite rule of P_1. It follows from $\langle t_i' \mid \phi_i \rangle \supseteq \langle t_n' \mid \phi_n \rangle$ that there exists a substitution γ_i' such that $t_i' \gamma_i' \in \langle t_i' \mid \phi_i \rangle$ and hence we have the reduction sequence $t_i' \gamma_i' \to_\mathcal{R} t_{i+1}' \gamma_{i+1}' \to_\mathcal{R} \cdots \to_\mathcal{R} t_n' \gamma_n'$. Repeating this process infinitely many times, we obtain an infinite execution path $t_0' \gamma_0 \to_\mathcal{R} t_1' \gamma_1 \to_\mathcal{R} \cdots \to_\mathcal{R} t_i' \gamma_i \to_\mathcal{R} t_{i+1}' \gamma_{i+1} \to_\mathcal{R} \cdots \to_\mathcal{R} t_n' \gamma_n (= t_i' \gamma_i') \to_\mathcal{R} t_{i+1}' \gamma_{i+1}' \to_\mathcal{R} \cdots$. □

5 Starvation Verification Under Process Fairness

When we do not assume *process fairness*, the AITS (i.e., $\mathcal{R}_1$) in Example 1.1 is not starvation free for P_1. On the other hand, as mentioned before, the AITS is starvation free for P_1 under process fairness. In this section, we propose a method to, in constructing APR pre-proofs, ignore a leaf with $\langle \mathsf{error} \mid \phi \rangle \Rightarrow \mathsf{success}$ such that repeating a path to the parent of the leaf implies an unfair execution path.

First, we formally define *process fairness* for concurrent LCTRSs.

Definition 5.1 (process fairness). *Let $\mathcal{R}$ be a concurrent LCTRS. An infinite execution path $t_1 \to_\mathcal{R} t_2 \to_\mathcal{R} \cdots$ is said to be* process fair *if for each i-th process P_i, rewrite rules for P_i are applied infinitely often, whenever rules for P_i are applicable infinitely often; otherwise, the path is called* process unfair.

Let us consider the finite APR disproof considered in Example 3.2 again. The disproof includes the three paths to leaves with $\langle \mathsf{error} | \ldots \rangle \Rightarrow \mathsf{success}$, e.g.,

(1) $\mathsf{state}(\mathsf{noncrit}_1, \mathsf{noncrit}_2, 1, 0) \to_{\tilde{\mathcal{R}}_1^\vee}$ (2) $\mathsf{state}(\mathsf{wait}_1, \mathsf{noncrit}_2, 1, 1)$
$\to_{\tilde{\mathcal{R}}_1^\vee}$ (5) $\mathsf{state}(\mathsf{wait}_1, \mathsf{wait}_2, 1, 2) \to_{\tilde{\mathcal{R}}_1^\vee}$ (9) $\mathsf{state}(\mathsf{wait}_1, \mathsf{crit}_2, 0, 3)$
$\to_{\tilde{\mathcal{R}}_1^\vee}$ (12) $\mathsf{state}(\mathsf{wait}_1, \mathsf{noncrit}_2, 1, 4) \to_{\tilde{\mathcal{R}}_1^\vee}$ (16) error

As explained in Sect. 4, the path (2), (5), (9), (12) causes starvation along the infinite path (5), (9), (12), (5), (9), (12), ..., where (2) and (12) are identified. At the nodes (5) and (12), process P_1 can be transitioned, and thus, the infinite path is process unfair. In addition, the node (12) has the one child (16). This means that the infinite path is the only one including (12). Thus, in considering process-fairness, we can ignore the node (16) which is a witness of the demonical invalidity. From the viewpoint of an APR pre-proof $((N, a, r, p), \xi)$, the above reduction is characterized as a path $v_i, v_{i+1}, \ldots, v_n, v$ ($0 \leq i < n - 1$) such that

- $a(v) = (\langle \mathsf{error} | \phi \rangle \Rightarrow \mathsf{success})$ (i.e., $r(v) = \mathsf{dis}$),
- $a(v_j) = (\langle t_j | \phi_j \rangle \Rightarrow \mathsf{success})$, and $(t'_j \triangleleft c_j) = t_j$ for all $i \leq j \leq n$,
- $\langle t'_i | \phi_i \rangle \supseteq \langle t'_n | \phi_n \rangle$,
- $\phi_n \implies (c_n - c_i = n - i)$ is valid, and
- a rewrite rule for P_1 is applicable to $\langle t_j | \phi_j \rangle$ for some $j \in \{i, \ldots, n-2\}$.

The leaf v is said to be *process unfair*. Note that v_n has exactly one child v (Proposition 3.4). In the example above, we have that $n = 4$, $v_1 = (5)$, $v_n = (12)$, $v = (16)$, $\phi_n = \mathsf{true}$, $c_n = 4$, and $c_1 = 1$.

Theorem 5.2. *Let $\mathcal{P} = (\mathcal{T}, \xi)$ be a finite APR pre-proof of an APR problem $\langle s_0 \triangleleft 0 | \phi_0 \rangle \Rightarrow \mathsf{success}$ w.r.t. $\mathcal{R}^\vee$. Assume that every open leaf v of $\mathcal{P}$ is either bud or process unfair. Then, $\mathcal{R}$ is starvation free for P_1 under process fairness.*

Proof. Let $\mathcal{T} = (N, a, r, p)$. Since the case where $\mathcal{P}$ is an APR proof is trivial, we consider the remaining case where there exists a process-unfair node v in $\mathcal{T}$. We assume w.l.o.g. that there exists exactly one process-unfair leaf v. Let u be the parent of v and t' a term such that $a(u) = (\langle t | \psi \rangle \Rightarrow \mathsf{success})$ and $(t' \triangleleft c') = t$. Let u' be an ancestor of u such that u' is a witness making v process-unfair. By removing counters from nodes, we obtain an APR proof $\mathcal{P}' = ((N \setminus \{v\}, a', r', p'), \xi')$ such that $a'(u) = (\langle t' | \psi \rangle \Rightarrow \mathsf{success})$, both $r'(u)$ and $p'(u)$ are undefined, $\xi'(u) = u'$, and for any node $v' \in N \setminus \{v, u\}$, $a'(v') = (\langle s'' | \phi' \rangle \Rightarrow \mathsf{success})$, $r'(v') = r(v')$, $p'(v') = p(v')$, and $\xi'(v') = \xi(v')$. where $a(v') = (\langle s' | \phi' \rangle \Rightarrow \mathsf{success})$ and $(s'' \triangleleft c') = s'$. Assume that there exists a reduction sequence of $\mathcal{R}$ that causes starvation for P_1 under process fairness. Then, the reduction sequence includes infinitely many instances of t' and is along an infinite path obtained from the path from u' to u by appending it infinitely many times. Since v is a process-unfair leaf, the reduction sequence is process unfair w.r.t. P_1. This contradicts the assumption that the reduction sequence is process fair. Therefore, $\mathcal{R}$ is starvation free under process fairness. □

Note that for starvation freedom under process fairness, not all unfair execution paths along paths of the APR pre-proof in Theorem 5.2 have to be ignored.

Example 5.3. Consider the modified LCTRS $\tilde{\mathcal{R}}_1^\forall$ in Example 3.2 and the finite APR pre-proof in Fig. 6. The APR pre-proof has the three leaves (16), (20), and (21) associated with dis: $r((16)) = r((20)) = r((21)) = $ dis. The three leaves (16), (20), and (21) are process unfair, and all other leaves are bud. Therefore, by Theorem 5.2, $\mathcal{R}_1$ is starvation free for P_1 under process fairness.

When we do not consider process fairness, we may stop the construction of an APR pre-proof immediately if we generate a leaf v with $r(v) = $ dis. Conversely, in considering process fairness, we do not stop the construction even if such a leaf is found. We try to construct a finite APR pre-proof, and once we obtain a finite pre-proof, we check whether each open leaf is either bud or process unfair. If so, we conclude that $\mathcal{R}$ is starvation free for P_1 under process fairness, and otherwise, we fail to prove starvation freedom for P_1 under process fairness.

6 Conclusion

In this paper, for the APR-based framework [13] for starvation freedom, we showed a disproof criterion for starvation freedom and extended the APR framework to starvation freedom under process fairness. We have implemented the proof system c-DCC in a prototype of Crisys2, an interactive proof system based on constrained rewriting induction [8,18].[5] The implemented tool Crisys2cdcc succeeded in proving the APR problem in Examples 5.3 under process fairness: The execution time is 0.21 s on a machine running MacOS 14.5 on Apple M2 8 cores with 24GB memory; Z3 (ver. 4.12.6) [19] was used as an external SMT solver. Note that the tool succeeded in disproving the APR problem in Example 3.2 in 0.13 s.

As described in [13], by considering, e.g., state as a predicate, LCTRSs representing AITSs must be seen as constrained Horn clauses (CHC, for short), while LCTRSs model functional or imperative programs with, e.g., function calls and global variables (see, e.g., [9]). In future work, the approach in [13] and this paper should be compared with the application of CHC-based approaches in, e.g., [11] from the theoretical and/or empirical viewpoint.

To improve the power of proving APR problems, a generalization method for APR problems is necessary as for other inductive proving methods such as *rewriting induction* [23] (cf. [8]). A future direction of this research is to develop such a technique.

References

1. Audemard, G., Bertoli, P., Cimatti, A., Korniłowicz, A., Sebastiani, R.: A SAT based approach for solving formulas over Boolean and linear mathematical propositions. In: Voronkov, A. (ed.) CADE 2002. LNCS (LNAI), vol. 2392, pp. 195–210. Springer, Heidelberg (2002). https://doi.org/10.1007/3-540-45620-1_17

---

[5] https://www.trs.css.i.nagoya-u.ac.jp/crisys/rp2024/.

2. Baader, F., Nipkow, T.: Term Rewriting and All That. Cambridge University Press, Cambridge (1998). https://doi.org/10.1145/505863.505888
3. Baier, C., Katoen, J.: Principles of Model Checking. MIT Press, Cambridge (2008)
4. Brotherston, J.: Cyclic proofs for first-order logic with inductive definitions. In: Beckert, B. (ed.) TABLEAUX 2005. LNCS (LNAI), vol. 3702, pp. 78–92. Springer, Heidelberg (2005). https://doi.org/10.1007/11554554_8
5. Ciobâcă, Ș, Lucanu, D., Buruiana, A.: Operationally-based program equivalence proofs using LCTRSs. J. Log. Algebraic Methods Program. **135**, 1–22 (2023). https://doi.org/10.1016/j.jlamp.2023.100894
6. Ciobâcă, Ș, Lucanu, D.: A coinductive approach to proving reachability properties in logically constrained term rewriting systems. In: Galmiche, D., Schulz, S., Sebastiani, R. (eds.) IJCAR 2018. LNCS (LNAI), vol. 10900, pp. 295–311. Springer, Cham (2018). https://doi.org/10.1007/978-3-319-94205-6_20
7. Fernández, M.: Programming Languages and Operational Semantics - A Concise Overview. Undergraduate Topics in Computer Science. Springer, Heidelberg (2014). https://doi.org/10.1007/978-1-4471-6368-8
8. Fuhs, C., Kop, C., Nishida, N.: Verifying procedural programs via constrained rewriting induction. ACM Trans. Comput. Log. **18**(2), 14:1–14:50 (2017). https://doi.org/10.1145/3060143
9. Kanazawa, Y., Nishida, N.: On transforming functions accessing global variables into logically constrained term rewriting systems. In: Niehren, J., Sabel, D. (eds.) Proceedings of the 5th International Workshop on Rewriting Techniques for Program Transformations and Evaluation. Electronic Proceedings in Theoretical Computer Science, vol. 289, pp. 34–52. Open Publishing Association (2019)
10. Kanazawa, Y., Nishida, N., Sakai, M.: On representation of structures and unions in logically constrained rewriting. IEICE Technical Report SS2018-38, the Institute of Electronics, Information and Communication Engineers, vol. 118, no. 385, pp. 67–72 (2019). In Japanese
11. Kobayashi, N., Nishikawa, T., Igarashi, A., Unno, H.: Temporal verification of programs via first-order fixpoint logic. In: Chang, B.-Y.E. (ed.) SAS 2019. LNCS, vol. 11822, pp. 413–436. Springer, Cham (2019). https://doi.org/10.1007/978-3-030-32304-2_20
12. Kojima, M., Nishida, N.: On reducing non-occurrence of specified runtime errors to all-path reachability problems of constrained rewriting. In: Ciobaca, S., Nakano, K. (eds.) Informal Proceedings of the 9th International Workshop on Rewriting Techniques for Program Transformations and Evaluation, pp. 1–16 (2022). https://easychair.org/publications/preprint/TM7q
13. Kojima, M., Nishida, N.: From starvation freedom to all-path reachability problems in constrained rewriting. In: Hanus, M., Inclezan, D. (eds.) PADL 2023. LNCS, vol. 13880, pp. 161–179. Springer, Cham (2023). https://doi.org/10.1007/978-3-031-24841-2_11
14. Kojima, M., Nishida, N.: Reducing non-occurrence of specified runtime errors to all-path reachability problems of constrained rewriting. J. Log. Algebraic Methods Program. **135**, 1–19 (2023). https://doi.org/10.1016/j.jlamp.2023.100903
15. Kojima, M., Nishida, N.: A sufficient condition of logically constrained term rewrite systems for decidability of all-path reachability problems with constant destinations. J. Inf. Process. **32**, 417–435 (2024). https://doi.org/10.2197/ipsjjip.32.417
16. Kojima, M., Nishida, N., Matsubara, Y.: Transforming concurrent programs with semaphores into logically constrained term rewrite systems. In: Riesco, A., Nigam, V. (eds.) Informal Proceedings of the 7th International Workshop on Rewriting Techniques for Program Transformations and Evaluation, pp. 1–12 (2020)

17. Kop, C., Nishida, N.: Term rewriting with logical constraints. In: Fontaine, P., Ringeissen, C., Schmidt, R.A. (eds.) FroCoS 2013. LNCS (LNAI), vol. 8152, pp. 343–358. Springer, Heidelberg (2013). https://doi.org/10.1007/978-3-642-40885-4_24
18. Kop, C., Nishida, N.: Automatic constrained rewriting induction towards verifying procedural programs. In: Garrigue, J. (ed.) APLAS 2014. LNCS, vol. 8858, pp. 334–353. Springer, Cham (2014). https://doi.org/10.1007/978-3-319-12736-1_18
19. de Moura, L., Bjørner, N.: Z3: an efficient SMT solver. In: Ramakrishnan, C.R., Rehof, J. (eds.) TACAS 2008. LNCS, vol. 4963, pp. 337–340. Springer, Heidelberg (2008). https://doi.org/10.1007/978-3-540-78800-3_24
20. Nishida, N., Winkler, S.: Loop detection by logically constrained term rewriting. In: Piskac, R., Rümmer, P. (eds.) VSTTE 2018. LNCS, vol. 11294, pp. 309–321. Springer, Cham (2018). https://doi.org/10.1007/978-3-030-03592-1_18
21. Ohlebusch, E.: Advanced Topics in Term Rewriting. Springer, Heidelberg (2002). https://doi.org/10.1007/978-1-4757-3661-8
22. Peterson, G.L.: Myths about the mutual exclusion problem. Inf. Process. Lett. **12**(3), 115–116 (1981). https://doi.org/10.1016/0020-0190(81)90106-X
23. Reddy, U.S.: Term rewriting induction. In: Stickel, M.E. (ed.) CADE 1990. LNCS, vol. 449, pp. 162–177. Springer, Heidelberg (1990). https://doi.org/10.1007/3-540-52885-7_86
24. Rosu, G., Serbanuta, T.: An overview of the K semantic framework. J. Log. Algebraic Methods Program. **79**(6), 397–434 (2010). https://doi.org/10.1016/j.jlap.2010.03.012
25. Ştefănescu, A., Ciobâcă, Ş, Mereuta, R., Moore, B.M., Şerbănuță, T.F., Roşu, G.: All-path reachability logic. In: Dowek, G. (ed.) RTA 2014. LNCS, vol. 8560, pp. 425–440. Springer, Cham (2014). https://doi.org/10.1007/978-3-319-08918-8_29
26. Stefanescu, A., Ciobâcă, Ş., Mereuta, R., Moore, B.M., Serbanuta, T., Rosu, G.: All-path reachability logic. Log. Methods Comput. Sci. **15**(2) (2019). https://doi.org/10.23638/LMCS-15(2:5)2019
27. Winkler, S., Middeldorp, A.: Completion for logically constrained rewriting. In: Kirchner, H. (ed.) Proceedings of the 3rd International Conference on Formal Structures for Computation and Deduction. Leibniz International Proceedings in Informatics, vol. 108, pp. 30:1–30:18. Schloss Dagstuhl–Leibniz-Zentrum für Informatik (2018). https://doi.org/10.4230/LIPIcs.FSCD.2018.30

Automata and Complexity

Rollercoasters with Plateaus

Duncan Adamson[1], Pamela Fleischmann[2], and Annika Huch[2(✉)]

[1] University of St Andrews, St Andrews, UK
duncan.adamson@st-andrews.ac.uk
[2] Kiel University, Kiel, Germany
{fpa,ahu}@informatik.uni-kiel.de

Abstract. In this paper we investigate the problem of detecting, counting, and enumerating (generating) all maximum length plateau-k-rollercoasters appearing as a subsequence of some given word (sequence, string), while allowing for plateaus. We define a plateau-k-rollercoaster as a word consisting of an alternating sequence of (weakly) increasing and decreasing *runs*, with each run containing at least k *distinct* elements, allowing the run to contain multiple copies of the same symbol consecutively. This differs from previous work, where runs within rollercoasters have been defined only as sequences of distinct values. Here, we are concerned with rollercoasters of *maximum* length embedded in a given word w, that is, the longest rollercoasters that are a subsequence of w. We present algorithms allowing us to determine the longest plateau-k-rollercoasters appearing as a subsequence in any given word w of length n over an alphabet of size σ in $O(n\sigma k)$ time, to count the number of plateau-k-rollercoasters in w of maximum length in $O(n\sigma k)$ time, and to output all of them with $O(n)$ delay after $O(n\sigma k)$ preprocessing. Furthermore, we present an algorithm to determine the longest common plateau-k-rollercoaster within a set of words in $O(Nk\sigma)$ where N is the product of all word lengths within the set.

Keywords: k-Rollercoaster · Plateaus · Enumeration · (Longest Common) Subsequences · Scattered Factors

1 Introduction

Subsequences, also known as scattered factors, of words are heavily studied with various motivations during the last decades: longest increasing or decreasing (contiguous) subsequence [1,14,15], longest common subsequence [11,12,18] and shortest common supersequence [5,9,13], string-to-string-correction [17], and questions related to bioinformatics (e.g. [16] and the references therein). The first two problems can be combined into the notion of *rollercoasters* which are - roughly speaking - sequences such that increasing and decreasing contiguous subsequences alternate. One is interested in the longest rollercoaster as a subsequence of a given sequence (cf. Biedl et al. [2,3]). More formally, a *run* is a maximal contiguous subsequence of a given sequence that is strictly increasing or strictly decreasing and, as introduced in [4], a *k-rollercoaster* is defined as a sequence over the real numbers such that every run is of length at least $k \geq 3$. For instance, the sequence $r = (4, 5, 7, 6, 3, 2, 7, 8)$ is a 3-rollercoaster since the runs $(4, 5, 7)$, $(7, 6, 3, 2)$, and $(2, 7, 8)$ are increasing resp. decreasing runs of length at least 3. Regarding the problem of finding the longest rollercoaster in a given sequence, the longest rollercoaster of $(5, 3, 6, 7, 1, 6, 4, 1)$ is $(5, 6, 7, 6, 4, 1)$. Notice that not all sequences contain a rollercoaster witnessed by $(5, 3, 5, 3)$ - all runs are only of length 2. Thus, for a subsequence s being a rollercoaster there has to exist a $k \geq 3$ such that s is a k-rollercoaster.

Rollercoasters over the real numbers as alphabet are heavily studied. Biedl et al. [2,3] introduced and solved the rollercoaster problem parametrised in the number $\ell \in \mathbb{N}$ of different letters in the given sequence by presenting an $O(\ell \log(\ell))$-time algorithm. Furthermore, they showed that a word of length n does contain a rollercoaster of length at least $\lceil \frac{n}{2} \rceil$ for $n > 7$. In their constructive proof they also gave an algorithm that computes a rollercoaster of this length in $O(n \log(n))$. For a word that is a permutation of $\{1, \ldots, n\}$ they gave an $O(n \log(\log(n)))$ solution. Gawrychowski et al. [8] later improved the runtime of the algorithm such that k-rollercoasters can be found in time $O(\ell k^2)$ for sequences with ℓ distinct letters. This result shows that longest rollercoasters are related to longest increasing subsequences, that are extensively studied, e.g., [1,6,14,15]. Moreover, in [7], Fujita et al. solved the *longest common rollercoaster problem* that is to find the longest common rollercoaster contained as a subsequence in two given words by providing two algorithms. The first runs in $O(nmk)$ time and space, where n, m are respectively the lengths of the two words. The second runs in $O(rk \log^3 m \log(\log(m)))$ time and $O(rk)$ space where $r = O(mn)$ is the number of pairs (i, j) of matching positions in the two words that need to contain the same set of letters.

Our Contribution. Generalising these ideas, we focus our work on rollercoasters with *plateaus*. We refer to a plateau in a rollercoaster as a factor of the rollercoaster containing only a single type of letter. For example, the rollercoaster $(1, 1, 2, 2, 3)$ contains the plateaus $(1, 1)$ and $(2, 2)$, while still being a 3-rollercoaster. We restrict our definition of k-rollercoaster to require that each maximal increasing and decreasing run contains k *unique* letters, rather than simply having length k. Therefore, $(1, 1, 2, 2, 3)$ is a plateau-3-rollercoaster, however despite containing a weakly increasing subsequence of length 5, it is not a plateau-5-rollercoaster. We focus on three algorithmic questions regarding plateau-k-rollercoaster that are related to reachability: first, we give an algorithm that determines the length of the longest plateau-k-rollercoaster appearing as a subsequence of a given word of length n, and therefore the minimum number of deletions to reach a maximum length plateau-k-rollercoaster from a given word of length n in $O(n\sigma k)$ time. We additionally provide an algorithm for outputting all maximum length plateau-k-rollercoasters that can be reached from an input word with at most linear delay relative to the length of the input word. Finally, we give an algorithm to compute the longest common plateau-k-rollercoaster within a set given words in $O(Nk\sigma)$ time where N is the product of all word lengths within the set.

Structure of the Work. In Sect. 2, the basic definitions are presented. In the following section we present the algorithm for computing the longest plateau-rollercoaster within a word and one for enumerating all longest plateau-rollercoasters. Section 4 considers the problem of finding the longest common plateau-rollercoasters within a set of words.

2 Preliminaries

Let $\mathbb{N}$ be the set of all natural numbers, $\mathbb{N}_0 = \mathbb{N} \cup \{0\}$, $[n] = \{1, \ldots, n\}$, $[m, n] = \{m, m+1, \ldots, n\}$, and $[n]_0 := [n] \cup \{0\}$ for $m, n \in \mathbb{N}$ with $m \leq n$. An *alphabet* Σ is a non-empty finite set whose elements are called *letters*. We assume w.l.o.g. $\Sigma = \{1, 2, \ldots, \sigma\}$ for some $\sigma \in \mathbb{N}$, with the usual order on $\mathbb{N}$. A *word* over an alphabet Σ is a finite sequence of letters from Σ, with the length of a word w, denoted $|w|$, being the number of letters in w. We define the *empty word* ε as the word containing no symbols, i.e., $|\varepsilon| = 0$. Let Σ^* denote the set of all words over Σ and Σ^n the set of all words of length $n \in \mathbb{N}$. Given a word $w \in \Sigma^n$, and integer $i \in [n]$, we denote by $w[i]$ the ith symbol in w, and therefore $w = w[1]w[2]\cdots w[n]$. Similarly, given a pair of integers $i, j \in [n]$ such that $i \leq j$, we denote by $w[i, j]$ the word $w[i]w[i+1]\cdots w[j]$ and set $w[j, i] = \varepsilon$ for $j > i$. We denote by $\mathrm{alph}(w)$ the set of unique letters in w, giving $\mathrm{alph}(w) = \{w[i] \mid i \in [|w|]\}$ The word $u \in \Sigma^*$ is called a *factor* of w if there exist $x, y \in \Sigma^*$ such that $w = xuy$. In the case $x = \varepsilon$, we call u a *prefix* of w and a *suffix* if $y = \varepsilon$. A word u is a *subsequence* (also known as a scattered factor) of w if there exists some set of indices $i_1, i_2, \ldots, i_{|u|} \in [n]$ such that $i_1 < i_2 < \cdots < i_{|u|}$ and $u = w[i_1]w[i_2]\ldots w[i_{|u|}]$. By w^R we denote the reversed word, i.e., $w^R = w[|w|]w[|w|-1]\cdots w[1]$.

Informally, a plateau-run is a (weakly) increasing or decreasing word. For instance, $w = 123345556$ is a plateau-6-run since firstly the letters in w are increasing. The subsequence 123456 is the longest strictly increasing run in w. Moreover, w contains the 5-plateau runs 12334555 and 23345556 as well as several shorter plateau-runs. Notice that none of these plateau-runs is maximal. The word 544133465 contains the maximal decreasing 3-plateau-run 5441 and the maximal increasing 4-plateau-run 13346.

Definition 1. *A factor $u = w[i, j]$ of $w \in \Sigma^n$ is a plateau-run if, for all $\ell \in [|u|-1]$, we either have $u[\ell] \leq u[\ell+1]$ or $u[\ell] \geq u[\ell+1]$. The orientation of a run is $\uparrow$ (increasing) if $u[\ell] \leq u[\ell+1]$, or $\downarrow$ (decreasing) if $u[\ell] \geq u[\ell+1]$ for every $\ell \in [|u|-1]$. Such a plateau-run is called maximal if $w[i-1] > u[1]$ and $u[|u|] > w[j+1]$ or $w[i-1] < u[1]$ and $u[|u|] < w[j+1]$ resp. (notice that if u is a prefix or suffix of w the related constraints on the maximality are omitted). A plateau-run u is a plateau-k-run for some $k \in \mathbb{N}$ if we have $|\mathrm{alph}(u)| \geq k$.*

Given a variable $\xi \in \{\uparrow, \downarrow\}$, we use $\overline{\xi}$ to denote the opposite orientation, i.e., $\overline{\uparrow} = \downarrow$ and $\overline{\downarrow} = \uparrow$.

Remark 2. Notice that one obtains the classical *run* introduced in [2,3] by changing $\leq$ and $\geq$ in Definition 1 into $<$ and $>$.

We define rollercoasters in our setting as a class of words containing alternating (weakly) increasing and decreasing factors that are neither left- nor right-extendable and all contain at least 3 distinct letters. For instance, 12345435667 is a plateau-3-rollercoaster consisting of the maximal plateau-3-runs 12345, 543, 35667 while 1234554567 is not a plateau-3-rollercoaster since 554 is a maximal run but not a plateau-3-run.

Definition 3. *A word $w \in \Sigma^*$ is a plateau-k-rollercoaster for $k \in \mathbb{N}, k \geq 3$ if every maximal plateau-run r in w is a plateau-k-run.*

Remark 4. Note that the required maximality of the runs within a plateau-rollercoaster implies that their orientation is alternating, i.e., if the ith run of w is an increasing run, then w's $(i+1)^{th}$ run is decreasing and vice versa.

Remark 5. First, notice that plateau-k-rollercoasters are an extension of classical rollercoasters by allowing plateau-k-runs instead of classical runs. In contrast to the classical rollercoaster, in plateau-k-rollercoasters the runs may overlap in more than one letter. Consider, for instance, the plateau-4-rollercoaster $w = 12223444321112345$ with the plateau-runs $r_1 = 12223444$, $r_2 = 44432111$, and $r_3 = 1112345$. In order to decompose a rollercoaster into its runs we define the concatenation $x_1 \overline{\cdot} x_2$ by $x_1 p^{-1} x_2$ where p is the longest prefix of x_2 which is also a suffix of x_1, for $x_1, x_2 \in \Sigma^*$, and $p^{-1} x_2 = y$ iff $x_2 = py$.

We extend the notion of $(k,h)_w$-rollercoasters as introduced by Biedl et al. [2,3] by not counting the length of the last run but the number of letters that are part of the strictly increasing or decreasing subsequence of the last run. This notion will be mainly used within the algorithmic constructions of rollercoasters. Here we follow the idea of forming a rollercoaster by successively appending letters and tracking whether we completed a k-run (which is done by the variable h), i.e., h counts the number of distinct letters in the last run and if this number reached k it is not counted further since any proper extension of the run will be a k-run.

Definition 6. *For $\xi \in \{\uparrow, \downarrow\}$, $k \in \mathbb{N}$ and $h, \ell \in [k]$, we say that a word w is a plateau-$(k,h)_\xi$-rollercoaster if the following properties hold for the decomposition $w = r_1 \overline{\cdot} \ldots \overline{\cdot} r_x$, $x \in \mathbb{N}$ of w into runs:*

1. *The last plateau-run r_x has orientation ξ.*
2. *For all $i \in [x-1]$, r_i is a plateau-k-run. The last plateau-run r_x is a plateau-h-run for $h \in [k]$ with h maximal.*

Remark 7. Note that a plateau-$(k,k)_\xi$-rollercoaster for $\xi \in \{\uparrow, \downarrow\}$ is a classic plateau-k-rollercoaster. Further, note that a plateau-$(k,1)_\uparrow$-rollercoaster is either a unary word or also a plateau-$(k,k)_\downarrow$-rollercoaster since neighboured runs do overlap in their respective ends/beginnings within unary factors (analogously for plateau-$(k,1)_\downarrow$-rollercoaster).

The word $r = 12234322$ is not a plateau-4-rollercoaster since its last run does only contain three distinct letters. Since the last run has orientation $\downarrow$, r is a plateau-$(4,3)_\downarrow$-rollercoaster. Further, consider $w = 43321$ and $k = 3$ which is a plateau-$(3,3)_\downarrow$-rollercoaster but also a plateau-$(3,1)_\uparrow$-rollercoaster since w's last letter not only belongs to the decreasing run but also starts a new increasing run of of length 1 itself.

For the remainder of this work, we are interested in plateau-k-rollercoasters that appear as subsequences of some given word w, utilising plateau-$(k,h)_\xi$-rollercoasters as a major tool in our algorithms. For these algorithmic results we use the standard computational model RAM with logarithmic word-size (see, e.g., [10]), i.e., we follow a standard assumption from stringology, that w is the input word for our algorithms with $\Sigma = \mathrm{alph}(w) = \{1, 2, \ldots, \sigma\}$.

3 Counting and Enumerating Plateau-k-Rollercoasters

In this section we present our results regarding detecting, counting, and enumerating the set of plateau-k-rollercoasters that appear as a subsequence within a given word $w \in \Sigma^n$ for some $n \in \mathbb{N}$. Thus, we are extending the classical rollercoaster problem - given a word $w \in \Sigma^*$, determine the longest subsequence of w which is a rollercoaster - to the plateau-rollercoaster scenario. For better readability, we say that a rollercoaster r is in a word w, if r is indeed a rollercoaster and additionally a subsequence of w.

We start with an algorithm for determining the longest plateau-k-rollercoaster in a given word $w \in \Sigma^n$. To do so, we introduce two sets of tables. First, the set of tables $L_w^{k,h,\xi}$ (**L**ongest plateau-$(k,h)_\xi$-rollercoaster), with $L_w^{k,h,\xi}[i]$ denoting the length of the longest plateau-$(k,h)_\xi$-rollercoaster in $w[1,i]$ ending at position i, noting that this may be different from the length of the longest plateau-$(k,h)_\xi$-rollercoaster in $w[1,i]$. We abuse our notation by using $L_w^{k,k,\xi}[i]$ to denote the length of any proper plateau-k-rollercoaster ending with a ξ-run at $w[i]$. Secondly, we have the $n \times \sigma$ table P where $P_w[i,x]$ contains the index i' such that $i' \in [1,i]$ where $w[i'] = x$ and, $\forall j \in [i'+1, i]$, $w[j] \neq x$.

Remark 8. Notice that P_w can be constructed in time $O(n\sigma)$ for a given $w \in \Sigma^n$.

For $w = 871264435161$, we have $L_w^{3,3,\uparrow}[9] = 7$ since the longest plateau-$(3,3)_\uparrow$-rollercoaster ending in $w[9]$ is 8712445. As a second example with an incomplete run in the end, we have $L_w^{3,2,\downarrow}(w)[8] = 7$ witnessed by the plateau-$(3,2)_\downarrow$-rollercoaster 8712443 ending in $w[8]$. In the following we present full exemplary tables for P_w and $L_w^{3,h,\xi}$ for $h \in [3]$, $\xi \in \{\downarrow, \uparrow\}$ (Table 1). Its longest plateau-3-rollercoasters are given by 8712644311 and 8712644356. Furthermore, note that $L_w^{k,1,\xi}$ and $L_w^{k,k,\bar{\xi}}$ do either represent unary plateau-rollercoasters or contain equal values (cf. Remark 7).

Further, we give some values of the rollercoaster table of w:

- $R_w^3[(11, \uparrow, 3)] = R_w^3[(12, \downarrow, 3)] = 1$, and
- $R_w^3[(3, \downarrow, 2)] = 2$ since both 81 and 71 are plateau-$(3,2)_\downarrow$-rollercoasters that end in $w[3] = 1$.

Table 1. The values of P_w and $L_w^{k,h,\xi}$ (the bold values mark the longest rollercoaster within w).

n\σ	1	2	3	4	5	6	7	8
1	-	-	-	-	-	-	-	1
2	-	-	-	-	-	-	2	1
3	3	-	-	-	-	-	2	1
4	3	4	-	-	-	-	2	1
5	3	4	-	-	-	5	2	1
6	3	4	-	6	-	5	2	1
7	3	4	-	7	-	5	2	1
8	3	4	8	7	-	5	2	1
9	3	4	8	7	9	5	2	1
10	10	4	8	7	9	5	2	1
11	10	4	8	7	9	11	2	1
12	12	4	8	7	9	11	2	1

$i \in [n]$	1	2	3	4	5	6	7	8	9	10	11	12
$L_w^{3,1,\downarrow}$	1	1	1	1	5	5	6	5	7	1	10	1
$L_w^{3,1,\uparrow}$	1	1	3	3	3	4	5	8	4	9	4	10
$L_w^{3,2,\downarrow}$	0	2	2	2	2	6	7	7	4	8	3	9
$L_w^{3,2,\uparrow}$	0	0	0	4	4	4	5	5	9	0	9	0
$L_w^{3,3,\downarrow}$	0	0	3	3	3	4	5	8	4	9	4	**10**
$L_w^{3,3,\uparrow}$	0	0	0	0	5	5	6	5	7	0	**10**	0

Here we can see that the longest plateau-3-rollercoasters that can be reached via deletions from w is of length 10 and are given by 8712644311 and 8712644356. Furthermore, the relation between $L_w^{k,1,\xi}$ and $L_w^{k,k,\overline{\xi}}$ (cf. Remark 7) gets perfectly visible since the respective rows do either represent unary plateau-rollercoasters or contain equal values. Using dynamic programming, we can compute the length of the longest common plateau-k-rollercoaster that ends in position i in a word w.

Lemma 9. *Given a word $w \in \Sigma^n$ and $i, k, h \in [n]$, then $L_w^{k,h,\xi}[i]$ can be determined in $O(\sigma)$ time from $L_w^{k,h',\xi'}[j]$, for all $\xi' \in \{\uparrow, \downarrow\}, h' \in [k], j \in [i-1]$.*

Corollary 10. *Given a word $w \in \Sigma^n$, the value of $L_w^{k,h,\xi}[i]$ can be determined in $O(nk\sigma)$ time for every $\xi \in \{\uparrow, \downarrow\}, h \in [1, k], i \in [1, n]$.*

Theorem 11. *Given a word $w \in \Sigma^n$, we can determine the length of the longest plateau-k-rollercoaster in $O(n\sigma k)$ time.*

Proof. From Corollary 10, we can construct the table $L_w^{k,h,\xi}$ in $O(n\sigma k)$ time. By definition, the longest rollercoaster is the value $\max_{i \in [1,n]} L_w^{k,\xi}[i]$, which may be determined in $O(n)$ time from $L_w^{k,h,\xi}$, giving the stated time complexity. □

Now, we consider the problems of counting and enumerating the set of maximum-length plateau-rollercoasters in w. To do so, we define for a word w the *k-rollercoaster table*, R_w. Informally, R_w can be thought of as an extension of $L_w^{k,h,\xi}$, storing not only the longest plateau-$(k, h)_\xi$-rollercoasters ending at each position, but also the number of such rollercoasters.

Definition 12 (Rollercoaster Table). *Given a word $w \in \Sigma^n$, for some $n \in \mathbb{N}$, and $k \in \mathbb{N}$, the* rollercoaster table *of w and k is the table R_w^k of size*

$n \times 2 \times k$, indexed by the triples $i \in [1,n]$, $\xi \in \{\uparrow, \downarrow\}$, and $h \in [1,k]$, where $R_w^k[i, \xi, h]$ is the number of subsequences s which are longest plateau-$(k,h)_\xi$-rollercoasters ending at position i in w, i.e. $w[i]$ is the last symbol of s.

The following theorem provides an algorithm to compute R_w^k based on the table $L_w^{k,h,\xi}$.

Theorem 13. *There exists an algorithm computing R_w^k for a given input word $w \in \Sigma^n$ and $k \in \mathbb{N}$ in $O(n\sigma k)$ time.*

Proof. We prove this statement using a dynamic programming approach. As a base case, note that $R_w^k[i, \xi, 1]$ is equal to 1 for every $i \in [n]$, $\xi \in \{\uparrow, \downarrow\}$, corresponding to the plateau-$(k,1)_\xi$-rollercoaster associated to position i, which may be either an increasing or decreasing run. In general, the value of $R_w^k[i, \xi, h]$ can be computed by one of six summations, depending on the values of ξ and h.

If $h \in [2, k-1]$, then $R_w^k[i, \uparrow, h]$ is equal to the sum of the number of plateau-$(k, h-1)_\uparrow$-rollercoasters of length $L_w^{k,h-1,\uparrow}[i] - 1$ that end on a symbol smaller than $w[i]$, plus the number of plateau-(k,h)-rollercoasters of length $L_w^{k,h,\uparrow}[i] - 1$ and ending with the symbol $w[i]$. Define the array $\mathrm{PP}_w^{k,h,\xi}$ of length n by

$$\mathrm{PP}_w^{k,h,\uparrow}[i] = \{P_w[i-1,x] \mid x \in [1, w[i]-1], L_w^{k,h-1,\uparrow}[P_w[i-1,x]] = L_w^{k,h,\uparrow}[i] - 1\}$$

for all $i \in [n]$. Then, we get

$$R_w^k[i, \uparrow, h] = \left(\sum_{i' \in PP_w^{k,h,\uparrow}[i]} R_w^k[i', \uparrow, h-1] \right) +$$

$$\begin{cases} R_w^k[P_w[w[i], i-1], \uparrow, h] & L_w^{k,h,\uparrow}[i] - 1 = L_w^{k,h,\uparrow}[P_w[w[i], i-1]] \\ 0 & L_w^{k,h,\uparrow}[i] - 1 \neq L_w^{k,h,\uparrow}[P_w[w[i], i-1]] \end{cases}.$$

An analogous summation can be derived for computing $R_w^k[i, \downarrow, h]$ by defining

$$PP_w^{k,h,\downarrow}[i] = \{P_w[i-1,x] \mid x \in [w[i]+1, \sigma], L_w^{k,h-1,\downarrow}[P_w[i-1,x]] = L_w^{k,h,\downarrow}[i] - 1\}.$$

If $h = k$ then the value of $R_w^k[i, \uparrow, k]$ is equal to the number of plateau-$(k, k-1)_\uparrow$-rollercoasters of length $L_w^{k,k,\uparrow}[i] - 1$ ending before position i with any symbol smaller than $w[i]$, plus the number of plateau-$(k,k)_\uparrow$-rollercoasters of length $L_w^{k,k,\uparrow}[i] - 1$ ending before position i with any symbol less than or greater than $w[i]$. Note the first set corresponds to the rollercoasters in the set $\mathrm{PP}_w^{k,k-1,\uparrow}[i]$, while the second is equal to the set

$$\mathrm{PP}_w^{k,h,\uparrow}[i]' = \{P_w[i-1,x] \mid x \in [1,w[i]], L_w^{k,h-1,\uparrow}[P_w[i-1,x]] = L_w^{k,h,\uparrow}[i] - 1\},$$

leading to

$$R_w^k[i, \uparrow, k] = \left(\sum_{i' \in PP_w^{k,k,\uparrow}[i]} R_w^k[i', \uparrow, k-1] \right) + \left(\sum_{i' \in PP_w^{k,k,\uparrow}[i]'} R_w^k[i', \uparrow, k] \right).$$

Again, an analogous summation may be derived for $R_w^k[i,\downarrow,k]$ by replacing the array $PP_w^{k,k,\uparrow}$ by $PP_w^{k,k,\downarrow}$ and $(PP_w^{k,k,\uparrow})'$ by

$$PP_w^{k,k,\downarrow}[i]' = \{P_w[i-1,x] \mid x \in [w[i],\sigma], L_w^{k,h-1,\downarrow}[P_w[i-1,x]] = L_w^{k,h,\downarrow}[i] - 1\}.$$

Finally, the value of $R_w^k[i,\xi,1]$ is exactly equal to the size of $\max\{1, R_w^k[i,\bar{\xi},k]\}$ by definition.

To determine the complexity of this computation, note that we can, as a base case, set the value of $R_w^k[1,\xi,1]$ to 1, and of $R_w^k[1,\xi,h]$ to 0, for every $h \in [2,k]$. Now, note that the sets $PP_w^{k,h,\uparrow}[i]$, $PP_w^{k,h,\downarrow}[i]$, $PP_w^{k,k,\uparrow}[i]'$, and $PP_w^{k,k,\downarrow}[i]'$ can each be computed in $O(\sigma)$ time assuming the values of $L_w^{k,h',\uparrow}[i']$ have been computed for every $h \in [1,k]$ and $i' \in [1,n]$. Then, using the above summations, we can compute the value of $R_w^k[i,\xi,h]$ in a further $O(\sigma)$ time, assuming that $R_w^k[i',\xi',h']$ has been computed for every $i' \in [1,i-1], \xi' \in \{\uparrow,\downarrow\}$, and $h' \in [1,k]$. As there are nk entries in the table R_w^k, we have a total time complexity of $O(nk\sigma)$. □

Regarding the problems of counting and enumerating plateau-rollercoasters, we need show that the longest common plateau-rollercoaster can only be found at a unique position within w.

Proposition 14. *Given a word $w \in \Sigma^n$ and pair of plateau-rollercoasters s and s' such that $s = w[i_1]w[i_2]\ldots w[i_m]$ and $s' = w[j_1]w[j_2]\ldots w[j_m]$ where m is the length of the longest plateau-rollercoaster that is a subsequence of w, either $s \neq s'$ or $(i_1, i_2, \ldots, i_m) = (j_1, j_2, \ldots, j_m)$.*

Proof. If $(i_1, i_2, \ldots, i_m) = (j_1, j_2, \ldots, j_m)$ then clearly $s = s'$. Assuming that $(i_1, i_2, \ldots, i_m) \neq (j_1, j_2, \ldots, j_m)$, we have that $s = s'$ if and only if $w[i_c] = w[j_c], \forall c \in [1,m]$. Let $t \in [1,m]$ be the value such that $i_1, \ldots, i_{t-1} = j_1, \ldots, j_{t-1}$ and $i_t \neq j_t$. If $s = s'$ then $w[i_t] = w[j_t]$. If $i_t < j_t$ then $w[i_1]\ldots w[i_t]w[j_t]\ldots w[j_m]$ must be a plateau-rollercoaster of length $m+1$, contradicting the assumption that m is the length of the longest plateau-rollercoaster in w. Analogously, if $i_t > j_t$, $w[j_1]w[j_2]\ldots w[j_t]w[i_t]w[i_{t+1}]\ldots w[i_m]$ must be a plateau-rollercoaster. Hence the claim holds. □

Lemma 15. *Let $w \in \Sigma^n$ for $n \in \mathbb{N}$ be a word where the length of the longest plateau-rollercoaster in w is $m \in \mathbb{N}$. Then, the total number of plateau-rollercoasters in w of length m is given by*

$$\sum_{i \in [1,n], \xi \in \{\uparrow,\downarrow\}} R_w^k[i,\xi,0].$$

Proof. Following Lemma 14, any plateau-rollercoaster of maximum length m in w must be unique. Therefore, given any pair of indices $i, i' \in [1,n]$, any m-length plateau-rollercoaster ending at $w[i]$ must be distinct from any such plateau-rollercoaster ending at $w[i']$. Hence the above statement holds. □

In order to enumerate the set of plateau-rollercoasters in a given word w, we need one additional auxiliary structure, the *next element graph* of w, denoted

NEG(w). Here every node represents a position of w, precisely v_i represents $w[i]$ for all $i \in [|w|]$ and unlabelled edges are not included.

Definition 16. *For a given word $w \in \Sigma^n$ for $n \in \mathbb{N}$, define* NEG(w) = (V, E) *as the edge-labeled, directed graph with $V = \{v_1, v_2, \ldots, v_n\}$, $E = \{(v_i, v_j) | i \in [n-1], j \in [i+1, n]\}$, and the edge labelling function $\ell : E \to \{\uparrow, \to, \downarrow\}$ with*

$$\ell(e) = \begin{cases} \uparrow, & \text{if } w[i] < w[j] \land \forall j' \in [i+1, j-1] : w[i] > w[j'] \lor w[j'] > w[j], \\ \to, & \text{if } w[i] = w[j] \land i = P_w[j-1, w[j]], \\ \downarrow, & \text{if } w[i] > w[j] \land \forall j' \in [i+1, j-1] : w[i] < w[j] \lor w[j'] < w[j]. \end{cases}$$

Lemma 17. *Given a word $w \in \Sigma^n$,* NEG(w) *can be constructed in $O(n\sigma)$ time (cf. Algorithm 1).*

Algorithm 1. Construct*NEG*

Input: $w \in \Sigma^n$, $k \in \mathbb{N}$, Rollercoaster table R_w^k
Output: graph $NEG(w)$

```
1  Initialise L_w^{k,h,ξ}, ∀ξ ∈ {↑,↓}, h ∈ [1,k]
2  Compute P_w
3  Set V = {v_1, v_2, ..., v_n}
4  Set E = ∅      // edges are triples (v_i, v_j, ξ)
5  // iteration over vertices to add the edges (v_1 has no incoming edges)
6  for i ∈ [2, n]
7      p_↑ = P_w[i-1, w[i]]   // initial bound on which ↑-labelled edges are blocked by existing
                                  edges, we do not need to explicitly construct the set
8      for x ∈ {w[i] - 1, ..., 1}
9          if P_w[i-1, x] > p_↑   // if P_w[i-1, x] after every symbol in w
10             E = E ∪ {(v_{P_w[i-1,x]}, v_i, ↑)}
11             p_↑ = max(p_↑, P_w[i-1, x])
12     p_↓ = P_w[i-1, w[i]]   // We now do the same for the edges labelled ↓
13     for x ∈ [w[i] + 1, σ]
14         if P_w[i-1, x] > p_↓   // if P_w[i-1, x] after every symbol in w
15             E = E ∪ {(v_{P_w[i-1,x]}, v_i, ↓)}
16             p_↓ = max(p_↓, P_w[i-1, x])
17     // add labelled → from v_{P_w[i-1,w[i]]} to v_i
18     E = E ∪ {(v_{P_w[i-1,w[i]]}, v_i, →)}
19 return (V, E)
```

First, we show that we can associate a path in NEG(w) to unique subsequences of w.

Lemma 18. *Let $w \in \Sigma^n$ for some $n \in \mathbb{N}$. Moreover, let $i_1, i_2, \ldots, i_\sigma$ be the set of position such that $w[i_q] = q$ and for all $i' \in [1, i_q - 1], w[i'] \neq q$, i.e., i_q is the left most occurrence of the letter q in w. Every path of length j in* NEG(w) *starting at any vertex v_{i_q} corresponds to a unique subsequence of w.*

Now, we prove that every maximum length plateau-rollercoaster in w corresponds to a path in NEG(w).

Lemma 19. *Given a word $w \in \Sigma^n$ for $n \in \mathbb{N}$ and a maximal plateau-rollercoaster u in w, there exists exactly one path $(i_1, i_2, \ldots, i_{|u|})$ in* NEG(w) *such that $u = w[i_1]w[i_2]\ldots w[i_{|u|}]$.*

Proof. Let i_1 be the value such that for all $i \in [1, i_1 - 1]$ we have $w[i] \neq u[1]$ and $u[1] = w[i_1]$, and i_2 be the value such that for all $j \in [i_1 + 1, i_2 - 1]$, we have $u[2] \neq w[j]$. As u is maximal in w, there must be the edge (v_{i_1}, v_{i_2}) in NEG(w), as otherwise there exists some $i' \in [i_1 + 1, i_2 - 1]$ such that either $w[i_1] \leq w[i'] \leq w[i_2]$ or $w[i_1] \geq w[i'] \geq w[i_2]$, contradicting the assumption that u is a maximum length plateau-rollercoaster in w. Repeating this argument for every index $i_j \in \{i_1, i_2, \ldots, i_{|u|}\}$ proves the statement. □

In order to use NEG(w), we label each vertex v_i with two lists, $L_{i,\uparrow}, L_{i,\downarrow}$, of length k.

Definition 20. *Let $w \in \Sigma^n$ for $n \in \mathbb{N}$. Define for all $i \in [n]$ and $h \in [k]$, $L_{i,\uparrow}[h]$ as the number of suffixes of plateau-rollercoasters in w of maximum length containing an increasing run where $w[i]$ is the hth unique element in the run. Let $m_{i,j,\uparrow}$ denote the length stored in $R^k_{w^R}[n - i, \uparrow, j] > 0$ if for all $m' \in [m_{i,j,\uparrow}, n]$, we have $R^k_{w^R}[n - i, m', \uparrow, j] = 0$. Define $L_{i,\downarrow}[h]$ analogously.*

With these additional data structures, we use NEG(w) to enumerate the set of all maximum length plateau-rollercoasters in a recursive manner. At a high level, the idea is to work backwards, using the tables $L^{k,h,\xi}_w$ to determine which positions in w correspond to the final positions in maximum length plateau-k-rollercoasters in w. Once the last symbol has been determined, we use the edges in NEG(w) to determine which preceding positions can be used to construct valid runs ending at the appropriate position. We repeat this until we have determined a valid plateau-rollercoaster, which we then output. By using the plateau-rollercoaster table, we can check efficiently if a given position is valid: $R^k_w[i, \xi, h]$ gives us the number partial plateau-$(k, h)_\xi$-rollercoaster ending at i, and therefore the number of prefixes of length $L^{k,h,\xi}_w[i]$ ending at i. Before, we can prove the main result of this section, we need one lemma relating maximum length plateau-rollercoaster and P_w.

Lemma 21. *Given $w \in \Sigma^n$, $k \in \mathbb{N}$, and a maximum length plateau-rollercoaster $s = w[i_1]w[i_2]\ldots w[i_m]$ in w, then $i_m = P_w[n, w[i_m]]$.*

Theorem 22. *Given a word $w \in \Sigma^n$ for some $n \in \mathbb{N}$ and $k \in \mathbb{N}$, where m is the length of the longest plateau-k-rollercoaster in w, there exists an algorithm outputting all plateau-k-rollercoasters in w of length m with $O(n)$ delay after $O(n\sigma k)$ preprocessing.*

Proof. Before we begin the enumeration, as our preprocessing step, we compute the tables $L^{k,h,\xi}_w$ for every $h \in [1, k]$, P_w, the plateau-rollercoaster table R^k_w and the next element graph NEG(w) = (V, E), requiring $O(n\sigma k)$ time. Further, we use m to denote the length of the longest plateau-k-rollercoaster in w.

To determine the first plateau-rollercoaster, let i_m be some index such that $i_m = P_w[n, x]$, for some $x \in \Sigma$, and $L^{k,k,\xi}_w[i_m] = m$, for some $\xi \in \{\uparrow, \downarrow\}$. By construction, there must exist some plateau-k-rollercoaster ending with a ξ-run at i_m and further, the preceding position i_{m-1} must satisfy either:

- $(v_{i_{m-1}}, v_{i_m}, \xi) \in E$, and $L_w^{k,k,\xi}[i_m] = L_w^{k,h,\xi}[i_{m-1}] + 1$, for some $h \in \{k-1, k\}$, or,
- $(v_{i_{m-1}}, v_{i_m}, \rightarrow) \in E$, and $L_w^{k,k,\xi}[i_m] = L_w^{k,k,\xi}[i_{m-1}] + 1$.

We can determine the value of i_{m-1} by checking each incoming edge (v_j, v_{i_m}, ξ) and the edge $(v_j, v_{i_m}, \rightarrow)$ in at most $O(\sigma)$ time. We assume, without loss of generality, that we choose the value of i_{m-1} satisfying for all $j \in [i_{m-1}+1, i_m - 1]$ either:

- $(v_{i_{m-1}}, v_{i_m}, \xi) \notin E$, $(v_{i_{m-1}}, v_{i_m}, \rightarrow) \notin E$, $(v_{i_{m-1}}, v_{i_m}, \xi) \in E$, and $L_w^{k,h',\xi}[j] + 1 < L_w^{k,k,\xi}[i_m]$, for all $h' \in \{k-1, k\}$, or
- $(v_{i_{m-1}}, v_{i_m}, \rightarrow) \in E$ and $L_w^{k,k,\xi}[j] + 1 < L_w^{k,k,\xi}[i_m]$.

In general, once the value of i_j has been determined, we determine the value of i_{j-1} as follows. Let us assume that i_j is the hth unique element of a ξ-run. If $h \geq k$, then i_{j-1} must be some index such that either:

- $(v_{i_{j-1}}, v_{i_j}, \xi) \in E$ and $L_w^{k,h',\xi}[j-1] + 1 = L_w^{k,k,\xi}[j]$, for some $h' \in \{k-1, k\}$, or,
- $(v_{i_{j-1}}, v_{i_j}, \rightarrow) \in E$, and $L_w^{k,k,\xi}[j-1] + 1 = L_w^{k,k,\xi}$.

If $h \in [2, k-1]$, then i_{j-1} must satisfy either:

- $(v_{i_{j-1}}, v_{i_j}, \xi) \in E$ and $L_w^{k,h-1,\xi}[j-1] + 1 = L_w^{k,h,\xi}[j]$, or,
- $(v_{i_{j-1}}, v_{i_j}, \rightarrow) \in E$ and $L_w^{k,h,\xi}[j-1] + 1 = L_w^{k,h,\xi}$.

Finally, if $h = 1$, then i_{j-1} must satisfy

- either $(v_{i_{j-1}}, v_{i_j}, \overline{\xi}) \in E$ and $L_w^{k,h',\xi}[j-1] + 1 = L_w^{k,k,\xi}[j]$, for some $h' \in \{k-1, k\}$,
- or $(v_{i_{j-1}}, v_{i_j}, \rightarrow) \in E$, and $L_w^{k,h',\xi}[j-1] + 1 = L_w^{k,1,\xi}$, for some $h' \in \{k-1, k\}$.

In all cases, we can determine this in $O(\sigma)$ time by checking each incoming edge to v_{i_j} in order, terminating once a valid index i_{j-1} has been found. Further, we assume that i_{j-1} is the largest index satisfying this condition and further that we check each candidate index from largest to smallest.

To get the delay in the first output, observe that no index is checked as a candidate to be added to the plateau-rollercoaster more than once. Therefore, as there are at most n possible checks, the total delay is $O(n)$.

To determine the next plateau-rollercoaster after outputting the plateau-rollercoaster $s = w[i_1]w[i_2]\ldots w[i_m]$, we require an auxiliary set of m counters, $c_1, c_2, \ldots, c_m$, where c_j counts the number of plateau-rollercoasters that have been output so far with the suffix $w[i_j]w[i_{j+1}]\ldots w[i_m]$. Note that if $c_j < R_w^k[i_j, \xi, h]$, where $w[i_j]$ is the hth (or at least h^{th} if $h = k$) unique symbol in a ξ-run in s, then there exists some plateau-rollercoaster with the suffix $w[i_j]w[i_{j+1}]\ldots w[i_m]$ that has not been output and otherwise every plateau-rollercoaster with this suffix has been output. Now, let j be the value such that $c_j < R_w^k[i_j, \xi_j, h_j]$ and $c_{j'} = R_w^k[i_{j'}, \xi_{j'}, h_{j'}]$, for every $j' \in [1, j-1]$ where i_f is the h_f^{th} unique symbol in a ξ_f run in s. We construct the new plateau-rollercoaster by first finding the new value i'_{j-1} in the same manner as for the first plateau-rollercoaster, with the additional constraint that $i'_{j-1} < i_{j-1}$. Once i'_{j-1} has been determined, we find the value of $i_{j'}$, for every $j' \in [j-2]$ in the same manner as above, without any additional restriction on the maximum value. Again, by choosing the largest possible value for $i'_{j'}$ each time, we ensure that at most $O(n)$ comparisons are needed, and thus the delay between output is at most $O(n)$. Once the new set of indices has been determined, we output the new plateau-rollercoaster $s = w[i'_1]w[i'_2]\ldots w[i'_{j-1}]w[i_j]w[i_j+1]\ldots w[i_m]$, and update the value of $c_{j'}$ to 1, if $j' < j$, or to $c_{j'} + 1$ if $j' \geq j$.

To show correctness, observe that each output corresponds to a unique path in $\mathrm{NEG}(w)$, and thus, by Lemma 18, a unique plateau-rollercoaster. Further, as we output $R^k_w[i_j, \xi, h]$ plateau-rollercoasters with the suffix $w[i_j]w[i_{j+1}], \ldots$, $w[i_m]$, when $w[i_j]$ is the hth element on a ξ-run, we must output every possible such plateau-rollercoaster, completing the proof. □

With this proof we conclude the section on searching for the longest plateau-k-rollercoaster within a word and move our attention to the problem on finding the longest common plateau-k-rollercoaster within a set of words.

4 Longest Common Plateaux Rollercoasters

In this section, we provide tools for finding the longest common plateau-k-rollercoaster s in a set of words $\mathcal{W} = \{w_1, w_2, \ldots, w_m\}$, i.e., s is a plateau-k-rollercoaster in every w_i, $i \in [m]$ and there does not exist a longer such plateau-k-rollercoaster. As with finding the longest single plateau-rollercoaster, there may be multiple such plateau-rollercoasters. We prove this using the same tools as in Sect. 3, namely the lists $L^{k,h,\xi}_{w_i}$, the tables P_{w_i} and rollercoaster tables $R^k_{w_i}$, for every $w_i \in \mathcal{W}, \xi \in \{\uparrow, \downarrow\}$ and $h \in [k]$. Our primary result in this section is an $O(Nk\sigma)$ time algorithm for finding the longest common plateau-k-rollercoaster in a given set of words $\mathcal{W}$, integer k, defined over an alphabet of size σ, where $N = \prod_{w \in \mathcal{W}} |w|$. First, we formally define longest common plateau-k-rollercoasters.

Definition 23. *Given a finite set $\mathcal{W} \subset \Sigma^*$, a plateau-$k$-rollercoaster s is common to $\mathcal{W}$ if s is a plateau-k-rollercoaster of every $w \in \mathcal{W}$. A longest common plateau-k-rollercoaster of the set $\mathcal{W}$ is a plateau-k-rollercoaster s which is common to $\mathcal{W}$ and any plateau-k-rollercoaster s' where $|s'| > |s|$ is not common to $\mathcal{W}$.*

For the remainder of this section, let $\mathcal{W} = \{w_1, \ldots, w_n\} \subset \Sigma^*$ for $n \in \mathbb{N}$ and $k \in \mathbb{N}$. We solve the longest common plateau-rollercoaster-problem using dynamic programming. The main tool we use is the set of tables $\mathrm{LCR}^{k,h,\xi}_\mathcal{W}$.

Definition 24. *Let $h \in [k]$, and $\xi \in \{\uparrow, \downarrow\}$. Define the $|w_1| \times \cdots \times |w_n|$ matrix $\mathrm{LCR}^{k,h,\xi}_\mathcal{W}$ such that $\mathrm{LCR}^{k,h,\xi}_\mathcal{W}[i_1, \ldots, i_m]$ contains the length of the longest common plateau-$(k,h)_\xi$-rollercoaster of the words $w_1[1, i_1], w_2[1, i_2], \ldots, w_m[1, i_m]$ that includes $w_1[i_1], w_2[i_2], \ldots, w_m[i_m]$.*

The last condition of the definition already contains the conclusion that the longest common plateau-rollercoasters is empty if it ends in different letters. We compute the values of $LCR_\mathcal{W}^{k,h,\xi}$ similarly to the values of $L_w^{k,h,\xi}$ in Sect. 3. For notational brevity, we introduce the table $\mathcal{P}_\mathcal{W}$ analogously to P_w. Notice that $\mathcal{P}_\mathcal{W}$ may be computed (analogously to Sect. 3) in $O(N\sigma)$ time, where $N = \prod_{w\in\mathcal{W}} |w|$.

Definition 25. *The table $\mathcal{P}_\mathcal{W}$ of size $|w_1| \times |w_2| \times \cdots \times |w_m| \times \sigma$ is defined by $\mathcal{P}_\mathcal{W}[i_1, i_2, \ldots, i_m, x] = (P_{w_1}[i_1, x], P_{w_2}[i_2, x], \ldots, P_{w_m}[i_m, x])$.*

For the above mentioned reason, we consider, without loss of generality, only index tuples $(i_1, \ldots, i_m)$ with $w_j[i_j] = w_{j'}[i_{j'}]$ for all $j, j' \in [m]$. The following lemma gives the computation for $\mathrm{LCR}_\mathcal{W}^{k,h,\xi}$ for all $h \in [k]$ and $\xi \in \{\uparrow, \downarrow\}$.

Lemma 26. *Given $h \in [k]$ and $(i_1, \ldots, i_m)$ appropriate, we have*

$$\mathrm{LCR}_\mathcal{W}^{k,k,\uparrow}[i_1, \ldots, i_m] =$$
$$1 + \max\left\{\max_{x \in [1, w_1[i_1]-1]}\{\mathrm{LCR}_\mathcal{W}^{k,k-1,\uparrow}[\mathcal{P}_\mathcal{W}[i_1-1,,\ldots,i_m-1,x]\},\right.$$
$$\left.\max_{x \in [1, w_1[i_1]]}\{\mathrm{LCR}_\mathcal{W}^{k,k,\uparrow}[\mathcal{P}_\mathcal{W}[i_1-1,,\ldots,i_m-1,x]\}\right\},$$

if either $LCR_\mathcal{W}^{k,k-1,\uparrow}[\mathcal{P}_\mathcal{W}[i_1-1,\ldots,i_m-1,x]) > 0$ for some $x \in [w_1[i_1]-1]$ or $\max_{x \in [1,w_1[i_1]]}(LCR_\mathcal{W}^{k,k,\uparrow}[\mathcal{P}_\mathcal{W}[i_1-1,\ldots,i_m-1,x]) > 0$ for some $x \in [w_1[i_1]]$,

$$\mathrm{LCR}_\mathcal{W}^{k,h,\uparrow}[i_1, \ldots, i_m] =$$
$$1 + \max\left\{\max_{x \in [1, w_1[i_1]-1]}\{\mathrm{LCR}_\mathcal{W}^{k,h-1,\uparrow}[\mathcal{P}_\mathcal{W}[i_1-1,\ldots,i_m-1,x]\},\right.$$
$$\left.\mathrm{LCR}_\mathcal{W}^{k,h,\uparrow}[\mathcal{P}_\mathcal{W}[i_1-1,\ldots,i_m-1,w_1[i_1]]\right\}$$

if either $LCR_\mathcal{W}^{k,h-1,\uparrow}[\mathcal{P}_\mathcal{W}[i_1-1,i_m-1,x]] > 0$ for some $x \in [w_1[i_1]-1]$, or $LCR_\mathcal{W}^{k,h,\uparrow}[\mathcal{P}_\mathcal{W}[i_1-1,\ldots,i_m-1,w_1[i_1]] > 0$ for some $x \in [w_1[i_1]]$, and 0 otherwise, and finally

$$\mathrm{LCR}_\mathcal{W}^{k,1,\uparrow}[i_1, \ldots, i_m] = \mathrm{LCR}_\mathcal{W}^{k,k,\downarrow}[i_1, \ldots, i_m],$$

if the respective plateau-$(k, 1)_\xi$-rollercoasters are not unary.
The values for $\xi =\downarrow$ can be computed analogously.

Now, we present the main result of this section, the computation of the longest common plateau-k-rollercoaster of a given finite set $\mathcal{W} \subset \Sigma^*$.

Theorem 27. *The set of tables $\mathrm{LCR}_\mathcal{W}^{k,h,\xi}$ and thus the length of the longest common plateau-k-rollercoaster, can be computed, for every $h \in [k], \xi \in \{\uparrow, \downarrow\}$, in $O(Nk\sigma)$ time, where $N = \prod_{w\in\mathcal{W}} |w|$.*

Proof. Lemma 26 provide the outlines for a recursive approach to computing the tables $LCR_{\mathcal{W}}^{k,h,\xi}$. As a base case, the values of $LCR_{\mathcal{W}}^{k,h,\xi}[i_1, \ldots, i_m]$ can be set to 0 for any set $i_1, \ldots, i_m$ where $w_j[i_j] \neq w_\ell[i_\ell]$ for some pair $j, \ell \in [m]$, requiring $O(Nk)$ time. Similarly, we can determine if the value of $LCR_{\mathcal{W}}^{k,1,\xi}[1, \ldots, 1]$ is 1, if $w_1[1] = \cdots = w_m[1]$, or 0 otherwise, for both values of $\xi \in \{\uparrow, \downarrow\}$. Finally, we set the value of $LCR_{\mathcal{W}}^{k,h,\xi}[1, \ldots, 1]$ to 0 for every $h \in [2, k]$.

In the general case, in order to compute the value of $LCR_{\mathcal{W}}^{k,h,\xi}[i_1, \ldots, i_m]$, let us assume that the values of $\text{LCR}_{\mathcal{W}}^{k,h',\xi}[i'_1, \ldots, i'_m]$ has been computed for every $i'_1 \in [i_1], \ldots, i'_m \in [i_m]$, $h' \in [k]$, and $\xi' \in \{\uparrow, \downarrow\}$, other than $(i'_1, \ldots, i'_m) = (i_1, \ldots, i_m)$. Further, we assume that, if $h = 1$, the value $\text{LCR}_{\mathcal{W}}^{k,k,\xi}[i_1, \ldots, i_m]$ has already been computed, noting that the value $\text{LCR}_{\mathcal{W}}^{k,1,\xi}[i_1, \ldots, i_m]$ does not depend upon $\text{LCR}_{\mathcal{W}}^{k,1,\xi}[i_1, \ldots, i_m]$. From Lemma 26, $\text{LCR}_{\mathcal{W}}^{k,h,\xi}[i_1, \ldots, i_m]$ can be computed in $O(\sigma)$ time using one of the the given formulae.

As there are N entries in $\text{LCR}_{\mathcal{W}}^{k,h,\xi}$, $2k$ tables $\text{LCR}_{\mathcal{W}}^{k,1,\xi}, \ldots, \text{LCR}_{\mathcal{W}}^{k,k,\xi}$, and the complexity of computing each entry is $O(\sigma)$ time, the total complexity of computing every table is $O(Nk\sigma)$.

The length can be obtained immediately from $\text{LCR}_{\mathcal{W}}^{k,h,\xi}$. □

5 Conclusion

Within this work we introduced and investigated the notion of plateau-k-rollercoaster as a natural extension to k-rollercoaster relaxing the strictly in/decreasing runs to weakly in/decreasing runs. First, we gave an $O(n\sigma k)$-algorithm to determine the longest plateau-k-rollercoaster for an n-length word over an σ-letter alphabet. Extending this idea, we introduced a *rollercoaster table* which allows for enumerating all longest plateau-k-rollercoaster of a word of length n with $O(n)$ delay after $O(n\sigma k)$ preprocessing. Second, we presented an algorithm to search for the longest common plateau-k-rollercoaster within a set of words. Via a dynamic programming approach, the longest common plateau-k-rollercoaster can be computed in $O(Nk\sigma)$ time where N is the product of all word lengths within the set. For further research, one might proceed with algorithmical studies on, e.g., the shortest common supersequence of a set of words that is a plateau-k-rollercoaster.

References

1. Aldous, D., Diaconis, P.: Longest increasing subsequences: from patience sorting to the Baik-Deift-Johansson theorem. Bull. Am. Math. Soc. **36**(4), 413–432 (1999)
2. Biedl, T., et al.: Rollercoasters and caterpillars. In: ICALP 2018. LIPIcs, vol. 107, pp. 18:1–18:15. Schloss Dagstuhl - Leibniz-Zentrum für Informatik (2018)
3. Biedl, T., et al.: Rollercoasters: Long sequences without short runs. SIAM J. Discret. Math. **33**(2), 845–861 (2019)

4. Biedl, T., Chan, T.M., Derka, M., Jain, K., Lubiw, A.: Improved bounds for drawing trees on fixed points with L-shaped edges. In: Frati, F., Ma, K.-L. (eds.) GD 2017. LNCS, vol. 10692, pp. 305–317. Springer, Cham (2018). https://doi.org/10.1007/978-3-319-73915-1_24
5. Fraser, C.B., Irving, R.W.: Approximation algorithms for the shortest common supersequence. Nord. J. Comput. **2**(3), 303–325 (1995)
6. Fredman, M.L.: On computing the length of longest increasing subsequences. Discret. Math. **11**(1), 29–35 (1975)
7. Fujita, K., Nakashima, Y., Inenaga, S., Bannai, H., Takeda, M.: Longest common rollercoasters. In: Lecroq, T., Touzet, H. (eds.) SPIRE 2021. LNCS, vol. 12944, pp. 21–32. Springer, Cham (2021). https://doi.org/10.1007/978-3-030-86692-1_3
8. Gawrychowski, P., Manea, F., Serafin, R.: Fast and longest rollercoasters. Algorithmica **84**(4), 1081–1106 (2022)
9. Jiang, T., Li, M.: On the approximation of shortest common supersequences and longest common subsequences. SIAM J. Comput. **24**(5), 1122–1139 (1995)
10. Kärkkäinen, J., Sanders, P., Burkhardt, S.: Linear work suffix array construction. J. ACM **53**(6), 918–936 (2006)
11. Kosowski, A.: An efficient algorithm for the longest tandem scattered subsequence problem. In: Apostolico, A., Melucci, M. (eds.) SPIRE 2004. LNCS, vol. 3246, pp. 93–100. Springer, Heidelberg (2004). https://doi.org/10.1007/978-3-540-30213-1_13
12. Kutz, M., Brodal, G.S., Kaligosi, K., Katriel, I.: Faster algorithms for computing longest common increasing subsequences. J. Discrete Algorithms **9**(4), 314–325 (2011)
13. Räihä, K., Ukkonen, E.: The shortest common supersequence problem over binary alphabet is np-complete. Theor. Comput. Sci. **16**, 187–198 (1981)
14. Romik, D.: The Surprising Mathematics of Longest Increasing Subsequences, vol. 4. Cambridge University Press, Cambridge (2015)
15. Schensted, C.: Longest increasing and decreasing subsequences. Can. J. Math. **13**, 179–191 (1961)
16. Sun, X., Woodruff, D.P.: The communication and streaming complexity of computing the longest common and increasing subsequences. In: SODA 2007, pp. 336–345. SIAM (2007)
17. Wagner, R., Fischer, M.: The string-to-string correction problem. JACM **21**(1), 168–173 (1974)
18. Yang, I., Huang, C., Chao, K.: A fast algorithm for computing a longest common increasing subsequence. Inf. Process. Lett. **93**(5), 249–253 (2005)

Quantum Automata and Languages of Finite Index

Andrea Benso[1], Flavio D'Alessandro[2,3](✉), and Paolo Papi[2]

[1] Dipartimento di Matematica e Informatica "U. Dini", Università di Firenze, 50134 Florence, Italy
andrea.benso@unifi.it

[2] Dipartimento di Matematica "G. Castelnuovo", Sapienza Università di Roma, 00185 Rome, Italy
{flavio.dalessandro,paolo.papi}@uniroma1.it

[3] Department of Mathematics, Boğaziçi University, Bebek, 34342 Istanbul, Turkey

Abstract. This paper continues the study of measure-once finite quantum automata building on work by Bertoni et al. and Blondel et al. We investigate conditions ensuring that, given a language recognized by such a device and a language generated by a context-free grammar of finite index or by a matrix context-free grammar, it is decidable whether or not they have a nonempty intersection.

Keywords: Quantum automata · Context-free languages · Algebraic groups · Decidability

1 Introduction

In this paper, we investigate the decidability of a problem for quantum automata introduced in [5,6] named *Intersection Problem*. Quantum automata were introduced at the beginning of 2000's in [21] as a new model of language recognizer (cf. [4]). The starting point of that investigation is a result of Blondel, Jeandel, Koiran, and Portier who showed in [7] that, given a "measure-once" quantum automaton $\mathcal{Q}$ (in the sense of [21]), it is decidable whether the language recognized by $\mathcal{Q}$ has a non trivial intersection with the free monoid generated by the input alphabet of $\mathcal{Q}$ (*Emptiness problem*). It is worth remarking that these problems are strictly related to some computational issues concerning matrices. Precisely, a crucial step in the construction of the decision procedure for both the Emptiness and Intersection Problem amounts to effectively compute the Zariski closure of a finitely generated group of matrices over $\mathbb{Q}$. For that task,

---

The second and the third author have been partially supported by Sapienza Ateneo 2023 Project "Representation Theory and Applications". The second author has been also partially supported by TUBITAK Project 2221 (The Scientific and Technological Research Council of Turkey). The authors acknowledge their membership to the National Group for Algebraic and Geometric Structures, and their Applications (GNSAGA–INdAM).

an appropriate algorithm has been developed in [11] by Derksen, Jeandel, and Koiran. Before introducing the Intersection Problem, we find then convenient to recall some developments of the research on that aspect. Two striking results have been obtained recently in connection with the algorithm by Derksen, Jeandel, and Koiran. In 2022 Nosan, Pouly, Schmitz, Shirmohammadi, and Worrell, following an alternative approach for such computation, have obtained in [23] a bound on the degree of the polynomials that define the Zariski closure of the set and showed that the latter can be computed in elementary time. The second was obtained in 2023 by Hrushovski, Ouaknine, Pouly, and Worrell for the study of the effective computation of polynomial invariants for affine programs [16]. Therein, it is exhibited an algorithm that, given a finite set of square matrices of the same size with coefficients in $\mathbb{Q}$, computes the Zariski closure of the semigroup generated by the set. It should be pointed out that the decidability issues for matrix semigroups is a challenging problem. Even in small size, a well-known result by Paterson shows that the problem of freeness and that of membership are both undecidable for matrix semigroups over $\mathbb{Z}$ of size 3×3 (see [8]). In contrast, in 2017, a surprising result of Potapov and Semukhin [26] shows that the membership is decidable for non-singular matrix integer semigroups of size 2×2.

We now give a formal presentation of our work. A *finite quantum automaton* (in the sense of [21]) is a quadruple

$$\mathcal{Q} = (s, \varphi, P, \lambda), \tag{1}$$

where $s \in \mathbb{R}^n$ is a row-vector of unit Euclidean norm, P is a projection of $\mathbb{R}^n$, and φ is a morphism

$$\varphi : \Sigma^* \longrightarrow O_n, \tag{2}$$

of the free monoid Σ^* into the group O_n of orthogonal $n \times n$-matrices in $\mathbb{R}^{n \times n}$, and the *threshold* λ has value in $\mathbb{R}$.

We are mainly interested in effective properties which require the quantum automaton to be effectively given. We thus consider the model of *rational quantum automaton*, i.e., where all the coefficients of the components of (1) and λ are rational numbers. For a real threshold λ, the languages recognized by $\mathcal{Q}$ with strict and nonstrict threshold λ are

$$|\mathcal{Q}_>| = \{w \in \Sigma^* \mid ||s\varphi(w)P|| > \lambda\}, \quad |\mathcal{Q}_\geq| = \{w \in \Sigma^* \mid ||s\varphi(w)P|| \geq \lambda\},$$

where $||\cdot||$ denotes the Euclidean norm of vectors. Let $\mathcal{L}$ be a family of effectively defined languages. The problem we tackle is the following:

$(L, \mathcal{Q})$ Intersection

INPUT: a language L in a family $\mathcal{L}$ of languages and a rational quantum automaton $\mathcal{Q}$.
QUESTION: does $L \cap |\mathcal{Q}_>| = \emptyset$ hold?
As already pointed out, the decidability of the above-mentioned problem has been proven in [7] in the specific case $L = \Sigma^*$. This result contrasts with the corresponding problems (strict and non strict, respectively) for probabilistic finite

automata. Indeed for this class of automata, the Emptiness problems w.r.t. $|\mathcal{Q}_>|$ and $|\mathcal{Q}_\geq|$ are both undecidable (see [25], Thm 6.17). Also it contrasts with the undecidability of the Emptiness problem of $|\mathcal{Q}_\geq|$ for quantum automata [7]. It is also proven in [19] that both the problems (strict and non strict) remain undecidable for the "measure many" model of quantum automata of Kondacs and Watrous [20].

In order to describe the main contributions of the paper, it may be useful to recall some lines of the technique used in the proof of the Intersection Problem in [6]. For this purpose, following the same strategy of [7], we first observe that the Intersection Problem is equivalent to the inclusion

$$\varphi(L) \subseteq |\mathcal{Q}_\leq|. \tag{3}$$

Since the function $M \to ||sMP||$ – where M is an arbitrary matrix in $\mathbb{R}^{n \times n}$, and s, P are components of (1) – is continuous, it is sufficient to prove that, for all matrices M in the Euclidean closure $\mathbf{Cl}(\varphi(L))$ of $\varphi(L)$, the condition $||sMP|| \leq \lambda$ holds (see Proposition 3). Such condition is then effectively tested by running, in parallel, two semialgorithms. The first tests the inclusion $\varphi(L) \subseteq |\mathcal{Q}_>|$ and it is readily seen to be semidecidable. The second, which is the main crux, is to prove that (3) is semidecidable. The semidecidability of the latter procedure is achieved by expressing the property (3) in first-order logic of the field of reals, which, in turn, consists to effectively compute a representation of $\mathbf{Cl}(\varphi(L))$ in terms of a semialgebraic set (a more general property than algebraic, which allows one to cope with the fact that algebraic sets are not closed w.r.t. the operation of product of sets). Afterwards, one applies on the constructed formula the Tarski-Seidenberg quantifier elimination to verify whether (3) holds true or not.

It turns out that, in the case the language L is context-free, the semialgebraicity of $\mathbf{Cl}(\varphi(L))$, and its effective computation are reduced to those of an algebro-combinatorial object that is associated with the grammar G generating L, and with every variable A of G: the *monoid of cycles of A*. Such monoid, denoted M_A, corresponds (up to a technical detail) to the matrix image, under taking the morphism φ of (1), of the language of cycles in G associated with A (see Eq. (8)). The reduction process mentioned above is made possible by resorting to two ingredients: the fact that the Euclidean closure $\mathbf{Cl}(M_A)$ is an algebraic set (because of a deep result of Algebraic Geometry, namely Theorem 1) and because of a suitably defined effective structuring of the derivations of G ([2], Sects. 4.1, 4.2, 6.1). In this theoretical setting, the work in [6] and the new results of this work provide the following picture of the Intersection Problem.

- If L is a context-free language, then $\mathbf{Cl}(\varphi(L))$ is semialgebraic. Moreover, if, for every variable A of G, $\mathbf{Cl}(M_A)$ is effectively computable, then $\mathbf{Cl}(\varphi(L))$ is so as well (Proposition 4).
- If, for every variable A of G, M_A is finitely generated, then $\mathbf{Cl}(\varphi(L))$ is effectively computable. This is guaranteed by applying Proposition 4 and the results [11,16,23] mentioned above. This, in particular, covers the cases

of context-free linear languages, bounded semi-linear languages and a new class of languages, proposed in this paper, generated by a class of restricted matrix context-free grammars which non trivially extend the previous ones (Proposition 5).
- The effective computation of $\mathbf{Cl}(\varphi(L))$ is achieved for a subfamily of finite index context-free languages, called *monoidal*, by considering the property of (algebraic) irreducibility on the Zariski closure of some monoids of cycles canonically associated with G (Proposition 6).

The new results of the paper are the following. In Sect. 5.1, we extend the main contribution of [6] – i.e., the statement of decidability of the Intersection Problem for linear context-free languages and for bounded semilinear languages – to a class of *matrix context-free grammar of finite index*. Precisely, whenever the quantum automaton is rational, such result (Proposition 5) allows one, on one hand, to recover both the previous two cases as immediate corollaries of it. On the other, it extends the decidability of the problem for non context-free languages of exponential growth. In the proof a suitable decomposition of the derivations in these grammars is used. Such a decomposition is based upon some combinatorial tools proposed by Choffrut, Varricchio and the second author in [9] for the study of rational relations.

In Sect. 5.2, we investigate the Intersection Problem for languages defined by *finite index context-free grammars*. These are grammars G where each word w in the language generated by G can be obtained by a derivation δ, whose index is uniformly bounded, i.e., the number of variables in each sentential form of δ is bounded by an integer not depending on w. A remarkable result by Ginsburg and Spanier [13] (see also Nivat [22]) provides a characterization of such languages in terms of composition of grammars: precisely, each language of this type is generated by a grammar G given by the composition

$$G = \mathcal{G}_1 \circ \mathcal{G}_2 \circ \cdots \circ \mathcal{G}_k, \tag{4}$$

of families $\mathcal{G}_i$, $1 \leq i \leq k$, of linear context-free grammars (Theorem 2). Here, we prove a statement of decidability for a subclass of these grammars called *monoidal*. Informally speaking, a grammar G of the form (4) is monoidal if the terminal productions of G are of the form $X \to \varepsilon$ and the remaining productions are unit ones or of the form $X \to uXv$, with X being a variable of some grammar of $\mathcal{G}_i$, $1 \leq i \leq k$ (see Definition 6).

We prove that, regardless of the length k of the composition (4), if the Zariski closure of the monoids of cycles of the linear grammars of the lowest level $\mathcal{G}_k$ are irreducible, then one can effectively compute the set $\mathbf{Cl}(\varphi(L))$ (Proposition 6, Corollary 3). Extensions of such result to broader families of finite index context-free languages seem to be related to possible extensions of the algorithms [11, 16, 23] (cf Sect. 6, [α]). A full version of this paper can be found in [2].

2 Preliminaries on Algebraic and Semialgebraic Sets

We give the definition of algebraic set over the field of real numbers [1,12,24].

Definition 1. *A subset $\mathcal{A} \subseteq \mathbb{R}^n$ is algebraic (over the field of real numbers), if $\mathcal{A}$ is the zero set of an arbitrary subset $\mathcal{P}$ of polynomials of $\mathbb{R}[x_1, \ldots, x_n]$, i.e., for every vector $v \in \mathbb{R}^n$, $v \in \mathcal{A}$ if and only if, for every $p \in \mathcal{P}$, $p(v) = 0$.*

Note that by the Hilbert's basis theorem (see [12], Ch. 1, Sec. 1, Theorem 1), one may assume that the set $\mathcal{P}$ is finite. Even more, since we are dealing with algebraic sets over $\mathbb{R}$, $\mathcal{P}$ can be reduced to a singleton since the condition in Definition 1 can be replaced by the equation $\sum_{p \in \mathcal{P}} p(x)^2 = 0$.

One can check that the family of algebraic sets is closed under finite unions and intersections. However, it is not closed under complement and projection. The following more general class of subsets enjoys extra closure properties and is therefore more robust.

Definition 2. *A subset $\mathcal{A} \subseteq \mathbb{R}^n$ is semialgebraic (over the field of real numbers) if it satisfies one of the two equivalent conditions*

(i) $\mathcal{A}$ is the set of vectors satisfying a finite Boolean combination of predicates of the form $p(x_1, \ldots, x_n) > 0$ where $p \in \mathbb{R}[x_1, \ldots, x_n]$.
(ii) $\mathcal{A}$ is first-order definable in the theory of the structure whose domain are the reals and whose predicates are of the form $p(x_1, \ldots, x_n) > 0$ or $p(x_1, \ldots, x_n) = 0$ with $p \in \mathbb{R}[x_1, \ldots, x_n]$.

We now instantiate the definitions above to the setting of square matrices.

Definition 3. *A set $\mathcal{A} \subseteq \mathbb{R}^{n \times n}$ of matrices is algebraic (resp., semialgebraic) if considered as a set of vectors of dimension n^2, it is algebraic (resp., semialgebraic).*

Notational convention. *In connection with the $(L, \mathcal{Q})$ Intersection Problem, we will adopt the following terminology. A set $\mathcal{A}$ of matrices will be called effective algebraic (resp., effective semialgebraic) if the polynomials (resp., the formula) defining $\mathcal{A}$ can be algorithmically computed from the input of the problem, i.e., an effectively defined language L and a finite quantum automaton $\mathcal{Q}$.*

Given a subset E of matrices, we denote by $\mathbf{Cl}(E)$ the closure of E for the topology induced by the Euclidean norm. The following useful result holds.

Proposition 1 *([6], Corollary 2). The topological closure of the product of two sets of matrices included in a compact subspace is equal to the product of the topological closures of the two sets.*

Given a subset E of a group, we denote by $\langle E \rangle$ and E^* the subgroup and the submonoid generated by E, respectively. It's useful now to recall that, for an arbitrary set E of orthogonal matrices, $\mathbf{Cl}(E^*) = \mathbf{Cl}(\langle E \rangle)$. In particular $\mathbf{Cl}(E^*)$ is a group with respect to the same operation of product defined in E^* [7] (see also [6], Theorem 6). The next theorem shows that the Euclidean topological closure of a monoid of orthogonal matrices is algebraic.

Theorem 1 *([24], Ch. 3, Sec. 4, Theorem 5, [6], Theorem 7). Let E be a set of orthogonal matrices. Then $\mathbf{Cl}(\langle E \rangle)$ is a subgroup and it is the zero set of all polynomials $p[x_{1,1}, \ldots, x_{n,n}]$ satisfying the conditions $p(I) = 0$ and $p(eX) = p(X)$ for all $e \in E$. Furthermore, if the matrices in E have rational coefficients, the above condition may be restricted to polynomials with coefficients in $\mathbb{Q}$.*

Algebraic Irreducible Sets. We recall the property of irreducibility for algebraic sets. Let us consider the topological space obtained by equipping $\mathbb{R}^m$ with the *Zariski topology* (see [12], Ch. 1). Given a non-empty subset $\mathcal{Z}$ of $\mathbb{R}^m$, $\mathcal{Z}$ is said to be *irreducible* if $\mathcal{Z}$ cannot be written as the union of two closed, i.e. algebraic, subsets of $\mathcal{Z}$, distinct from $\emptyset$ and $\mathcal{Z}$ itself. It can be proven that a non-empty algebraic subset $\mathcal{Z}$ of $\mathbb{R}^m$ is irreducible if and only if its vanishing ideal is a prime ideal (cf [12], Prop. 1.1.12). Such sets play a role in the theoretical developments presented later (see Sect. 5.2). Therefore, we recall a definition and some results.

Definition 4. *Given non-empty algebraic sets $\mathcal{V} \subseteq \mathbb{R}^n$ and $\mathcal{W} \subseteq \mathbb{R}^m$, a map $f: \mathcal{V} \to \mathcal{W}$ is said to be regular (or a morphism of algebraic sets), if there exist $p_1, \ldots, p_m \in \mathbb{R}[x_1, \ldots, x_n]$ (where the x_i are indeterminates) such that, for all $\mathbf{x} = (x_1, \ldots, x_n)$ in $\mathcal{V}$, $f(\mathbf{x}) = (p_1(\mathbf{x}), \ldots, p_m(\mathbf{x}))$.*

Notation. We find useful to introduce some notation.
1) The Zariski closure of a set $\mathcal{A}$ will be denoted by $\overline{\mathcal{A}}$.
2) Given a subset $\mathcal{A}$ of the group O_m of orthogonal matrices (over $\mathbb{R}$) of size $m \times m$, we write $\mathcal{A} \in (\mathcal{H})$ if

$$\mathcal{A} = \bigcup_{i=1}^{k} \mathcal{A}_i, \tag{5}$$

where, for every $i = 1, \ldots, k$, $\mathcal{A}_i$ is an irreducible, effective algebraic set that contains the identity matrix I.

By using some standard arguments on irreducible sets, one can prove the following result (cf [2], Prop. 6).

Proposition 2. *Let $\mathcal{A}$ be a subset of O_{nk} with $\mathcal{A} \in (\mathcal{H})$ and let Ψ be a regular map $\Psi: O_{nk} \longrightarrow O_n$. If, for every $i = 1, \ldots, k$, $\Psi(\mathcal{A}_i)$ contains I, then $\overline{\Psi(\mathcal{A})}^* \in (\mathcal{H})$. In particular $\Psi(\mathcal{A})$ is effective algebraic.*

The Operation ($\oplus$). In the sequel, the symbol ($\oplus$) denotes the *sum* of square matrices $M_1, M_2, \ldots, M_k$ of the same size, whose result is the square block matrix

$$M_1 \oplus \cdots \oplus M_k = \begin{pmatrix} M_1 & 0 & \cdots & 0 \\ 0 & M_2 & \cdots & 0 \\ \vdots & \vdots & \vdots & \vdots \\ 0 & 0 & 0 & M_k \end{pmatrix} \tag{6}$$

These notations extend to subsets of matrices in the natural way. Observe that if the matrices are orthogonal, so is their sum. Such matrices form a subgroup of orthogonal matrices of size $kn \times kn$, where $n \times n$ is the size of $M_1, M_2, \ldots, M_k$.

3 Preliminaries on Context-Free Languages

We recall some essential notions and results on context-free languages and regular sets (see, for instance, [3,14,15]). Given a monoid M, a subset of M is said to be *regular* if it is finite or it is obtained from finite sets by using finitely many times the *regular operations*, i.e., the set union and the product of two sets and the star (*), this last operation associating with a set $\mathcal{X}$ the submonoid $\mathcal{X}^*$ generated by $\mathcal{X}$ in M. A *context-free grammar* G is a quadruple $\langle V, \Sigma, P, S \rangle$ where Σ is the alphabet of *terminal symbols*, V is the set of *variables* or *nonterminal symbols*, P is the set of *productions* (or *rules*), and S is the *axiom* of the grammar. A word over the alphabet Σ is called *terminal*. As usual, the nonterminal symbols are denoted by uppercase letters as, for instance, A, B. An arbitrary production of the grammar is written as $A \to \alpha$, where α is a finite sequence of variables and terminals. The *derivation* relation of G is denoted by $\stackrel{*}{\Rightarrow}$. The *language generated by G* is the subset of terminal words $L(G) = \{w \in \Sigma^* : S \stackrel{*}{\Rightarrow} w\}$.

We say that a context-free language L is *of index k*, if there exist a context-free grammar G with $L = L(G)$ and an integer $k \in \mathbb{N}$ such that the following property holds: for every word $w \in L(G)$, there exists a derivation $\delta = (S \stackrel{*}{\Rightarrow} w)$ such that the number of occurrences of variables in each sentential form of δ does not exceed k. The family of context-free languages of index k will be denoted by $Ind(k)$ and $\mathcal{L}(IND_{FIN})$ will denote the family of all the languages *of finite index*, i.e., $\mathcal{L}(IND_{FIN}) = \bigcup_{k \in \mathbb{N}} Ind(k)$.

We now recall a characterization, by Ginsburg and Spanier [13] (see also Nivat [22]), of languages of $\mathcal{L}(IND_{FIN})$, that will be used in Sect. 5.2. To this purpose, we first recall that, given alphabets Σ_1 and Σ_2, and a family $\mathcal{F}$ of languages of Σ_2^*, a $\mathcal{F}$-*substitution* is a morphism $\theta : \Sigma_1^* \to \Pi(\Sigma_2^*)$ from the free monoid Σ_1^* into the multiplicative monoid $\Pi(\Sigma_2^*)$ of subsets of the free monoid Σ_2^* such that, for every $x \in \Sigma_1$ $\theta(x) \in \mathcal{F}$. Let us define, recursively, the family $\{Qrt(k)\}_{k \in \mathbb{N}}$ of *quasi-rational languages of rank k* (or *bounded derivation languages of rank k*) as:

$$Qrt(k) = \begin{cases} LIN & \text{if } k = 1, \\ LIN \circ Qrt(k-1) & \text{if } k > 1, \end{cases} \quad (7)$$

where $LIN \circ Qrt(k-1)$ denotes the family of all the languages obtained as the images of linear languages, via $Qrt(k-1)$-substitutions, $k > 1$, i.e. $Qrt(k) = \{\theta(L) : L \in LIN, \, \forall x \in \Sigma_1 \, \theta(x) \in Qrt(k-1)\}$.

The family $Qrt(k)$ is equivalently formulated in terms of of *composition of grammars* (see [3], Sec. II.2). The following holds.

Theorem 2 *([13], Theorem 4.2, cf [3], Sec. VII.5, Theorem 5.2). Let L be a context-free language and $k \geq 1$. Then $L \in Ind(k)$ if and only if $L \in Qrt(k)$. Moreover, $L \in Qrt(k)$ if and only if there exists a grammar G with $L = L(G)$ such that G is a composition $\mathcal{G}_1 \circ \cdots \circ \mathcal{G}_k$ of k families of linear grammars.*

Example 1. Let $\mathcal{L} = \{a^n b^n : n \in \mathbb{N}\}^*$ be the language over the alphabet $\Sigma_2 = \{a, b\}$. Observe that $\mathcal{L}$ is the image of the language $\{\sigma^n : n \in \mathbb{N}\}$ over the alphabet $\Sigma_1 = \{\sigma\}$, under the substitution $\theta : \Sigma_1^* \to \Pi(\Sigma_2^*)$, with $\theta(\sigma) = \{a^n b^n :$

$n \in \mathbb{N}\}$. Hence $\mathcal{L}$ is in $Qrt(2)$. The language $\mathcal{L}$ is generated by the composition $G = \mathcal{G}_1 \circ \mathcal{G}_2$ of the linear grammars $\mathcal{G}_1 = \langle V_1, \Sigma_1, P_1, S \rangle$ and $\mathcal{G}_2 = \langle V_2, \Sigma_2, P_2, \sigma \rangle$ where, for the grammar $\mathcal{G}_1$, $V_1 = \{S\}$ and $P_1 = \{S \to \varepsilon, S \to \sigma S\}$, while, for the grammar $\mathcal{G}_2$, $V_2 = \{\sigma\}$ and $P_2 = \{\sigma \to \varepsilon,\ \sigma \to a\sigma b\}$.

4 Context-Free Languages and Semialgebraic Sets

We return to the decidability of the $(L, \mathcal{Q})$ Intersection problem defined in the Introduction. The bulk of the proof of the decidability of the problem is based on the following results [6].

Proposition 3. *Let $\mathcal{Q}$ be a rational quantum automaton. Let $L \subseteq \Sigma^*$ be a formal language such that the Euclidean closure $\mathbf{Cl}(\varphi(L))$ of $\varphi(L)$ is effective semialgebraic. It is recursively decidable whether or not $L \cap |\mathcal{Q}_>| = \emptyset$ holds.*

Thus, by Proposition 3, the main task amounts to the computation of $\mathbf{Cl}(\varphi(L))$ as a semialgebraic set of matrices. It turns out that the semialgebraicity of $\mathbf{Cl}(\varphi(L))$, and its effective computation are reduced to those of an algebro-combinatorial object that is canonically associated with the grammar G spanning L, and with every variable A of G: the *monoid of cycles of A*. With an arbitrary nonterminal A of G, associate the set M_A defined as

$$M_A = \{\varphi(u_1) \oplus \varphi(u_2)^T \mid A \stackrel{*}{\Longrightarrow} u_1 A u_2\}, \tag{8}$$

where $\varphi(u_2)^T$ is the transpose of $\varphi(u_2)$ and $(\oplus)$ denotes the operation defined in (6). M_A will be called the *monoid of cycles of A* since the following holds.

Lemma 1. *Let A be a nonterminal symbol of G. M_A is a monoid and its closure $\mathbf{Cl}(M_A)$ is an algebraic group.*

Proposition 4. *Let L be a context-free language. The set $\mathbf{Cl}(\varphi(L))$ is semialgebraic. Moreover, if for every $A \in V$, $\mathbf{Cl}(M_A)$ is effective algebraic, then $\mathbf{Cl}(\varphi(L))$ is effective semialgebraic.*

Corollary 1 *([6], Prop. 22). If L is a context-free linear language and the quantum automaton $\mathcal{Q}$ is rational, then, for every variable A, $\mathbf{Cl}(M_A)$ is effective algebraic, so that $\mathbf{Cl}(\varphi(L))$ is an effective semialgebraic set. Thus the $(L, \mathcal{Q})$ Intersection Problem is recursively decidable.*

5 Main Results

5.1 The Case of Finite Index Matrix Languages

Now we propose a first non trivial extension of the main results of decidability provided in [6]. Such an extension is based upon the machinery of Sect. 4, and some combinatorial tools developed by Choffrut, Varricchio and the second author in [9] for the study of rational relations. To this purpose, following [10],

Ch. 1, Sec. 1.1, we first recall that a *matrix context-free grammar* (which we will henceforth simply call *matrix grammar*) is a tuple $G = \langle V, \Sigma, M, S \rangle$, where V, Σ, and M are respectively the finite sets of nonterminals, terminals, and matrix (of rules), and $S \in V$ is the start symbol. Each matrix $m \in M$ is a finite sequence $m = (p_1, \ldots, p_s)$ where each p_i, $1 \leq i \leq s$, is a context-free production from V to $(V \cup \Sigma)^*$. We denote by P_M the set of all the context-free productions that appear in the matrices of M. For an arbitrary $m \in M$, we define the *1-step derivation in G*, $x \Rightarrow_m y$, with $x, y \in (V \cup \Sigma)^*$ if $x = x_0 \Rightarrow_{p_1} x_1 \Rightarrow_{p_2} \cdots \Rightarrow_{p_s} x_s = y$, where $\Rightarrow_{p_i}$ is the standard context-free derivation relation. This is equivalent to say that $x = x_0, x_s = y$, and, for every $i = 1, \ldots, s$, $x_{i-1} = x'_{i-1} X_i x''_{i-1}$, $x_i = x'_{i-1} \alpha_i x''_{i-1}$, where $p_i = (X_i \to \alpha_i)$, and for some $x'_{i-1}, x''_{i-1} \in (V \cup \Sigma)^*$. In other words, a 1-step derivation in the matrix grammar G corresponds to a s-step derivation in the context-free grammar $\langle V, \Sigma, P_M, S \rangle$, i.e., using, one by one, all the context-free productions of m. Let M^* be the set of the finite sequences of matrices and $\alpha \in M^*$. If $\alpha = \varepsilon$, then $x \Rightarrow_\alpha x, x \in V^*$; if one has

$$\alpha = m_1 m_2 \cdots m_s \quad \text{and} \quad x_0 \underset{m_1}{\Rightarrow} x_1 \underset{m_2}{\Rightarrow} \cdots x_{s-1} \underset{m_s}{\Rightarrow} x_s,$$

with $x_i \in (V \cup \Sigma)^*$, $m_j \in M$, $1 \leq j \leq s$, $0 \leq i \leq s$, then we write $x_0 \Rightarrow_\alpha x_s$. The *language generated* by G is $L(G) = \{w \in \Sigma^* \mid S \Rightarrow_\alpha w\}$. The following definition is instrumental.

Definition 5. *A matrix grammar $G = (V, \Sigma, M, S)$ is said to be* restricted of index k, *or simply* restricted, *if the set of nonterminals of G has the form $V = \{S\} \cup V'$, where $S \notin V'$ and $V' = V_1 \cup \cdots \cup V_k$, where $V_i \cap V_j = \emptyset$, for all $i \neq j$, and a matrix of rules of G can be only of one of the following forms*

- $(X_1 \to u_1 Y_1 v_1, \ldots, X_k \to u_k Y_k v_k)$, *for some $X_i, Y_i \in V_i$, $u_i, v_i \in \Sigma^*$, with $1 \leq i \leq k$.*
- *For every $X_i \in V_i$ and $1 \leq i \leq k$, $(X_1 \to \varepsilon, \ldots, X_k \to \varepsilon)$.*
- $(S \to X_1 \cdots X_k)$, *for some $X_i \in V_i$, with $1 \leq i \leq k$.*

Example 2. Let $L = \{a^n b^n c^n : n \in \mathbb{N}\}$ be the language over the alphabet $\Sigma = \{a, b, c\}$. Let G be the restricted matrix grammar with the matrices of rules $m_1 = (S \to ABC)$, $m_2 = (A \to aA, B \to bB, C \to cC)$, and $m_3 = (A \to \varepsilon, B \to \varepsilon, C \to \varepsilon)$. Note that the other components of G can be deduced from these matrices. One easily checks that $L = L(G)$. Indeed, each word $a^n b^n c^n, n \geq 0$, can be generated by using first the matrix m_1, then the matrix m_2 $n-1$ times and finally the matrix m_3. Hence $L \subseteq L(G)$. The reverse inclusion is proved similarly.

The main result of this section is the following.

Proposition 5. *Let L be a language generated by a restricted grammar and let Q be a quantum automaton. Then one has:*

i. $\mathrm{Cl}(\varphi(L))$ is semialgebraic.

ii. If the quantum automaton $\mathcal{Q}$ is rational, then $\mathbf{Cl}(\varphi(L))$ is effective semialgebraic and the $(L, \mathcal{Q})$ Intersection problem is decidable.

As a straightforward corollary of Proposition 5 for the matrix grammars of index $k = 1$, we get Corollary 1. Moreover, the language of Example 2 is an instance of the following more general result.

To this purpose, we recall that a *bounded semilinear* language is a finite union of *linear* languages which are languages of the form

$$L = \{u_1^{n_1} \cdots u_k^{n_k} \mid (n_1, \ldots, n_k) \in R\} \tag{9}$$

for some fixed words $u_i \in \Sigma^*$ for $i = 1, \ldots, k$ and $R \subseteq \mathbb{N}^k$ is a linear set, i.e., there exists $\mathbf{v}_0, \mathbf{v}_1, \ldots, \mathbf{v}_\ell \in \mathbb{N}^k$. such that $R = \{\mathbf{v}_0 + \lambda_1 \mathbf{v}_1 + \cdots + \lambda_\ell \mathbf{v}_\ell \mid \lambda_1, \ldots, \lambda_\ell \in \mathbb{N}\}$.

Lemma 2. *Every bounded semi-linear language can be generated by a restricted matrix grammar.*

Proposition 5 and Lemma 2 provide a new proof of the following.

Corollary 2 *([6], Prop. 21). Let L be a bounded semilinear language. Then its closure $\mathbf{Cl}(\varphi(L))$ is semialgebraic. Moreover, if the quantum automaton $\mathcal{Q}$ is rational, the $(L, \mathcal{Q})$ Intersection problem is decidable.*

Example 2 (continued). Let $\mathcal{Q} = (s, \varphi, P, \lambda)$ be a finite quantum automaton where $\varphi : \Sigma^* \longrightarrow O_n$ is the morphism (2) associated with $\mathcal{Q}$. Recall that $\mathbf{Cl}(\mathcal{A})$ and $\overline{\mathcal{A}}$ denote, respectively, the Euclidean and the Zariski closure of a set $\mathcal{A}$. The set $\mathbf{Cl}(\varphi(L))$ is then equal to

$$\mathbf{Cl}(\{\varphi(a)^n \varphi(b)^n \varphi(c)^n : n \in \mathbb{N}\}) = \mathbf{Cl}(\Psi(\{\varphi(a)^n \oplus \varphi(b)^n \oplus \varphi(c)^n : n \in \mathbb{N}\})) \tag{10}$$

where $\oplus$ is the operation defined in (6) and Ψ is the map defined, for every $X, Y, Z \in O_n$, as $\Psi(X \oplus Y \oplus Z) := XYZ$. By Proposition 1, the right-side term of (10) rewrites as $\Psi(\mathbf{Cl}(\{\varphi(a)^n \oplus \varphi(b)^n \oplus \varphi(c)^n : n \in \mathbb{N}\}))$, so that

$$\mathbf{Cl}(\varphi(L)) = \Psi(\mathbf{Cl}(\{g\}^*)) = \Psi(\mathbf{Cl}(\overline{\langle\{g\}\rangle})),$$

where $g = \varphi(a) \oplus \varphi(b) \oplus \varphi(c)$. By Theorem 1, $\mathbf{Cl}(\langle\{g\}\rangle) = \overline{\langle\{g\}\rangle}$ is algebraic. From this, by some standard properties of semialgebraic sets (see [2], Proposition 3), it follows that $\mathbf{Cl}(\varphi(L)) = \Psi(\overline{\langle\{g\}\rangle})$ is semialgebraic.

If the automaton $\mathcal{Q}$ is a rational, then, by applying Derksen's algorithm to $\{g\}$, one computes $\overline{\langle\{g\}\rangle}$ as an algebraic set, which, in turn, implies that $\Psi(\overline{\langle\{g\}\rangle})$ is effective semialgebraic.

Proposition 5 is an extension of Corollary 1 and Corollary 2. Indeed the following example provides languages, that are neither bounded, nor context-free.

Example 3. Let L_k, with $k \in \mathbb{N}$, be the language over $\Sigma = \{a, b\}$ given by $L_k = \{u^k : u \in \Sigma^*\}$. It is easily checked that L_k can be generated by the restricted grammar $G = \langle V, \Sigma, M, S \rangle$, where $V = \{S, X_1, \ldots, X_k\}$, and M is given by the matrices $(S \to X_1 \cdots X_k)$, $(X_1 \to \varepsilon, \ldots, X_k \to \varepsilon)$, and, for every $\sigma \in \Sigma$, $(X_1 \to \sigma X_1, \ldots, X_k \to \sigma X_k)$. Note that the very same argument used in Example 2 allows one to verify that $\mathbf{Cl}(\varphi(L))$ is semialgebraic and effective semialgebraic in the case that the quantum automaton is rational.

5.2 The Case of Monoidal Languages

We start with the following definition that is instrumental for this section.

Definition 6. *A grammar $G = \langle V, \Sigma, P, S \rangle$ is said to be monoidal of index k if it is the composition*
$$G = \mathcal{G}_1 \circ \mathcal{G}_2 \circ \cdots \circ \mathcal{G}_k,$$
of k families of linear grammars where the productions of G satisfy the following conditions:

- *if $(X \to u) \in P$, $u \in \Sigma^*$ (terminal production) then $u = \varepsilon$*
- *if $(X \to uYv) \in P$, where X, Y are distinct variables of a grammar in $\mathcal{G}_i$, $1 \leq i \leq k$, then $u = v = \varepsilon$.*

A language is said to be *monoidal* if it is generated by a monoidal grammar of index k, for some $k \in \mathbb{N}$. The main result of this section is the following.

Proposition 6. *Let (L, Q) be a pair where Q is a rational quantum automaton and L is a language generated by a monoidal grammar $G = \mathcal{G}_1 \circ \mathcal{G}_2 \circ \cdots \circ \mathcal{G}_k$. If the Zariski closure of the monoids of cycles of the grammars of $\mathcal{G}_k$ are irreducible, then $\mathbf{Cl}(\varphi(L))$ is effective semialgebraic.*

We will do the proof of Proposition 6 for $k \leq 2$, i.e., for a monoidal grammar $G = \mathcal{G}_1 \circ \mathcal{G}_2$ of index 2, since, as a consequence of (7), the general case can be treated similarly.

For the sequel, it's worth recalling that, for a set $\mathcal{A}$ of matrices, $\mathcal{A} \in (\mathcal{H})$ stands for $\mathcal{A}$ satisfies Eq. (5) of Sect. 2. The following lemmata hold.

Lemma 3. *Let $G = \mathcal{G}_1$ be a monoidal grammar of index 1 and $L = L(G)$. If the Zariski closure of all the monoids of cycles of G are irreducible, effective algebraic sets, then there exist a set $H \in (\mathcal{H})$ and a regular map Ψ with $\mathbf{Cl}(\varphi(L)) = \Psi(H)$. In particular $\mathbf{Cl}(\varphi(L))$ is effective semialgebraic and contains I.*

Let $\mathcal{G}_1 = \langle V_1, \Sigma_1, P_1, S \rangle$ be the first grammar of the composition $G = \mathcal{G}_1 \circ \mathcal{G}_2$. Let $\Sigma_1^* \times \Sigma_1^*$ be the multiplicative monoid of binary relations on Σ_1^* and let $\Pi(O_n \oplus O_n)$ be the multiplicative monoid of subsets of $O_n \oplus O_n$. Define now the morphism between the two monoids
$$\zeta : \Sigma_1^* \times \Sigma_1^* \longrightarrow \Pi(O_n \oplus O_n)$$
as: for every $(u, v) \in \Sigma_1^* \times \Sigma_1^*$, with $u = A_1 \cdots A_s$, $v = B_1 \cdots B_t$, $A_i, B_i \in \Sigma_1$
$$\zeta(u, v) = \varphi(L_{A_1}) \cdots \varphi(L_{A_s}) \oplus \varphi(L_{B_1})^T \cdots \varphi(L_{B_t})^T \tag{11}$$
where, for every $i = 1, \ldots, t$, $\varphi(L_{B_i})^T$ denotes the set of matrices
$$\{m^T \ : \ m \in \varphi(L_{B_i})\}. \tag{12}$$

Lemma 4. *For every $A \in V_1$, there exists an effective computable regular subset $\mathcal{R}$ of $\Sigma_1^* \times \Sigma_1^*$ such that $M_A = \zeta(\mathcal{R})$.*

Lemma 5. *For every* $A \in V_1$, $\mathbf{Cl}(M_A) \in (\mathcal{H})$. *In particular,* $\mathbf{Cl}(M_A)$ *is effective algebraic.*

Proof. By Lemma 4, for every $A \in V_1$, there exists an effective computable regular subset $\mathcal{R}$ of $\Sigma_1^* \times \Sigma_1^*$ such that $M_A = \zeta(\mathcal{R})$. We will prove the following slightly more general claim:

Claim. *Let $\mathcal{R}$ be a regular subset of $\Sigma_1^* \times \Sigma_1^*$. Then $\mathbf{Cl}(\zeta(\mathcal{R}))$ contains I and $\mathbf{Cl}(\zeta(\mathcal{R})) = \Psi(H)$ for some $H \in (\mathcal{H})$ and for some regular map Ψ. Further, if $\mathcal{R} = \mathcal{R}^*$, then $\mathbf{Cl}(\zeta(\mathcal{R})) \in (\mathcal{H})$.*

We begin to show the claim for a finite set $\mathcal{R}$. We may first suppose $\mathcal{R} = \{r\}$, where $r = (\alpha, \beta)$ with $\alpha = A_1 A_2 \cdots A_s$, $\beta = B_1 \cdots B_t$, $A_i, B_i \in \Sigma_1$, $s, t \geq 0$. We assume also that α and β have the same length, i.e., $s = t$. By (11), the fact that ζ is a monoid morphism and Proposition 1, one gets

$$\mathbf{Cl}(\zeta(\mathcal{R})) = \mathbf{Cl}(\varphi(L_{A_1})) \cdots \mathbf{Cl}(\varphi(L_{A_s})) \oplus \mathbf{Cl}(\varphi(L_{B_1}))^T \cdots \mathbf{Cl}(\varphi(L_{B_s}))^T,$$

taking into account that, for the set (12), one has $\mathbf{Cl}(\varphi(L_{A_i})^T) = \mathbf{Cl}(\varphi(L_{A_i}))^T$, with $1 \leq i \leq s$. By Lemma 3, the previous formula gives

$$\mathbf{Cl}(\zeta(\mathcal{R})) = \Psi(H_1) \cdots \Psi(H_s) \oplus \Psi(G_1)^T \cdots \Psi(G_s)^T,$$

where Ψ is a regular map and, for every $i = 1, \ldots, s$, $H_i, G_i \in (\mathcal{H})$. Since each $\Psi(H_i)$ and $\Psi(G_i)$ of the last formula, contain I so holds for $\mathbf{Cl}(\zeta(\mathcal{R}))$. Set $H = H_1 \oplus \cdots \oplus H_s \oplus G_1 \oplus \cdots \oplus G_s$ and observe that, by a closure property of irreducible sets ([2], Sec. 2.4, Prop. 5), $H \in (\mathcal{H})$. Then $\mathbf{Cl}(\zeta(\mathcal{R})) = \Psi'(H)$, where Ψ' is the regular map defined as $\Psi'(X_1 \oplus \cdots \oplus X_s \oplus Y_1 \oplus \cdots \oplus Y_s) = \Psi(X_1) \cdots \Psi(X_s) \oplus \Psi(Y_1)^T \cdots \Psi(Y_s)^T$. Observe that, if $s > t$, then we can write the word $\beta = B_1 \cdots B_t$ as $\beta = B_1 \cdots B_s$, with $B_j = \varepsilon$, for $j = t+1, \ldots, s$. Afterwards, set $L_{B_j} = \{\varepsilon\}$ and $\varphi(L_{B_j}) = \{I\}$, $t+1 \leq j \leq s$. Then one proceeds to prove the claim, by repeating the argument above for $s = t$. This proves that the claim for $\mathcal{R} = \{r\}$. Finally, the claim for an arbitrary finite set $\mathcal{R}$ is obtained by applying Case (i) below.

We now prove the claim according to the following cases:

(i) $\mathcal{R} = \mathcal{R}_1 \cup \mathcal{R}_2$, (ii) $\mathcal{R} = \mathcal{R}_1 \mathcal{R}_2$, (iii) $\mathcal{R} = \mathcal{R}_1^*$,

where

$$\mathbf{Cl}(\zeta(\mathcal{R}_1)) = \Psi_1(H_1), \quad \mathbf{Cl}(\zeta(\mathcal{R}_2)) = \Psi_2(H_2), \tag{13}$$

with $H_1, H_2 \in (\mathcal{H})$ and Ψ_1, Ψ_2 being regular maps.

(i) By (13), starting from Ψ_1 and Ψ_2, one has $\mathbf{Cl}(\zeta(\mathcal{R})) = \Psi(H_1 \cup H_2)$, for a suitably defined regular map Ψ, so that the claim comes from $H_1 \cup H_2 \in (\mathcal{H})$.

(ii) By Proposition 1 and the fact that ζ is a monoid morphism, (13) yields $\mathbf{Cl}(\zeta(\mathcal{R})) = \Psi_1(H_1)\Psi_2(H_2)$. From this, since $I \in \Psi_1(H_1), \Psi_2(H_2)$ then $I \in \mathbf{Cl}(\zeta(\mathcal{R}))$. Now, observe that, by closure properties of semialgebraic sets ([2], Sect. 2.3) and by a closure property of irreducible sets ([2], Sect. 2.4, Prop. 5), the condition $H_1, H_2 \in (\mathcal{H})$ implies $H_1 \oplus H_2 \in (\mathcal{H})$. The claim follows from the fact that $\Psi_1(H_1)\Psi_2(H_2) = \Psi(H_1 \oplus H_2)$, where Ψ is the regular map defined as $\Psi(X \oplus Y) := \Psi_1(X)\Psi_2(Y)$.

(iii) Since $\mathcal{R} = R^*$ and the fact that ζ is a monoid morphism, then one has $Cl(\zeta(\mathcal{R})) = Cl(\zeta(R^*)) = Cl(\zeta(R)^*)$. From this, it trivially follows that $Cl(\zeta(\mathcal{R}))$ contains I. Now, since, by Proposition 1, for an arbitrary set $\mathcal{X}$,

$$Cl(\mathcal{X}^*) = Cl(Cl(\mathcal{X})^*), \qquad (14)$$

one has $Cl(\zeta(\mathcal{R})) = Cl(Cl(\zeta(R))^*)$. By Theorem 1, one has

$$Cl(Cl(\zeta(R))^*) = \overline{Cl(\zeta(R))}^*, \qquad (15)$$

where $\overline{Cl(\zeta(R))}^*$ is the Zariski closure of $Cl(\zeta(R)^*)$. Recall that, by (13), one has $Cl(\zeta(R)) = \Psi(H)$, where Ψ is a regular map and $H \in (\mathcal{H})$. This fact and (15) now implies $Cl(\zeta(\mathcal{R})) = \overline{\Psi(H)}^*$. Since $H \in (\mathcal{H})$, we can write $H = \bigcup_{i=1}^{k} H_i$ as a finite union of irreducible effective algebraic sets H_i, every one of which containing the identity matrix I. This implies $Cl(\zeta(\mathcal{R})) = \overline{(\Psi(H_1) \cdots \Psi(H_k))}^*$. Observe that in the formula above, we can write $\Psi(H_1) \cdots \Psi(H_k)$ as
$\widehat{\Psi}(H_1 \oplus \cdots \oplus H_k)$, where $\widehat{\Psi}$ is the regular map defined as $\widehat{\Psi}(X_1 \oplus \cdots \oplus X_k) = \Psi(X_1) \cdots \Psi(X_k)$. Hence we get $Cl(\zeta(\mathcal{R})) = \overline{(\widehat{\Psi}(H))}^*$, with $H = H_1 \oplus \cdots \oplus H_k$. Again, $H \in (\mathcal{H})$ as one easily checks. The claim follows by applying Proposition 2. This completes the proof. □

Proof of Proposition 6: As remarked before, it is enough to prove the claim for $k \leq 2$ since, by (7), the general case is treated with the same argument. Let $G = \langle V, \Sigma, P, S \rangle$ and observe that, for every $A \in V$, $Cl(M_A)$ is effective algebraic by hypothesis if $k = 1$ and, by Lemma 5, if $k = 2$. The claim now follows from Proposition 4. □

By the closure property of semialgebraic sets ([2], Sec. 2.3) and Proposition 3, an immediate corollary of Proposition 6 is the following.

Corollary 3. *If the pair (L, Q) satisfies the hypothesis of Proposition 6, then the (L, Q) Intersection problem is decidable.*

Example 1 (continued). We run the proofs of Propositions 2 and 6 on the language of Example 1. Since $\mathcal{L} = L^*$, with $L = \{a^n b^n : n \in \mathbb{N}\}$, then

$$Cl(\varphi(\mathcal{L})) = Cl(\varphi(L^*)) = Cl(\varphi(L)^*),$$

where φ is the morphism (2) of the quantum automaton. Recall that $Cl(\mathcal{A})$ and $\overline{\mathcal{A}}$ denote, respectively, the Euclidean and the Zariski closure of a set $\mathcal{A}$. By Proposition 1 and Theorem 1, one has $Cl(\varphi(L)^*) = \overline{\varphi(L)}^*$. This implies $Cl(\varphi(\mathcal{L})) = \overline{\Psi(G)}^*$, where

- $G = M^*$ is the monoid generated by the matrix $M = \varphi(a) \oplus \varphi(b) \in O_n \oplus O_n$,
- Ψ is the regular map defined as $\Psi(A \oplus B) := AB$.

We now prove that, if $\overline{G}$ is irreducible and effective algebraic, then $\overline{\Psi(G)^*}$ is in $(\mathcal{H})$ and thus $\mathbf{Cl}(\varphi(\mathcal{L}))$ is effective algebraic as well.

For this purpose, observe first that $\Psi : O_n \oplus O_n \to O_n$ is a regular and thus a continuous map. This implies that

$$\Psi(\overline{G}) \subseteq \overline{\Psi(G)},$$

which, in turn, implies

$$\overline{\Psi(\overline{G})} = \overline{\Psi(G)}. \tag{16}$$

Let $H_1 = \overline{\Psi(\overline{G})}$. By the fact that $\overline{G}$ is irreducible and Ψ is a regular map, then H_1 is still irreducible (cf [12], Remark 1.3.2). Moreover, being G a group, one has $I \in \Psi(G)$ so $I \in H_1$. Since G is finitely generated as a monoid, then, by using Derksen's algorithm, one can effectively calculate $\overline{G}$. Afterwards, following [11], Sec. 3.1, by using *Gröbner bases techniques*, since $\overline{G}$ is effective algebraic, then one can effectively compute H_1.

Let $H_2 = \overline{H_1 \cdot H_1}$. Observe first that $I \in H_1$ implies $I \in H_2$. Similarly to the previous case, one observes that, for arbitrary subsets $\mathcal{A}, \mathcal{B}$ of O_n

$$\overline{\overline{\mathcal{A}} \cdot \overline{\mathcal{B}}} = \overline{\mathcal{A} \cdot \mathcal{B}}. \tag{17}$$

By using the very same argument for H_1 one verifies that H_2 is an irreducible effective algebraic set. Moreover, by (16) and (17), one has $H_2 = \overline{\Psi(G)^2}$.

Finally, let us define recursively the family $\{H_i\}_{i \geq 1}$ of sets of O_n, where $H_0 = \{I\}$ and, for every $i \geq 1$, $H_i = \overline{H_{i-1} \cdot H_1}$. Getting an ascending chain

$$H_1 \subseteq H_2 \subseteq \cdots H_i \subseteq \cdots$$

of irreducible, effective algebraic sets, then there exists some $m \in \mathbb{N}$ such that

$$H_{m+1} = \overline{\Psi(G)^{m+1}} = \overline{\Psi(G)^m} = H_m. \tag{18}$$

Finally let $\mathcal{X} = \bigcup_{i \geq 1} H_i$. By (18), $\overline{\mathcal{X}} = H_1 \cup \cdots \cup H_m$. On the other hand, since, for every $i \geq 1$, $H_i = \overline{\Psi(G)^i}$, then it is checked that

$$\overline{\mathcal{X}} = H_1 \cup \cdots \cup H_m = \overline{\Psi(G)^*}.$$

Hence $\overline{\Psi(G)^*}$ is effective algebraic.

6 Concluding Remarks and Open Problems

[α]. One could extend Proposition 6 to a broader class of quantum automata. Such extension would require the computation of the Zariski closure of a group G of matrices that is *algebraically presented*, i.e. G admits a set of generators that is algebraic itself. It may be of interest to investigate conditions ensuring the extension of the algorithms of [11,16,23] for such groups.

[β]. It is open whether the (L, Q) Intersection problem is decidable for context-free languages. We conjecture that the answer to this problem is negative. An approach may be to reduce the problem to a so called *"birdie problem"* [27].

[γ]. The (L, Q) INTERSECTION problem is decidable for bounded semi-linear languages. It may be of interest to extend this result to the languages accepted by reversal bounded non deterministic counter machines [17]. This extension could exploit a characterization of such languages provided by Ibarra in [18] in terms of RLGMC-grammars (right-linear grammar with multi-counters). A RLGMC-grammar is a regular grammar that associates with each derivation a weight, i.e. a vector in the additive monoid $\mathbb{N}^d$. Then a derivation is considered successful if the corresponding weight belongs to the monoid generated by $(1, 1, \ldots, 1)$. The main crux is the construction of a formula (see Proposition 3) based on the computation of the Zariski closure of the intersection of two finitely generated monoids of matrices. The intrinsic difficulty to *control* the accumulation points of the set – i.e., to ignore those not coming from the effective functioning of the machine – makes such a computation a non trivial task.

Acknowledgements. The authors would like to thank the reviewer for the numerous comments that improve the paper. The authors have no competing interests to declare that are relevant to the content of this article.

Disclosure of Interests. The authors have no competing interests to declare that are relevant to the content of this article.

References

1. Basu, S., Pollack, R., Roy, M.-F.: Algorithms in Real Algebraic Geometry. Springer, Berlin (2003). https://doi.org/10.1007/3-540-33099-2
2. Benso, A., D'Alessandro, F., Papi, P.: Quantum finite automata and languages of finite index. Preprint arXiv:2406.13797v2 (2024)
3. Berstel, J.: Transductions and Context-Free Languages. Teubner, Stuttgart (1979)
4. Bertoni, A., Mereghetti, C., Palano, B.: Quantum computing: 1-way quantum automata. In: Ésik, Z., Fülöp, Z. (eds.) DLT 2003. LNCS, vol. 2710, pp. 1–20. Springer, Heidelberg (2003). https://doi.org/10.1007/3-540-45007-6_1
5. Bertoni, A., Choffrut, C., D'Alessandro, F.: Quantum finite automata and linear context-free languages: a decidable problem. In: Béal, M.-P., Carton, O. (eds.) DLT 2013. LNCS, vol. 7907, pp. 82–93. Springer, Heidelberg (2013). https://doi.org/10.1007/978-3-642-38771-5_9
6. Bertoni, A., Choffrut, C., D'Alessandro, F.: On the decidability of the intersection problem for quantum automata and context-free languages. Internat. J. Found. Comput. Sci. **25**, 1065–1081 (2014)
7. Blondel, V.D., Jeandel, E., Koiran, P., Portier, N.: Decidable and Undecidable Problems about Quantum Automata. SIAM J. Comput. **34**, 1464–1473 (2005)

8. Cassaigne, J., Nicolas, F.: On the decidability of semigroup freeness. RAIRO Theor. Informatics Appl. **46**, 355–399 (2012)
9. Choffrut, C., D'Alessandro, F., Varricchio, S.: Bounded rational languages of trace monoids. Theory Comput. Syst. **46**, 351–369 (2010)
10. Dassow, J., Paun, G.: Regulated Rewriting in Formal Language Theory. EATCS Monographs in Theoretical Computer Science, vol. 18, Springer, Heidelberg (1989)
11. Derksen, H., Jeandel, E., Koiran, P.: Quantum automata and algebraic groups. J. Symb. Comput. **39**, 357–371 (2005)
12. Geck, M.: An Introduction to Algebraic Geometry and Algebraic Groups. Oxford University Press, Oxford (2003)
13. Ginsburg, S., Spanier, E.H.: Derivation-bounded languages. J. Comput. System Sci. **2**(3), 228–250 (1968)
14. Ginsburg, S.: The Mathematical Theory of Context-Free Languages. Mc Graw-Hill, New York (1966)
15. Harrison, M.A.: Introduction to Formal Language Theory. Addison-Wesley Publishing Co., Reading (1978)
16. Hrushovski, E., Ouaknine, J., Pouly, A., Worrell, J.: On Strongest Algebraic Program Invariants. J. ACM **70**, 1–22 (2023)
17. Ibarra, O.H.: Reversal-bounded multicounter machines and their decision problems. J. ACM **25**, 116–133 (1978)
18. Ibarra, O.H.: Grammatical characterizations of NPDAs and VPDAs with counters. Theoret. Comput. Sci. **746**, 136–150 (2018)
19. Jeandel, E.: Indécidabilité sur les automates quantiques. Master's thesis. ENS Lyon (2002)
20. Kondacs, A., Watrous, J.: On the power of quantum finite state automata. In: Proceedings of the 38th Annual Symposium on Foundations of Computer Science, pp. 66–75 (1997)
21. Moore, C., Crutchfield, J.: Quantum automata and quantum grammars. Theoret. Comput. Sci. **237**, 275–306 (2000)
22. Nivat, M.: Transductions des langages de Chomsky. Ann. Inst. Fourier (Grenoble) **18**(1), 339–455 (1968)
23. Nosan, K., Pouly, A., Schmitz, S., Shirmohammadi, M., Worrell, J.: On the computation of the Zariski closure of finitely generated groups of matrices. In: Proceedings of ISSAC 2022, International Symposium on Symbolic and Algebraic Computation, pp. 129–138. ACM (2022). ISBN 978-1-4503-8688-3
24. Onishchik, A., Vinberg, E.: Lie Groups and Algebraic Groups. Springer, Berlin (1990). https://doi.org/10.1007/978-3-642-74334-4
25. Paz, A.: Introduction to Probabilistic Automata. Academic Press, New York (1971)
26. Potapov, I., Semukhin, P.: Decidability of the membership problem for 2×2 integer matrices. In: Proceedings of SODA 2017 ACM-SIAM Symposium on Discrete Algorithms, SIAM 2017, pp. 170–186 (2017). ISBN 978-1-61197-478-2
27. Ullian, J.S.: Partial Algorithm Problems for Context Free Languages. Inf. and Control. **11**, 80–101 (1967)

On Shortest Products for Nonnegative Matrix Mortality

Andrew Ryzhikov

Department of Computer Science, University of Oxford, Oxford, UK
ryzhikov.andrew@gmail.com

Abstract. Given a finite set of matrices with integer entries, the matrix mortality problem asks if there exists a product of these matrices equal to the zero matrix. We consider a special case of this problem where all entries of the matrices are nonnegative. This case is equivalent to the NFA mortality problem, which, given an NFA, asks for a word w such that the image of every state under w is the empty set. The size of the alphabet of the NFA is then equal to the number of matrices in the set. We study the length of shortest such words depending on the size of the alphabet. We show that for an NFA with n states this length can be at least $2^n - 1$ for an alphabet of size n, $2^{(n-4)/2}$ for an alphabet of size 3 and $2^{(n-2)/3}$ for an alphabet of size 2. We also discuss further open problems related to mortality of NFAs and DFAs.

Keywords: matrix mortality · NFA mortality · nonnegative matrices

1 Introduction

Given a finite set $\mathcal{A}$ of integer square matrices of the same dimension, the matrix mortality problem asks if the monoid generated by $\mathcal{A}$, that is, the set of all possible products of matrices from $\mathcal{A}$, contains the zero matrix. This problem is undecidable already for sets of 3×3 matrices [20], and is a subject of extensive research, see [6] for a survey of recent developments.

We consider a special case of matrix mortality where all matrices in $\mathcal{A}$ are nonnegative, that is, all their entries are nonnegative. We call this problem nonnegative matrix mortality. The theory of nonnegative matrices has applications in the theory of codes [4], Markov chains [28] and symbolic dynamics [16], and recently there has been a significant interest in reachability problems for sets of nonnegative matrices and monoids generated by them [9,21,29].

Clearly, for nonnegative matrix mortality it only matters which entries of the matrices in $\mathcal{A}$ are strictly positive and which are zero. In particular, this implies that nonnegative matrix mortality is decidable in polynomial space. Moreover, for a set $\mathcal{A}$ of nonnegative $n \times n$ matrices, the length of a shortest product from this set resulting in the zero matrix is at most $2^n - 1$. This easily follows from the interpretation of $\mathcal{A}$ as a non-deterministic finite (semi-)automaton (NFA), as explained, e.g., in [22]. Since our results, as well as most of the existing literature on nonnegative matrix mortality, use this correspondence, we provide its details.

We view a finite set $\mathcal{A}$ of nonnegative $n \times n$ matrices as an NFA with a state set $Q = \{1, 2, \ldots, n\}$. Each letter a_j of its alphabet Σ corresponds to a matrix $A_j \in \mathcal{A}$. To define the transition relation $\Delta : Q \times \Sigma \to 2^Q$, we take $\Delta(i, a_j)$ to be the set of all h such that the entry (i, h) in A_j is strictly positive. When the transition relation is clear from the context, we denote $\Delta(q, a)$ as $q \cdot a$. We homomorphically extend it to words over Σ, and then to subsets $S \subseteq Q$ as $S \cdot w = \{p \mid p \in q \cdot w \text{ for some } q \in S\}$. In thus constructed NFA (Q, Σ, Δ), with respect to matrix mortality, words correspond to products of matrices from $\mathcal{A}$: we have $Q \cdot w = \emptyset$ if and only if the product of matrices corresponding to w is the zero matrix. Such words w are called *mortal*. To obtain the mentioned upper bound of $2^n - 1$ (observed, e.g., in [13]), it is now enough to note that for a shortest mortal word $w = x_1 x_2 \ldots x_k$ the sets $Q \cdot x_1 x_2 \ldots x_i$ for different values of i are pairwise different subsets of Q, and hence $k \leq 2^n - 1$.

The length of a shortest mortal word of an NFA is called its *mortality threshold*. Let $\mathbf{mt}(n, m)$ denote the maximum mortality threshold of an NFA with n states and m letters admitting a mortal word. In this paper, we investigate the values $\mathbf{mt}(n, m)$ depending on m. As shown above, for every n and m, $\mathbf{mt}(n, m) \leq 2^n - 1$. We survey the known bounds on $\mathbf{mt}(n, m)$ in the next section, and in the rest of this section we mention some other relevant results.

The problem of checking if a given NFA admits a mortal word is PSPACE-complete, even when restricted to binary NFAs [14]. As usually, we call an NFA *binary* if its alphabet has size two, and *ternary* if it has size three.

In [15], the case where $\mathcal{A}$ is a set of nonnegative integer matrices of joint spectral radius at most one is considered. This case includes, in particular, all sets $\mathcal{A}$ of $\{0, 1\}$-matrices (that is, matrices whose entries belong to the set $\{0, 1\}$) with the property that the monoid generated by $\mathcal{A}$ contains only $\{0, 1\}$-matrices. Such sets $\mathcal{A}$ correspond to unambiguous NFAs, which are NFAs with the property that for every pair p, q of states and every word w there is at most one path from p to q labelled by w. It is proved in [15] that for sets of joint spectral radius at most one, the existence of a product resulting in the zero matrix can be checked in polynomial time, and for all m we have $\mathbf{mt}_{\rho \leq 1}(n, m) \leq n^5$, where $\mathbf{mt}_{\rho \leq 1}(n, m)$ is $\mathbf{mt}(n, m)$ restricted to such sets. The best known lower bound is quadratic in n [25]. Some more results on mortal words for unambiguous NFAs are presented in [5].

Our Contributions. We study the following question: given a set of m nonnegative $n \times n$ matrices, how long can their shortest product resulting in the zero matrix be? In Sect. 2 we survey known results. Section 3 contains our main contributions. Namely, for $m = n$, we show a lower bound of $2^n - 1$ (Theorem 4), matching the upper bound for arbitrary m. We prove it by constructing an NFA simulating a binary counter counting from $2^n - 1$ to zero. The case of constant m uses the same idea, but requires much more involved constructions to implement it, resulting in a lower bound of $2^{(n-4)/2}$ for $m = 3$ (Theorem 5) and a lower bound of $2^{(n-2)/3}$ for $m = 2$ (Theorem 6). We conclude the paper by discussing other approaches to NFA and DFA mortality in Sect. 4, and providing a list of open problems related to this setting in Sect. 5.

2 Existing Results and Their Adaptations

In this section, we explain existing results on $\mathbf{mt}(n,m)$. The first result about this value seems to be in the work [10], where it is proved that $\mathbf{mt}(n,m)$ is not bounded by any polynomial in n. Most of the known results are not stated in terms of matrix or NFA mortality, so we appropriately rephrase them. We also show that strong lower bounds on $\mathbf{mt}(n,m)$ can be easily obtained from existing results that are not obviously related to mortality.

2.1 Factorial Languages and Shortest Rejected Words

Clearly, mortal words for an NFA are precisely the words that do not label a path between any two states in the NFA. Equivalently, these are words that are not accepted by the NFA if we make each state both initial and final. Shortest such words were studied in [14,22], building upon an earlier work [8] about a more general question: what is the maximum length of a shortest word not accepted by an NFA with a given number of states? The result relevant to our work is as follows.

Theorem 1 (Theorem 16 in [14], rephrased). *For every n, there exists an NFA with n states over a ternary alphabet such that its shortest mortal word is of length at least $2^{n/75}p(n)$, where $p(n)$ is some polynomial.*

2.2 D1-Directing and Carefully Synchronizing Words

A word w is called *D1-directing* for an NFA if there exists a state q such that for every state p we have $p \cdot w = \{q\}$ [13]. An NFA is called *total*[1] if for every state q and every letter a we have $q \cdot a \neq \emptyset$. A state q in an NFA is called a *sink* state (called a trap state in [12]) if for every letter a we have $q \cdot a = \{q\}$. A word is D1-directing for a total NFA $\mathcal{A}$ with a sink state q if and only if it is mortal for the NFA obtained from $\mathcal{A}$ by removing q. We thus get the following.

Theorem 2 (Theorem 3 in [12]). *For every $n \geq 3$, $\mathbf{mt}(n, 2^n - 5) = 2^n - 1$.*

In Theorem 4 below we show that only n letters are enough to achieve the same lower bound, thus obtaining that for every $n \geq 1$, $\mathbf{mt}(n,n) = 2^n - 1$.

The most well-studied setting of D1-directing words for NFAs is that of DFAs, in which case such words are often called *carefully synchronizing*. Recall that a DFA is an NFA such that $|q \cdot a| \leq 1$ for every state q and every letter a.

---
[1] Such NFAs are often called complete, for example in [12,13]. However, in e.g. [19,23] the term "complete" refers to objects corresponding to matrix monoids that do not contain the zero matrix, which we find reasonable. We thus find the term "total" more appropriate, since an NFA is total if and only if the action of every its letter is a total binary relation. A total NFA is then necessarily complete (that is, does not admit a mortal word), but the opposite does not always hold, as illustrated by the following NFA with two letters: .

As usually, we use a partial function $\delta : Q \times \Sigma \rightharpoonup Q$ instead of $\Delta : Q \times \Sigma \to 2^Q$ to highlight this. A DFA admitting a carefully synchronizing word is also called *carefully synchronizing*. Strong lower bounds are known for the length of shortest carefully synchronizing words for DFAs, and these bounds can be transferred to shortest mortal words for NFAs as follows.

Let $\mathcal{A} = (Q, \Sigma, \delta)$ be a carefully synchronizing DFA. Consider an NFA $\mathcal{A}' = (Q, \Sigma \cup \{r\}, \Delta')$ obtained from $\mathcal{A}$ as follows: for every pair $q \in Q, a \in \Sigma$ such that $\delta(q, a)$ is undefined, take $\Delta'(q, a) = Q$. If $\delta(q, a)$ is defined, take $\Delta'(q, a) = \{\delta(q, a)\}$. Pick a state $p \in Q$ such that there exists a carefully synchronizing word w mapping all states to p. Take $\Delta(p, r) = \emptyset$ and $\Delta(q, r) = Q$ for all $q \neq p$. If w is a carefully synchronizing word for $\mathcal{A}$, then wr is a mortal word for $\mathcal{A}'$. Conversely, by construction, every mortal word for $\mathcal{A}'$ must contain a carefully synchronizing word for $\mathcal{A}$ as a factor. By applying this transformation to Theorems 14 and 16 from [7], we obtain the following.

Theorem 3 ([7]). *We have* $\mathbf{mt}(n, 4) = \Omega(\frac{2^{2n/5}}{n})$ *and* $\mathbf{mt}(n, 3) = \Omega(\frac{2^{n/3}}{n\sqrt{n}})$.

Note that sometimes adding a new letter r is not necessary in the construction described above, which seems to be the case for [7]. However, proving that requires analysing the whole construction for careful synchronization, which is quite involved. Since our results on NFA mortality are stronger than those in Theorem 3 even without the additional letter, we do not pursue this any further. We also remark that the applicability of this approach is limited by the upper bound on shortest carefully synchronizing words for n-state DFAs, which is $3^{(1+\epsilon)n/3}$ for every $\epsilon > 0$ [24]. This upper bound is asymptotically tight already for an alphabet of linear size [18,24,27].

3 NFA Mortality Lower Bounds

For all lower bounds in this section, the main idea is to construct an NFA simulating a binary counter. Namely, for every k we construct an NFA with the number of states linear in k such that, when reading a word w, a dedicated set $Q = \{q_1, \ldots, q_k\}$ of k states behaves in the following way. We say that a state q is *active* after the application of a word w if $q \in p \cdot w$ for some state p, otherwise we say that it is *inactive*. Initially, all states in Q are active. We interpret active and inactive states from Q as a number in the binary encoding. Namely, given a word $w \in \Sigma^*$, define $\text{bin}(w) = x_1 x_2 \ldots x_k$, where $x_i = 1$ if $q_i \in Q \cdot w$, and $x_i = 0$ otherwise. We treat $\text{bin}(w)$ as a natural number represented in the binary notation, with the most significant bit first. By construction of the NFA we guarantee that when reading a letter, this number is either decremented by at most one or is set again to the number consisting of all ones. The length of a shortest mortal word for such NFA is then at least $2^k - 1$.

For an alphabet whose size is equal to the number of states, this idea can be implemented in a fairly straightforward way with a k-state NFA. To implement it with a ternary and a binary alphabet, we need more advanced techniques. The main challenge is that at least one of the letters must have a different effect on the encoding of the number depending on its value. To achieve that, between every two decrements we shift the number to the right until it ends with a one, thus iteratively dividing it by two, and simultaneously add some additional marker bits to preserve the information about the number of shifts we performed. In particular, it is important to remark that in these constructions not every application of a letter decrements the value of the number that we are tracking. Namely, we will have one letter that actually decrements the value, and all other letters will perform auxiliary operations such as shifting the representation. These ideas will be explained in detail in due course.

Conventions for Figures. In all figures in this section, we use a vertical bold outgoing arrow to indicate that there is a transition from the state it originates from to all the states of the NFA. Similarly, a vertical bold incoming arrow indicates that there is a transition from each state of the NFA (except for states with a dashed outgoing arrow) to the state where the arrow ends. A dashed outgoing arrow from a state means that the image of this state is empty.

3.1 Linear-Size Alphabet is Enough for a Tight Lower Bound

We start by proving the following statement. As its consequence, together with the upper bound mentioned in the introduction, we get that $\mathbf{mt}(n,n) = 2^n - 1$.

Theorem 4. *For every integer $n \geq 1$, there exists an NFA with n states and n letters such that the length of its shortest mortal word is $2^n - 1$.*

Proof. For every positive n we construct an NFA $\mathcal{A} = (Q, \Sigma, \Delta)$ as follows. We take $Q = \{q_1, \ldots, q_n\}$ (thus we have $k = n$ in the informal explanations in the beginning of this section). Let $\Sigma = \{a_1, \ldots, a_n\}$. Define

$$\Delta(q_i, a_j) = \begin{cases} \emptyset & \text{if } i = j = n; \\ \{q_{i+1}, \ldots, q_n\} & \text{if } 1 \leq i = j < n; \\ Q & \text{if } 1 \leq j < i \leq n; \\ \{q_i\} & \text{if } 1 \leq i < j \leq n. \end{cases}$$

For $n = 5$, the action of a_2 is illustrated below:

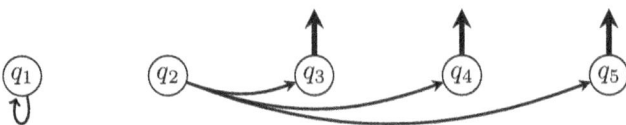

First we note that there exists a mortal word for $\mathcal{A}$. We construct it letter by letter. At each step we apply letter a_j, where j is such that state q_j is currently active, and states $q_{j+1}, \ldots, q_n$ are not active. In other words, j is the position of the last 1 in the representation of active states of Q as a number in binary. The length of thus constructed word is $2^n - 1$. This is in fact the only mortal word of this length, and to better illustrate its structure, let us describe how it can be constructed recurrently. Take $w_n = a_n$, and for all $1 \leq i \leq n-1$ take $w_i = w_{i+1} a_i w_{i+1}$. The word w_1 is then the shortest mortal word for $\mathcal{A}$. The shortest mortal word for the example above for $n = 5$ is

$$a_5 a_4 (a_5) a_3 (a_5 a_4 a_5) a_2 (a_5 a_4 a_5 a_3 a_5 a_4 a_5) a_1 (a_5 a_4 a_5 a_3 a_5 a_4 a_5 a_2 a_5 a_4 a_5 a_3 a_5 a_4 a_5),$$

where the partition of the word indicates the second appearance of w_{i+1} in each step of the recurrence $w_i = w_{i+1} a_i w_{i+1}$.

Now we show that this is the shortest mortal word for $\mathcal{A}$. Let w be a mortal word for $\mathcal{A}$. As mentioned above, we consider $\text{bin}(w')$ for prefixes w' of w. We start with $\text{bin}(\epsilon) = 2^n - 1$. By construction, reading a letter can decrement the value of the number by at most one: for every $a \in \Sigma$, we have $\text{bin}(w'a) \geq \text{bin}(w') - 1$. Indeed, let $a = a_j$ and let $\text{bin}(w') = x_1 \ldots x_n$ with $x_1, \ldots, x_n \in \{0,1\}$. If $x_{j+1} = \ldots = x_n$, then $\text{bin}(w'a) = x_1 \ldots x_{j-1} 0 1 1 \ldots 1$. If $x_j = 1$, we have $\text{bin}(w'a) = \text{bin}(w') - 1$, otherwise $\text{bin}(w'a) > \text{bin}(w')$. Now consider the case where $x_i = 1$ for some $i > j$. Then $\text{bin}(w'a) = 2^n - 1$ by construction. Hence, the length of a shortest mortal word for $\mathcal{A}$ is at least $2^n - 1$. □

3.2 Ternary Case Lower Bound

Now we implement a similar idea in a much more restricted setting of only three letters. We remark that combining the result from the previous subsection with a standard technique for reducing the size of the alphabet (for example, via a composition with a finite prefix code) is not enough to obtain a good lower bound. Indeed, going from a linear-size alphabet to an alphabet of size three requires appending to each state a tree with a linear number of inner states to simulate a linear number of choices (represented by the leaves of this tree). This results in a quadratic increase of the number of states. A possible improvement of this approach might be then to reuse the trees between different states in a clever way. However, we were not able to make it work, and our approach is instead to use shifts that directly change the binary representation of the number that we are tracking. A significant part of the construction is thus devoted to guaranteeing that this number cannot be decremented too much by applying a short word. The goal of this subsection is to prove the following.

Theorem 5. *For every even $n \geq 6$, there exists an NFA with n states and 3 letters such that the length of its shortest mortal word is at least $2^{(n-4)/2}$.*

Proof. We construct an NFA $\mathcal{A} = (P \cup Q \cup \{f\}, \{s, d, c\}, \Delta)$. Its set of states consists of a set of $k+1$ states $P = \{p_0, p_1, \ldots, p_k\}$, which we refer to as the left half, a set of k states $Q = \{q_1, \ldots, q_k\}$, which we refer to as the right half, and a special "flag" state f. We thus have $n = 2k + 2$.

As mentioned above, we will be tracking the values $\text{bin}(w')$ for prefixes w' of an input word w. However, since we only have three letters, we will have to use a more involved approach to decrementing these values by one. Namely, when applying letters to the NFA, the encoding of the tracked number will sometimes be shifted. In more detail, this encoding will be represented by a continuous segment of k states in the sequence $p_1, p_2, \ldots, p_k, q_1, q_2, \ldots, q_k$. Most other states in this sequence will be inactive, except for one state remembering by how much the representation was shifted. The main reason why we need to shift the number, and therefore why we need more states, is that, just like the action of a_j is different for different positions in the proof of Theorem 4, the action of one of the three available letters must have a different effect depending on the current position of the last 1 in the encoding of the number.

We first define the action of letter s that performs shifting. Its initial purpose is to make active precisely the set $Q \cup \{f\}$. After that, it will allow shifting $\text{bin}(w)$ to the right until its last digit is 1. If another shift is performed after that, the set $Q \cup \{f\}$ gets reactivated, so the whole process is reset to the beginning. The actions of letters d and c defined later will guarantee that the number of removed zeroes at the end of $\text{bin}(w)$ is remembered elsewhere, and hence the value of the tracked number is not lost.

Formally, we define

$$p_i \cdot s = \begin{cases} \{p_{i+1}\} & \text{if } 0 \le i < k; \\ \{q_1\} & \text{if } i = k; \end{cases} \qquad q_i \cdot s = \begin{cases} \{q_{i+1}\} & \text{if } 1 \le i < k; \\ Q \cup \{f\} & \text{if } i = k; \end{cases}$$

$$f \cdot s = \{f\}.$$

The action of letter s for $k = 4$ is illustrated below:

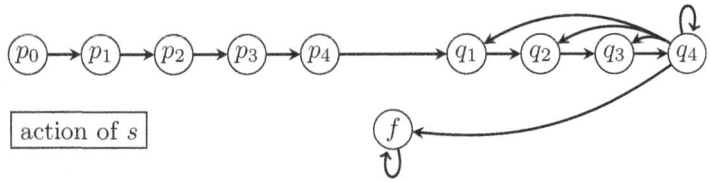

Next we define the action of letter c that checks that the representation of the number that we are tracking was shifted to the right half and removes the "marker" bit at the end of this representation. This "marker" bit is a detail that we have not mentioned before, but which is crucial for the construction. Most of the time, the encoding of the number we are tracking will end with an additional 1 that prevents s from changing the number without resetting the counting. However, once c is applied, which can be done only when the number is shifted to the right half, this bit is removed, and we can remove zeroes at the end of the tracked number by applying s. To stop the tracked number from decreasing fast, an application of c activates state p_0. After that, by construction, letter c cannot be applied as long as there are active states in the left half, which prevents us from using it again to remove more ones at the end of the tracked number.

The number of shifts performed by s after an application of c is then equal to the index of the unique active state among the states $p_0, p_1, \ldots, p_k$. The action of d (defined later) will take that into account when performing the decrement and returning the number to its proper representation.

Letter c also deactivates state f. This state is used to disallow using d too often: as we define later, if f is active, using d will reactivate all states.

Formally, we define

$$p_i \cdot c = P \cup Q \cup \{f\} \text{ for } 0 \leq i \leq k; \qquad q_i \cdot c = \begin{cases} \{q_i, p_0\} & \text{if } 1 \leq i < k; \\ \emptyset & \text{if } i = k; \end{cases}$$

$$f \cdot c = \emptyset.$$

The action of letter c is illustrated below:

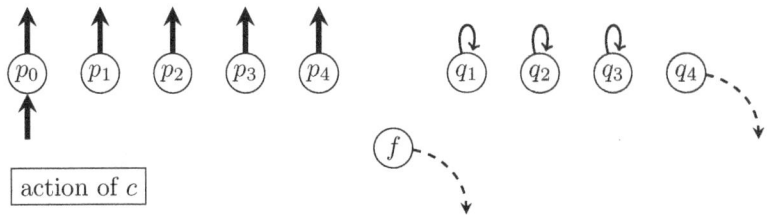

Finally, we define the action of letter d that decrements the tracked value by one and simultaneously shifts its representation by k positions to the left. We make sure that between any two applications of d there must be at least one application of c. It is guaranteed by the fact that d activates f, and if f is already active, an application of d makes all the states active. Hence, before d is applied, the representation of the tracked number is deconstructed into two parts: the number of performed shifts is remembered in the unique active state of the left half, and the rest of the number is in the right half. An application of d then subtracts one from the value of the number, and composes the representation of the number back to the proper binary representation. Besides that, the action of d adds back the "marker" bit at the end of the representation, which was removed by an application of c.

Formally, we define

$$p_i \cdot d = \begin{cases} P \cup Q \cup \{f\} & \text{if } i = 0; \\ \{q_1, \ldots, q_i\} \cup \{f\} & \text{if } 1 \leq i < k; \\ \{f\} & \text{if } i = k; \end{cases} \qquad q_i \cdot d = \begin{cases} \{p_i\} \cup \{f\} & \text{if } 1 \leq i < k; \\ \emptyset & \text{if } i = k; \end{cases}$$

$$f \cdot s = P \cup Q \cup \{f\}.$$

The action of d for some states is illustrated below. The transitions from p_1 and p_2 are omitted for the clarity of presentation.

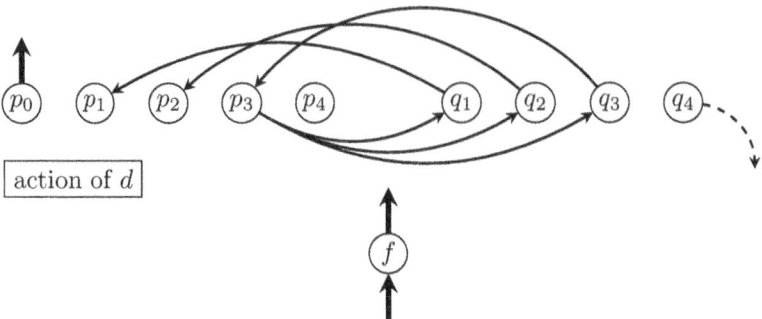

We have now fully defined $\mathcal{A}$. Let us first verify that there exists a mortal word for $\mathcal{A}$. Initially, we apply s until only states in $Q \cup \{f\}$ are active. Then we keep repeating the following process. Apply letter c, which in particular makes p_0 active. Then keep applying s until q_k becomes active. If s was applied h times, then p_h is active. When we apply d, we get that the set of active states is a subset of $p_{h+1}, \ldots, p_k, q_1, \ldots, q_h$, together with f. Moreover, by construction, the encoded value is decreased by one (ignoring the marker bit at the end). Repeat the described process until the encoded value is zero. This means that only f and the state corresponding to the marker bit at the end are active, and they can be made inactive by shifting them to the right and applying c.

The figure below illustrates a few first steps of such counting. The number that we are tracking, including the last marker bit if it is present, is highlighted.

f	p_0	p_1	p_2	p_3	p_4	q_1	q_2	q_3	q_4	
1	1	1	1	1	1	1	1	1	1	s^5
1	0	0	0	0	0	1	1	1	1	c
0	0	0	0	0	0	1	1	1	0	s
0	0	0	0	0	0	0	1	1	1	d
1	0	0	1	1	0	1	0	0	0	

Now we need to show a lower bound on the length of shortest mortal words for $\mathcal{A}$. Intuitively, any divergence from the mortal word described above results either in a full reset of the counting (via reactivating all the states) or in a number larger than the current number (if the representation in the right half contains zeroes at the end just before d is applied). Most of our argument is already outlined in the explanations of the actions of the letters. Formally, let w be a mortal word for $\mathcal{A}$. We can assume that the set of active states is $Q \cup \{f\}$, since this assumption will not make the mortal word longer. We look at the value of $\text{bin}(w')$ for all prefixes w' ending with cs. We claim that for two such subsequent prefixes w' and w'', where w' is shorter than w'', we have that $\text{bin}(w'') \geq \text{bin}(w') - 1$. We can assume that the whole set $Q \cup \{f\}$ does not get active again. Then after applying w' we can only apply s or d. If after some number of applications of s state q_k is not active and d is applied, then by construction the encoded number is increased. Now d cannot be applied again, so the increased number has to be shifted to the right half. If q_k is active, by the same reasoning we get $\text{bin}(w'') = \text{bin}(w') - 1$. We start counting from 2^{k-1} (cs removes the marker bit at the end), hence the length of w is at least 2^{k-1}. □

3.3 Binary Case Lower Bound

We now implement an idea similar to the idea for a ternary alphabet, but trading a letter for k more states. Again, we remark that standard techniques for reducing the size of the alphabet increase the number of states too much, so we have to introduce new techniques to directly simulate a counter using only two letters. We prove the following.

Theorem 6. *For every $n \geq 5$ such that $n = 3k + 2$ for some integer k, there exists an NFA with n states and 2 letters such that the length of its shortest mortal word is at least $2^{(n-2)/3}$.*

Proof. We construct an NFA $\mathcal{A} = (P \cup R \cup Q \cup \{f\}, \{s, d\}, \Delta)$ with

$$P = \{p_0, p_1, \ldots, p_{k-1}\}, R = \{r_0, r_1, \ldots, r_k\}, Q = \{q_1, \ldots, q_k\}.$$

Hence, we have $n = 3k+2$. We appropriately refer to P as the left part, R as the middle part and Q as the right part. The difference from the construction in the proof of Theorem 5 is that, intuitively, we get rid of letter c. To do so, we split the role of the states in P from the ternary case into two parts P and R. The new left part P now makes sure that the representation of the tracked number is shifted to the right part Q and that d is not applied too much, and the action of d on the middle part R performs the actual decrement. This time, no marker bit at the end of the representation is needed, but one of the states from $P \cup R$ will be active at all times, except for the very beginning of the counting process.

First we define the action of letter s, which has no conceptual differences from the action of s in the ternary case. Formally,

$$p_i \cdot s = \begin{cases} \{p_{i+1}\} & \text{if } 0 \leq i < k-1; \\ \{r_0\} & \text{if } i = k-1; \end{cases} \qquad r_i \cdot s = \begin{cases} \{r_{i+1}\} & \text{if } 0 \leq i < k; \\ \{q_1\} & \text{if } i = k; \end{cases}$$

$$q_i \cdot s = \begin{cases} \{q_{i+1}\} & \text{if } 1 \leq i < k; \\ Q \cup \{f\} & \text{if } i = k; \end{cases} \qquad f \cdot s = \{f\}.$$

The action of letter s for $k = 3$ is illustrated below:

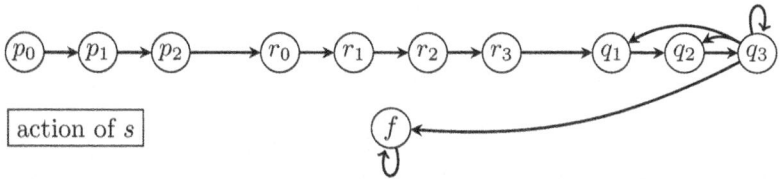

Now we define the action of letter d. As mentioned above, we split the roles of states from the set P from the ternary case between the sets P and R. State f now also plays a different role: if the counting is done correctly, f is only active in the very beginning of the counting process. After d is applied at least once, p_0 becomes active. From this moment onwards, there will always be at least one active state from the set $P \cup R$. The leftmost active state will remember by how much the representation of the tracked number is shifted. The action of d forces this representation to be shifted to the right part, otherwise all the states are reactivated. After this shift is performed, there is exactly one active state in R, and this state remembers how many zeroes there are at the end of the tracked number. Then an application of d decrements the value of the tracked number by one, just like in the ternary case (or, also like in the ternary case, makes the number larger if it was not shifted enough), and one state in P gets reactivated to preserve the information about the current shift. Formally,

$$p_i \cdot d = P \cup R \cup Q \cup \{f\} \quad \text{for } 0 \leq i \leq k-1;$$

$$r_i \cdot d = \begin{cases} \{p_i\} \cup \{q_1, \ldots, q_i\} & \text{if } 0 \leq i < k; \\ \emptyset & \text{if } i = k; \end{cases}$$

$$q_i \cdot d = \begin{cases} \{r_i\} & \text{if } 1 \leq i < k; \\ \emptyset & \text{if } i = k; \end{cases} \quad f \cdot d = \{p_0\}.$$

The action of d is illustrated below, with transitions from r_0, r_1, r_3 omitted:

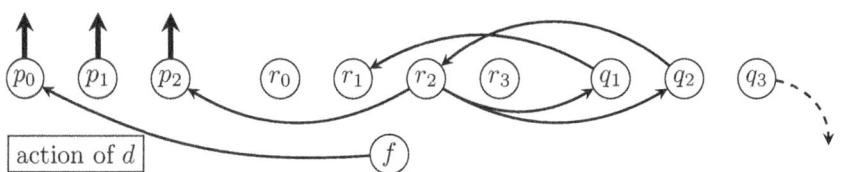

The proof that thus constructed NFA $\mathcal{A}$ admits a mortal word is very similar to that in Theorem 5: if all shifts are performed correctly (that is, if $Q \cup \{f\}$ never gets reactivated) and fully, then the tracked number decreases by one with each application of d, until it becomes zero and we can use d to kill the remaining active state from P. A few first steps are illustrated below, with the tracked number highlighted:

f	p_0	p_1	p_2	r_0	r_1	r_2	r_3	q_1	q_2	q_3	
1	1	1	1	1	1	1	1	1	1	1	s^7
1	0	0	0	0	0	0	0	1	1	1	d
0	1	0	0	0	1	1	0	0	0	0	s^4
0	0	0	0	0	1	0	0	0	1	1	d
0	0	1	0	0	0	1	0	1	0	0	

To prove a lower bound on the length of a shortest mortal word, we formally look at $\mathrm{bin}(w')$ for prefixes w' of an applied word w such that state p_0

is active after the application of w'. As in the ternary case, we argue that for two such subsequent prefixes w' and w'' we have that $\text{bin}(w'') \geq \text{bin}(w') - 1$ by construction. Thus, the length of a shortest mortal word is at least 2^k. □

4 Why do Automata Die Slowly?

The presented constructions, as well as all mentioned constructions from the literature, have purely combinatorial nature. In this brief section we discuss a possibility of a more algebraic view on automata mortality. We focus on a simpler case of strongly connected DFAs, since even for this case very little is known. A DFA is called *strongly connected* if for every pair p, q of its states there is a word mapping p to q.

In [1], another reachability property of strongly connected complete DFAs, namely the reset threshold, was linked to exponents of primitive matrices. A word is called *synchronizing* for a complete DFA $\mathcal{A}$ if it maps all states of $\mathcal{A}$ to the same state. The length of a shortest synchronizing word of $\mathcal{A}$ is then called its *reset threshold*. Let M be the sum of the matrices of all letters of $\mathcal{A}$. If $\mathcal{A}$ is synchronizing, M must be primitive. A matrix is *primitive* if some its power has only strictly positive entries. The smallest such power is called the *exponent* of a matrix. In [1] it is proved that the exponent of M is at most the reset threshold of $\mathcal{A}$ plus $n - 1$, where n is the number of states of $\mathcal{A}$. Thus, one obstacle for a complete DFA to synchronize fast is a large exponent of the corresponding matrix. A more advanced property of similar kind is proposed in [11].

The DFA mortality problem is sometimes framed as synchronization of a particular class of complete DFAs (namely, of complete DFAs with a single sink state), but we find that DFA mortality significantly differs from DFA synchronization and deserves independent treatment. One reason for that is the dependence on the size of the alphabet: adding more letters to a complete DFA usually helps to synchronize it faster, but, in contrast, it allows a DFA to "stall" for longer when applying a mortal word. Thus, the known series of complete DFAs with the largest reset threshold have two letters [1], but the series with the largest known mortality threshold have the number of letters equal to the number of states [25]. Perhaps more importantly, the results of [1] do not apply to DFA mortality, since after adding a sink state a DFA stops being strongly connected.

Mortality and synchronization can be viewed as very combinatorial phenomena: having a set of matrices, we ask for just one product of such matrices with a certain property, and hence have to make a lot of independent choices when constructing this product. Primitivity of a single matrix has much more algebraic flavour to it, since it concerns iterated multiplication of a single matrix.

We thus pose the following semi-formal question: *is there an algebraic property of a DFA telling us something about its mortality threshold?*

A simpler question is to ask for an algebraic property guaranteeing that a mortal word exists. For DFAs this is trivial, but for unambiguous NFAs, a proper superclass of DFAs, such property is precisely that the joint spectral radius of the corresponding set of matrices is strictly smaller than one [15]. For general NFAs one has to keep in mind that deciding the existence of a mortal word becomes PSPACE-complete, and hence one has to look for a "non-constructive" property.

The discovery of the connection between reset threshold and the exponent of a matrix in [1] was done by observing in exhaustive search experiments the similarity between the attainable large values of reset thresholds on the one hand and exponents of primitive matrices on the other hands. The present paper can be seen as a step towards a similar potential development for mortality. For small n, it is not hard to write down the sets of $n \times n$ matrices with large mortality thresholds provided in the previous section, in case one wants to check some conjectures connected to our question. To further assist with this goal, we now briefly survey what is known about DFA mortality, and provide a new simple family of binary DFAs with large mortality thresholds.

Denote by $\mathbf{mt_d}(n, m)$ the maximum mortality threshold of a DFA with n states and m letters admitting a mortal word. In [25] it is proved that for all m we have $\mathbf{mt_d}(n, m) \leq \frac{n(n+1)}{2}$, and that this upper bound is tight already for $m = n$. The binary case is more intriguing. To the best of our knowledge, the first series of n-state binary DFAs with quadratic mortality threshold is provided in Proposition 3.4 of [3], showing that $\mathbf{mt_d}(n, 2) \geq \frac{(n-1)^2}{4}$. Later on, such series with slightly larger mortality threshold were provided in [17] and [2], showing, respectively,

$$\mathbf{mt_d}(n, 2) \geq \frac{n^2 + 8n}{4} + \mathcal{O}(1) \quad \text{and} \quad \mathbf{mt_d}(n, 2) \geq \frac{n^2 + 10n}{4} + \mathcal{O}(1).$$

A few more series with simpler structure but with slightly smaller mortality threshold are provided in [26]. It is not known if $\mathbf{mt_d}(n, 2) = \frac{n^2}{4} + \mathcal{O}(n)$.

Below we provide a very simple series of binary DFAs with mortality threshold $\frac{n^2+8n}{4}+\mathcal{O}(1)$ which does not seem to appear in the literature. It can be seen as a simplification of the construction from [17] with approximately the same mortality threshold. Let $\mathcal{A} = (Q, \{a, b\}, \delta)$ be such that $Q = \{q_1, \ldots, q_k, p_1, \ldots, p_k\}$. Define

$$q_i \cdot a = \begin{cases} q_{i+1} & \text{if } 1 \leq i < k; \\ q_1 & \text{if } i = k; \end{cases} \qquad p_i \cdot a = \begin{cases} p_{i+1} & \text{if } 1 \leq i < k; \\ \emptyset & \text{if } i = k; \end{cases}$$

$$q_i \cdot b = \begin{cases} p_1 & \text{if } i = 1; \\ q_k & \text{if } i = 2; \\ q_{i-1} & \text{if } 2 < i \leq k; \end{cases} \qquad p_i \cdot b = q_1 \text{ for } 1 \leq i \leq k.$$

The fact that $\mathcal{A}$ has mortality threshold $\frac{n^2+8n}{4} + \mathcal{O}(1)$ follows, for example, from Lemma 1 in [2], since this is another example of adding a tail to an almost permutation automaton, formally defined in [2]. While most of the known series with large mortality thresholds have this structure, not all of them do: for example, DFAs from the series in [3] do not have a tail, and have a cycle going through all states instead.

An illustration for the case $k = 4$ is provided below, with the action of a depicted by solid arrows and the action of b by dashed arrows.

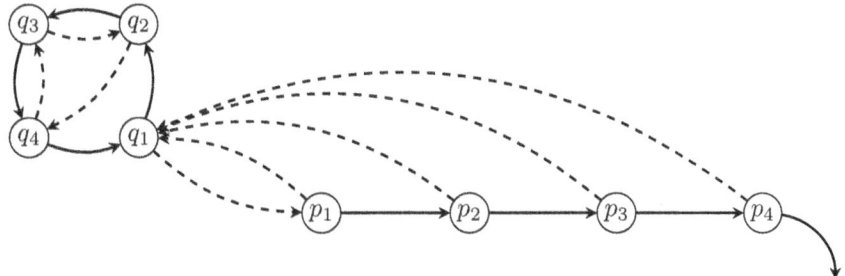

The unique shortest mortal word for this DFA is $a^k b a^k a b a^k (aaba^k)^{k-2}$. The intuition behind this word is that if everything is done "optimally", after killing (that is, applying a word that is undefined for a state) two states among $q_1, \ldots, q_k$, the structure of the cycles on them maintains that q_1 and q_k are inactive, and hence they have to be traversed in the process of killing every remaining active state from $q_1, \ldots, q_k$.

5 Conclusions and Open Problems

One of the most interesting questions that remains open is the precise behaviour of $\mathbf{mt}(n, m)$ and $\mathbf{mt_d}(n, m)$ for $m = 2$ and $m = 3$. As already discussed in Sect. 4, lower bounds on $\mathbf{mt_d}(n, 2)$ were investigated quite actively [2,3,17,26], and all these bounds are still of the form $\frac{n^2}{4} + \mathcal{O}(n)$. To the best of our knowledge, there is no plausible conjecture on the value of $\mathbf{mt_d}(n, 2)$ in the literature, and there are no known upper bounds better than $\frac{n(n+1)}{2}$ [25], a bound that holds for any size of the alphabet. It is in fact easy to see that this bound cannot be tight for $\mathbf{mt_d}(n, 2)$, but we were not able to obtain a significant improvement for it. However, we conjecture that it can be improved to at least the bound $\mathbf{mt_d}(n, 2) \leq \frac{7n^2}{16} + \mathcal{O}(n)$. As for bounds on $\mathbf{mt_d}(n, 3)$, we are not aware of any results. Since the dependence of $\mathbf{mt_d}(n, m)$ on m seems to be far from trivial, this is a natural first step to approach its investigation. Alternatively, one can ask for bounds on $\mathbf{mt_d}(n, \frac{n}{2})$.

The same questions can be asked about $\mathbf{mt}(n, m)$, and the situation with existing results is similar: we already surveyed known lower bounds in Sect. 2 and mentioned that $\mathbf{mt}(n, m) \leq 2^n - 1$. To the best of our knowledge, no better upper bounds for small m are known, and the same comments that we made about the precise dependence of $\mathbf{mt_d}(n, m)$ on m also apply to $\mathbf{mt}(n, m)$.

Mortality of unambiguous NFAs is even further from being understood. As mentioned in the introduction, we know that the mortality threshold for them is between $\frac{n(n+1)}{2}$ [25] and n^5 [15]. The lower bound is provided by a series of DFAs, and all known families of unambiguous NFAs with large mortality threshold are

DFAs or their co-deterministic counterparts (obtained by reversing the direction of each transition). Since unambiguous NFAs constitute a much more general class than DFAs, this is a curious situation that deserves further investigation.

Finally, it would be interesting to see any subclasses with good (that is, polynomial) behaviour of $\mathbf{mt}(n,m)$ besides the class of unambiguous NFAs [15].

Acknowledgments. The author greatly benefited from talking to (and sometimes at) Andrei Draghici. The comments of anonymous reviewers improved the presentation of the paper. The SynchroViewer software (github.com/marekesz/synchroviewer) was helpful for testing conjectures about shortest mortal words for DFAs. The author is supported by the European Research Council (ERC) under the European Union's Horizon 2020 research and innovation programme (Grant agreement No. 852769, ARiAT).

References

1. Ananichev, D.S., Volkov, M.V., Gusev, V.V.: Primitive digraphs with large exponents and slowly synchronizing automata. J. Math. Sci. **192**(3), 263–278 (2013). https://doi.org/10.1007/s10958-013-1392-8
2. Ananichev, D.S., Vorel, V.: A new lower bound for reset threshold of binary synchronizing automata with sink. J. Autom. Lang. Combin. **24**(2-4), 153–164 (2019). https://doi.org/10.25596/JALC-2019-153 https://doi.org/10.25596/JALC-2019-153
3. Béal, M.P., Crochemore, M., Mignosi, F., Restivo, A., Sciortino, M.: Computing forbidden words of regular languages. Fundam. Inform. **56**(1-2), 121–135 (2003). http://content.iospress.com/articles/fundamenta-informaticae/fi56-1-2-08
4. Berstel, J., Perrin, D., Reutenauer, C.: Codes and Automata, vol. 129. Cambridge University Press, Cambridge (2010)
5. Boccuto, A., Carpi, A.: On the length of uncompletable words in unambiguous automata. RAIRO - Theor. Inform. Appl. **53**(3–4), 115–123 (2019). https://doi.org/10.1051/ITA/2019002
6. Cassaigne, J., Halava, V., Harju, T., Nicolas, F.: Tighter undecidability bounds for matrix mortality, zero-in-the-corner problems, and more. CoRR abs/1404.0644 (2014). http://arxiv.org/abs/1404.0644
7. De Bondt, M., Don, H., Zantema, H.: Lower bounds for synchronizing word lengths in partial automata. Int. J. Found. Comput. Sci. **30**(1), 29–60 (2019). https://doi.org/10.1142/S0129054119400021
8. Ellul, K., Krawetz, B., Shallit, J.O., Wang, M.: Regular expressions: new results and open problems. J. Autom. Lang. Combin. **9**(2/3), 233–256 (2004). https://doi.org/10.25596/JALC-2004-233
9. Gerencsér, B., Gusev, V.V., Jungers, R.M.: Primitive sets of nonnegative matrices and synchronizing automata. SIAM J. Matrix Anal. Appl. **39**(1), 83–98 (2018). https://doi.org/10.1137/16M1094099
10. Goralčík, P., Hedrlín, Z., Koubek, V., Ryslinková, V.: A game of composing binary relations. RAIRO - Theor. Inform. Appl. **16**(4), 365–369 (1982). https://doi.org/10.1051/ITA/1982160403651
11. Gusev, V.V.: Lower bounds for the length of reset words in Eulerian automata. Int. J. Found. Comput. Sci. **24**(2), 251–262 (2013). https://doi.org/10.1142/S0129054113400108

12. Imreh, B., Imreh, C., Ito, M.: On directable nondeterministic trapped automata. Acta Cybern. **16**(1), 37–45 (2003). https://cyber.bibl.u-szeged.hu/index.php/actcybern/article/view/3608
13. Imreh, B., Steinby, M.: Directable nondeterministic automata. Acta Cybernet. **14**(1), 105–115 (1999)
14. Kao, J., Rampersad, N., Shallit, J.O.: On NFAs where all states are final, initial, or both. Theoret. Comput. Sci. **410**(47–49), 5010–5021 (2009). https://doi.org/10.1016/J.TCS.2009.07.049
15. Kiefer, S., Mascle, C.N.: On nonnegative integer matrices and short killing words. SIAM J. Discret. Math. **35**(2), 1252–1267 (2021). https://doi.org/10.1137/19M1250893
16. Lind, D., Marcus, B.: An Introduction to Symbolic Dynamics and Coding. Cambridge University Press, Cambridge (2021)
17. Martugin, P.V.: A series of slowly synchronizing automata with a zero state over a small alphabet. Inf. Comput. **206**(9–10), 1197–1203 (2008). https://doi.org/10.1016/J.IC.2008.03.020
18. Martyugin, P.V.: A lower bound for the length of the shortest carefully synchronizing words. Russ. Math. **54**(1), 46–54 (2010). https://doi.org/10.3103/S1066369X10010056
19. Mika, M., Szykuła, M.: The Frobenius and factor universality problems of the Kleene star of a finite set of words. J. ACM **68**(3) (2021). https://doi.org/10.1145/3447237
20. Paterson, M.: Unsolvability in 3×3 matrices. Stud. Appl. Math. **49**, 105–107 (1970)
21. Protasov, V.Y., Voynov, A.: Sets of nonnegative matrices without positive products. Linear Algebra Appl. **437**(3), 749–765 (2012)
22. Rampersad, N., Shallit, J.O., Xu, Z.: The computational complexity of universality problems for prefixes, suffixes, factors, and subwords of regular languages. Fund. Inform. **116**(1–4), 223–236 (2012). https://doi.org/10.3233/FI-2012-680
23. Restivo, A.: Some remarks on complete subsets of a free monoid. Quaderni ricerca scientifica CNR Roma **109**, 19–25 (1981)
24. Rystsov, I.K.: Asymptotic estimate of the length of a diagnostic word for a finite automaton. Cybernetics **16**(2), 194–198 (1980). https://doi.org/10.1007/BF01069104
25. Rystsov, I.K.: Reset words for commutative and solvable automata. Theoret. Comput. Sci. **172**(1–2), 273–279 (1997). https://doi.org/10.1016/S0304-3975(96)00136-3
26. Ryzhikov, A.: Mortality and synchronization of unambiguous finite automata. In: Mercaş, R., Reidenbach, D. (eds.) WORDS 2019. LNCS, vol. 11682, pp. 299–311. Springer, Cham (2019). https://doi.org/10.1007/978-3-030-28796-2_24
27. Ryzhikov, A., Shemyakov, A.: Subset synchronization in monotonic automata. Fund. Inform. **162**(2–3), 205–221 (2018). https://doi.org/10.3233/FI-2018-1721
28. Seneta, E.: Non-negative Matrices and Markov Chains. Springer, Heidelberg (2006)
29. Wu, Y., Zhu, Y.: Primitivity and Hurwitz primitivity of nonnegative matrix tuples: a unified approach. SIAM J. Matrix Anal. Appl. **44**(1), 196–211 (2023). https://doi.org/10.1137/22M1471535

Hardness of Busy Beaver Value BB(15)

Tristan Stérin[✉] and Damien Woods

Hamilton Institute and Department of Computer Science, Maynooth University,
Maynooth, Ireland
{tristan.sterin,damien.woods}@mu.ie
https://dna.hamilton.ie

Abstract. The busy beaver value $BB(n)$ is the maximum number of steps made by any n-state, 2-symbol deterministic halting Turing machine starting on blank tape. The busy beaver function $n \mapsto BB(n)$ is uncomputable and, from below, only 4 of its values, $BB(1) \ldots BB(4)$, are known to date. This leads one to ask: from above, what is the smallest BB value that encodes a major mathematical challenge? Knowing BB(4,888) has been shown by Yedidia and Aaronson [32] to be at least as hard as solving Goldbach's conjecture, with a subsequent improvement, as yet unpublished, to BB(27) [2,6]. We prove that knowing BB(15) is at least as hard as solving the following Collatz-related conjecture by Erdős, open since 1979 [10]: for all $n > 8$ there is at least one digit 2 in the base 3 representation of 2^n. We do so by constructing an explicit 15-state, 2-symbol Turing machine that halts if and only if the conjecture is false. This 2-symbol Turing machine simulates a conceptually simpler 5-state, 4-symbol machine which we construct first. This makes, to date, BB(15) the smallest busy beaver value that is related to a *natural* open problem in mathematics, bringing to light one of the many challenges underlying the quest of knowing busy beaver values.

1 Introduction

In the theory of computation, there lies a strange and complicated beast, the busy beaver function. This function tracks a certain notion of algorithmic complexity, namely, the maximum number of steps any halting algorithm of a given program-size may take. Although formulated in terms of Turing machines, the underlying notion of "maximum algorithmic bang for your buck" that it captures could be defined in any reasonable programming language.

The busy beaver function $n \mapsto BB(n)$ was introduced by Tibor Radó in 1962 and corresponds to the maximum number of steps made by a halting determinis-

---

T. Stérin and D. Woods—Hamilton Institute and Department of Computer Science, Maynooth University, Ireland. Research supported by European Research Council (ERC, grant agreement No 772766, Active-DNA project) under the European Union's Horizon 2020 research and innovation programme, European Innovation Council (EIC, DISCO, No 101115422); and Science Foundation Ireland (SFI) under Grant number 18/ERCS/5746. Research sponsored by prgm.dev.

tic Turing machine with n states and 2 symbols starting from blank input [24].[1] It was generalised by Brady [4,22] to machines with k symbols; $BB(n,k)$. Busy beaver functions, i.e. $n, k \mapsto BB(n,k)$, or $n \mapsto BB(n,k)$ for fixed $k \geq 2$, are not computable. Otherwise the halting problem on blank tape would be computable: take any machine with n states and k symbols, run it for $BB(n,k) + 1$ steps; if it has not halted yet, we know that it will never halt. They also dominate any computable function.

To date, only four non-trivial[2] busy beaver values are known: $BB(2) = 6$, $BB(3) = 21$, $BB(4) = 107$ and $BB(2,3) = 38$ [20,22]. Machines that halt without input after so many steps that they beat previously known record-holders, with the same n and k, are called busy beaver *champions* and they give lower bounds for $BB(n)$ or $BB(n,k)$. It was shown[3] in July 2024 that $BB(5) = 47{,}176{,}870$ [1,2,19]. It is known [14,20] that $BB(6) \geq 10 \uparrow\uparrow 15 = 10^{10^{10^{\cdot^{\cdot^{10}}}}}$, a tower of 15 powers of 10, well beyond 10^{80}, the estimated number of atoms in the observable universe. Also, $BB(2,6) \geq 10 \uparrow\uparrow (10 \uparrow\uparrow 10^{10^{315}})$ [15,20] and $BB(3,4) \geq \mathrm{Ack}(14)$, with the Ackerman number defined as $\mathrm{Ack}(n) = n \uparrow^n n$ [16,18]. These results highlight the sheer growth of the busy beaver function (it provably grows faster than any computable function [24]). However, lower bounds like this one do not give us a concrete sense of how hard it is to know busy beaver values, we just know that they quickly become "huge".

Can we give a more formal notion of *hardness* of finding busy beaver values? Indeed, for specific values of n and k, how hard is it to find $BB(n,k)$? A recent trend, that we follow here, seeks to formally relate values of $BB(n,k)$ to notoriously hard mathematical problems. For instance, one can imagine [5] designing a Turing machine that, starting from blank tape, will halt if and only if it finds a counterexample to Goldbach's conjecture (every even integer greater than 2 is the sum of two primes). Such a machine was actually built, using 4,888 states and 2 symbols [32]. This result implies that knowing the value of $BB(4{,}888)$ would allow us to computably decide Goldbach's conjecture: run the machine for $BB(4{,}888) + 1$ steps – which, as we've seen, must be an unbelievably huge number – and if it has halted before, then the conjecture is false otherwise it is true. The result was later claimed to be improved to a 27-state 2-symbol machine [2,6], which, subject to being proved correct, would be the smallest busy beaver value, before our work, that relates to a *natural* mathematical problem. Similar

---

[1] Radó originally used the notation $S(n)$. The more modern [2] notation $BB(n)$ can be a source of confusion since some authors, for example [12,14], use BB to mean Radó's Σ which counts the number of 1s on the final tape of a halting Turing machine [24].

[2] Trivial values include those where $n = 1$ since for all $k \geq 1$, $BB(1,k) = 1$.

[3]

	A (init)	B	C	D	E
0	1 R B	1 R C	1 R D	1 L A	Halt
1	1 L C	1 R B	0 L E	1 L D	0 L A

Current busy beaver champion [19] for machines with 5 states (A–E) and 2 symbols (0,1), it halts in 47,176,870 steps starting from all-0 input and state **A**, which gives the lowerbound $BB(5) \geq 47{,}176{,}870$. Proving that bound tight is the goal of the bbchallenge project [1]. Is the 5-state, 2 symbol (0,1) busy beaver. Marxen [19] showed it halts in 47,176,870 steps starting from all-0 input and state **A**, and the bbchallenge collective [1] proved that no machine halts in more steps, i.e. that $BB(5) = 47{,}176{,}870$.

efforts led to the construction of a 5,372-state 2-symbol machine that halts if and only if the Riemann hypothesis is false [32]; with a later claimed improvement to 744 states [2], hence knowing the value of BB(744) is at least as hard as solving the Riemann hypothesis.

Results and Discussion. Here, we continue the approach of relating small busy beaver values to hard mathematical problems towards obtaining insight into the location of the frontier between knowable and unknowable busy beaver values. Our particular approach gives upper bounds on the smallest counterexample to such problems. We do this by giving machines that search for counterexamples to an open, Collatz-related conjecture formulated in 1979 by Erdős:

Conjecture 1 (Erdős [10]). For all natural numbers $n > 8$ there is at least one digit 2 in the base 3 representation of 2^n.

In Sect. 1.1 we discuss existing literature on the conjecture as well as its relationship with the Collatz and weak Collatz conjectures [8,9,17,29].

The main technical contribution of this paper is to prove the following theorem:

Theorem 1. *There is an explicit 15-state 2-symbol Turing machine that halts if and only if Erdős' conjecture is false.*

That Turing machine, called $M_{15,2}$, is given in Fig. 4 and proven correct in Sect. 4. The proof shows that $M_{15,2}$ *simulates*, in a tight (linear time) fashion, an intuitively simpler machine with 5 states and 4 symbols, that we call $M_{5,4}$ (Fig. 3) and whose behaviour is proven correct in Sect. 3:

Theorem 2. *There is an explicit 5-state 4-symbol Turing machine that halts if and only if Erdős' conjecture is false.*

From these theorems, we get that knowing the value of BB(15) is at least as hard as solving Erdős' conjecture since BB(15) gives a finite—although particularly impractical—algorithmic procedure to decide the conjecture as it upper-bounds the set of values that need to be considered:

Corollary 3. *Erdős' conjecture is equivalent to the following conjecture over a finite set: for all $8 < n \leq \min(BB(15), BB(5,4))$ there is at least one digit 2 in the base 3 representation of 2^n.*

Finally, we also get an upper bound on the smallest counterexample to Erdős conjecture, if it exists:

Corollary 4. *Let $n \in \mathbb{N}$ be the smallest counterexample to Erdős' conjecture, that is the smallest positive integer $n > 8$ such that 2^n does not contain any digit 2 in base 3, if it exists. Then we have:* $BB(15) \geq n$ *and* $BB(5,4) \geq n$.

Discussion. Why are these theorems important? They give us more knowledge on the busy beaver function: we get that values as small as BB(15) are related to hard open problems in mathematics and thus correspondingly hard to know. Our results make BB(15) the smallest busy beaver value linked to an open problem in mathematics, a significant improvement on the BB(4,888) and BB(5,372) results [32] (that have a proof of correctness), and the BB(27) and BB(744) results (that as yet lack a published proof of correctness).

An even more drastic way to establish hardness is to find an n for which BB(n) is independent of some standard set of axioms such as Peano Arithmetic (PA), or ZFC. In that spirit, a 7,910-state machine whose halting problem is independent of ZFC was constructed [32] and this was later improved to 748 states [2,13] and further claimed to be improved to 643 states [25]. These machines explicitly looks for a contradiction in ZFC (such as a proof of $0 = 1$) which is, by Gödel's second incompleteness theorem, independent of ZFC. Aaronson [2] conjectures that BB(10) is independent of PA and BB(20) is independent of ZFC meaning that the frontier between knowable and unknowable busy beaver values could be as low as BB(10) if we limit ourselves to typical inductive proofs. With our work, any independence result on Erdős' conjecture would automatically transfer to busy beaver value BB(15) and higher.

As for Erdős' conjecture, Corollary 3 gives a finite bound on the number of counter-examples needed to check in order to prove or refute the conjecture. However, this is not a practical bound: first, we don't know its value, and second, it must be so astronomical as to not give a tractable procedure to prove or disprove the conjecture. Moreover, since we now know that BB(15) *embeds* Erdős' conjecture, and that it could embed other horrible beasts, finding BB(15) itself could indeed be much harder than answering the conjecture. However, it is not unthinkable that methods for automatically deciding whether certain Turing machines halt or not [1,11,19] could one day solve the halting problem of the machines constructed in this article (or future improved machines with less states) and thus solve Erdős' conjecture that way. This approach might highlight unforeseen links between theoretical computer science and number theory.

As we've noted, some claimed results on the busy beaver function come with proofs, but some do not [2]. We advocate for proofs, although we acknowledge the challenges in providing human-readable correctness proofs for small programs that are, or almost are, program-size optimal.[4] Here, the proof technique for correctness of our BB(15) candidate $M_{15,2}$ amounts to proving by induction that $M_{15,2}$ *simulates* (in a tight sense via Definition 8 and Lemma 10) another Turing machine $M_{5,4}$ (giving Theorem 1). $M_{5,4}$ exploits the existence of a tiny finite state transducer (FST), in Fig. 2, for multiplication by 2 in base 3. We directly prove $M_{5,4}$'s correctness by induction on its time steps (giving Theorem 2).

A Turing machine simulator, bbsim, was built in order to test our constructions[5]. The reader is invited to run the machines of this paper using the simulator.

---

[4] The situation can be likened to the hunt for small, and fast, universal Turing machines [31], or simple models like Post tag systems, with earlier literature often missing proofs of correctness, but some later papers using induction on machine configurations to do the job, for example refs [23,30].

[5] Simulator and machines available here: https://github.com/tcosmo/bbsim.

1.1 Erdős' Conjecture and its Relationship to the Collatz and Weak Collatz Conjectures

A first obvious fact about Conjecture 1 is that the bound $n > 8$ is set so as to exclude the three known special cases: 1, 4 and 256 whose ternary representations are respectively 1, 11 and 100111. Secondly, since having no digit 2 in ternary is equivalent to being a sum of distinct powers of three, Conjecture 1 can be restated as: for all $n > 8$, the number 2^n is not a sum of distinct powers of 3.

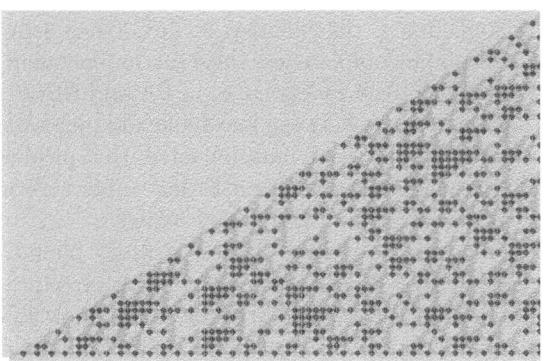

Fig. 1. The first 75 powers of 2 assembled in base 3 by the size-6 Wang tile set introduced in [7,27]. Reading left-to-right, each column of glues (colours) corresponds to a power of 2: beige glues represent ternary digit 0, green glues 1 and red glues 2. For instance, the rightmost column encodes (from top to bottom) $110210021020202011202012000020112001021021222022_3 = 2^{75}$. The complexity of the patterns illustrates the complexity of answering Erdős conjecture which amounts to asking if each glue-column (except for the few first) has ≥ 1 red glue. (Color figure online)

The conjecture has been studied by several authors. Notably, Lagarias [17] showed a result stating that, in some sense, the set of powers of 2 that omits the digit 2 in base three, is small. In [9], the authors showed that, for p and q distinct primes, the digits of the base q expansions of p^n are equidistributed on average (averaging over n) which in our case suggests that digits $0, 1, 2$ should appear in equal proportion in the base 3 representation of 2^n. Dimitrov and Howe [8] showed that 1, 4 and 256 are the only powers of 2 that can be written as the sum of at most twenty-five distinct powers of 3.

The Collatz conjecture states that iterating the Collatz function T on any $x \in \mathbb{N}$ eventually yields 1, where $T(x) = x/2$ if x is even and $T(x) = (3x+1)/2$ if x is odd. The weak Collatz conjecture states that if $T^k(x) = x$ for some $k \geq 1$ and x a natural number then $x \in \{0, 1, 2\}$.

Although solving the weak Collatz conjecture, given current knowledge, would not directly solve Erdős' conjecture, intuitively, Erdős' conjecture seems to be the simpler problem of the two. Indeed, Tao [29] justifies calling Erdős' conjecture a "toy model" problem for the weak Collatz conjecture by giving the number-theoretical reformulation that there are no integer solutions to

$2^n = 3^{a_1} + 3^{a_2} + \cdots + 3^{a_k}$ with $n > 8$ and $0 \leq a_1 < \cdots < a_k$, which in turn seems like a simplification of a statement equivalent to the weak Collatz conjecture, also given in [29] (Conjecture 3; Reformulated weak Collatz conjecture).

The three conjectures have been encoded using the same size-6 Wang tile set by Cook, Stérin and Woods [7,27]. As Fig. 1 shows, their construction illustrates the complexity of the patterns occurring in ternary representations of powers of 2 which gives a sense of the complexity underlying Erdős' conjecture. Beyond making complex, albeit pretty, pictures there is deeper connection here: the small tile set of [7,27] can be shown to simulate the base 3, **mul2**, Finite State Transducer that we introduce in Sect. 3, Fig. 2, and our small Turing machines use it to look for counterexamples to Erdős' conjecture. The tile set also simulates the *inverse* of the **mul2** FST (which computes the operation $x \mapsto x/2$ in ternary) which was used to build a 3-state 4-symbol non-halting machine that runs the Collatz map on any ternary input [21] and finally, it also simulates the *dual* of that FST (which computes the operation $x \mapsto 3x + 1$ in binary) which can be used to simultaneously compute the Collatz map both in binary and ternary [28]. So these three closely-related FSTs are all encoded within that small tile set, that in turn encodes the three conjectures.

Future Work. This paper opens a number of avenues for future work.

We've given a finite bound on the first counterexample to Erdős' conjecture, an obvious future line of work is to improve that cosmologically large bound by shrinking the program-size of our 15-state machine. An avenue for reducing the number of states needed to encode Erdős' conjecture is to study small Turing machines and look for machines with the expected behaviour "in the wild"[6].

It would be interesting to design small Turing machines that look for counterexamples to the weak Collatz conjecture. That way we could relate the fact of knowing busy beaver values to solving that notoriously hard conjecture. We have already found 124-state 2-symbol, and 43-state 4-symbol, machines that look for such counterexamples[7] but we would like to further optimise them (i.e. reduce their number of states) before formally proving that their behaviour is correct.

There are certain other open problems amenable to BB-type encodings. For instance, the Erdős-Mollin-Walsh conjecture states that there are no three consecutive powerful numbers, where $m \in \mathbb{N}$ is *powerful* if $m = a^2 b^3$ for some $a, b \in \mathbb{N}$. It would not be difficult to give a small Turing Machine, that enumerates unary encoded natural numbers, i.e. candidate m and a, b values, and runs the logic to check for counterexamples. Algorithms that bounce back and forth on a tape marking off unary strings have facilitated the finding of incredibly small universal Turing machines [23,26,31], so stands a chance to work well here too, although we wouldn't expect it to beat our main results in terms of program-size.

In fact, any problem expressible as a Π_1 sentence [2] should be relatable to a value of BB in the same way that we did for Erdős' conjecture, although many

---

[6] Such an effort has been done for 5-state Turing machines [1].

[7] These machines are also available here: https://github.com/tcosmo/bbsim.

problems would presumably yield unsuitably large program-size. It is also worth noting that some problems seem out of reach of the busy beaver framework: for instance the Collatz conjecture: to date, there are no known ways for an algorithm to recognise whether or not an arbitrary trajectory is divergent, making it seemingly impossible for a program to search for a divergent counterexample[8]. Aaronson [2] gives open problems on the structure of BB functions.

2 Definitions: Busy Beaver Turing Machines

Let $\mathbb{N} = \{0, 1, 2 \dots\}$ and $\mathbb{Z} = \{\dots, -1, 0, 1, \dots\}$. We consider Turing machines that are deterministic, have a single bi-infinite tape with tape cells indexed by $\mathbb{Z}$, and a finite alphabet of k tape symbols that includes a special 'blank' symbol (we use # for $M_{5,4}$ and b for $M_{15,2}$). For readability, we use bold programming-style names, for example **check_halt**. We use '(init)' to denote the initial state. Each state has k transitions, one for each of the k tape symbols that might be read by the tape head, and a transition is either (1) the halting instruction 'Halt' or (2) a triple: write symbol, tape head move direction (L or R), and next state. In the busy beaver setting used throughout this paper, we start machines in their initial state on all-blank tape and, at each time step, according to what symbol is read on the tape, the specified transition is performed, until 'Halt' is encountered (if ever).

Let TM(n, k) be the set of such n-state, k-symbol Turing machines. Given a machine $M \in$ TM(n, k), let $s(M)$ be the number of transitions it executes before halting, including the final Halt instruction[9] and let $s(M) = \infty$ if M does not halt. Then, BB(n, k) is defined [2] by

$$\mathrm{BB}(n, k) = \max_{M \in \mathrm{TM}(n,k),\ s(M) < \infty} s(M)$$

In other words, BB(n, k) is the number of steps by a machine in TM(n, k) that runs the longest without halting. By convention, BB(n) = BB$(n, 2)$, this being the most classic and well-studied busy beaver function.

Some conventions: A busy beaver candidate is a Turing machine for which we don't currently *know* whether it halts or not on blank input. A busy beaver contender is a Turing machine that halts on blank input. A busy beaver champion is a Turing machine that halts on blank input in more steps than any other *known* machine with the same number of states and symbols.

A Turing machine configuration is given by: the current state, the tape contents, and an integer tape head position. We sometimes write configurations in the following condensed format: **state_name**, $\cdots$ * * * *$\underline{*}$* * * * $\ldots$ with * any tape symbol of the machine and _ for tape head position. Let c_1 and c_2 two

---

[8] Said otherwise: given current knowledge there are no reason to believe that the set of counterexamples to the Collatz conjecture is recursively enumerable.

[9] As in [2], we took the liberty of not defining a symbol to write and a direction to move the tape head to when the halting instruction is performed as this does not change the number of transitions that the machine executed.

configurations of some machine M and let $k \in \mathbb{N}$, we write $c_1 \vdash_M^k c_2$ to mean that M transitions from c_1 to c_2 in k steps. We write $\vdash$ (without M) if M is clear from the context.

3 Five State, Four Symbol Turing Machine

In this section we prove Theorem 2. To do this we define, and prove correct, a 5-state 4-symbol Turing machine that searches for a counterexample to Erdős' conjecture, the machine is given in Fig. 3. The construction begins with the **mul2** finite state transducer (FST) in Fig. 2.

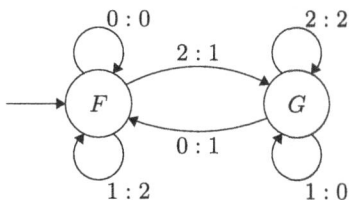

Fig. 2. The **mul2** Finite State Transducer (FST) that multiplies a reverse-ternary represented number (base-3 written in reverse digit order) by 2. For example, the base 10 number 64_{10} is 2101_3 in base 3, which we represent in reverse-ternary with a leading zero to give the input 10120, which in turn yields the FST output 20211; the reverse-ternary of 128_{10}. Transition arrows are labelled $r : w$ where r is the read symbol and w is the write symbol.

A similar FST, and its 'dual', can be used to compute iterations of the Collatz map [28] (Appendix B). The fact that there is an FST that multiplies by 2 in base 3 is not surprising, since there is one for any affine transformation in any natural-number base ≥ 1 [3]. However, in this section, we will exploit **mul2**'s small size. We begin by proving its behaviour is correct (by reverse-ternary we mean the base-3 representation written in reverse digit order):

Lemma 5. Let $x \in \mathbb{N}$ and let $w = w_1 \ldots w_n 0 \in \{0, 1, 2\}^n 0$ be its reverse-ternary representation, with $n \geq 0$, and a single leading 0. Then, on input w the **mul2** FST outputs $\gamma = \gamma_1 \gamma_2 \ldots \gamma_{n+1} \in \{0, 1, 2\}^{n+1}$ which represents $y = 2x$ in reverse-ternary.

Proof. We give an induction on ternary word length n, with the following induction hypothesis: given a word $w = w_1 \ldots w_n 0$ that represents $x \in \mathbb{N}$ (as in the lemma statement), if the FST reads from state F then the operation $x \mapsto 2x$ is computed in reverse-ternary, and if the FST reads from state G then the operation $x \mapsto 2x + 1$ is computed in reverse-ternary.

For the base case, when $n = 0$ we have $w = 0$ and, when started from state F the FST outputs 0 and when started from state G the FST outputs 1 which corresponds to respectively applying $x \mapsto 2x$ and $x \mapsto 2x + 1$.

Let's assume that the induction hypothesis holds for n and consider the base 3 word $w = w_1 \ldots w_{n+1} 0 \in \{0,1,2\}^{n+1} 0$ that represents some $x \in \mathbb{N}$. We first handle state F. There are three cases for the value of the least significant digit $w_1 \in \{0,1,2\}$. If $w_1 = 0$, from the FST state F the first transition will output 0 and return to state F. Then, from state F, by applying the induction hypothesis to the length-n word $w_2 \ldots w_{n+1}$, which represents the number $(x - w_1)/3 = x/3$ in base 3, we get that the FST output (on $w_2 \ldots w_{n+1}$) is the representation of $2\frac{x}{3}$.

Then, by including the first output 0, the complete output of the FST represents the number $0 + 3 \cdot 2\frac{x}{3} = 2x$ which is what we wanted. Similarly, if $w_1 = 1$ the FST outputs the representation of the number $2 + 3 \cdot 2\frac{x-1}{3} = 2x$ since the FST outputs 2 when reading a 1 in state F, or if $w_1 = 2$ the FST outputs the representation of $1 + 3 \cdot (2\frac{x-2}{3} + 1) = 2x$ since it moves to state G after the first transition. Likewise, if we start the FST in state G the FST outputs: $1 + 3 \cdot 2\frac{x}{3} = 2x + 1$ if $w_0 = 0$, or $0 + 3 \cdot (2\frac{x-1}{3} + 1) = 2x + 1$ if $w_0 = 1$, or $2 + 3 \cdot (2\frac{x-2}{3} + 1) = 2x + 1$ if $w_0 = 2$. In all cases we get the result. □

	mul2 F	mul2 G (init)	find 2	rewind	check halt
0	0 R mul2_F	1 R mul2_F	0 L find_2	0 L rewind	1 R rewind
1	2 R mul2_F	0 R mul2_G	1 L find_2	1 L rewind	2 R rewind
2	1 R mul2_G	2 R mul2_G	2 L rewind	2 L rewind	Halt
#(blank)	# L find_2	1 R mul2_F	# L check_halt	# R mul2_F	0 R rewind

Fig. 3. 5-state 4-symbol Turing machine $M_{5,4}$ that halts if and only if Erdős' conjecture is false. The initial state of the machine is **mul2_G**, denoted '(init)'. The blank symbol is # and, since this is a busy-beaver candidate, the initial tape is empty: $\ldots \#\#\underline{\#}\#\# \ldots$ (tape head underlined). Example 6 shows the initial 333 steps of $M_{5,4}$. States **mul2_F** and **mul2_G** implement states F and G of the "mul2" FST in Fig. 2, that multiplies a reverse-ternary number by 2. The other states check whether the result is a counterexample to, or one of the three special cases of, Erdős' conjecture.

Intuitively, the 5-state 4-symbol Turing machine $M_{5,4}$ in Fig. 3 works as follows. Starting from all-# tape and state **mul2_G**, $M_{5,4}$ constructs successive natural number powers of 2 in base 3 (in fact in reverse-ternary) by iterating the **mul2** FST, which is embedded in its states **mul2_F** and **mul2_G**. Then, for each power of two, $M_{5,4}$ checks that there is at least one digit 2 using state **find_2**. If at least one digit 2 is found, then, using state **rewind** the machine goes back to the start (left) and iterates on to the next power of 2. If no digit 2 is found, such as in the three known special cases $1_{10} = 1_3$, $4_{10} = 11_3$ and $256_{10} = 100111_3$ (giving the condition $n > 8$ in Erdős' conjecture), then a counter is incremented on the tape (using state **check_halt**) and if this counter goes beyond value three the machine halts. The machine halts, iff we have found a counterexample to Erdős' conjecture, a fact we prove formally below in Theorem 2.

Example 6. Here, we highlight 11 out of the first 334 configurations of $M_{5,4}$. Five of the first 16 configurations are shown on the left (read from top to bottom), and 6 out of configurations 17 to 334 are shown on the right (see Footnote 5). From step 5 onwards, the content of the tape is of the form $c\#w_1\ldots w_n\#$ with $c, w_i \in \{0, 1, 2\}$, where c represents a single-symbol counter keeping track of the 3 special cases of Erdős conjecture and $w_1 \ldots w_n$ is the reverse-ternary representation of a power of two. For instance, in the final configuration below we have 1101011202221 as the reverse-ternary representation of $2^{20} = 122021101011_3$:

	mul2_G,	...###**#**###...	$\vdash^1$	rewind,	...#1**#**11#...
$\vdash^5$	rewind,	...##0**#**1#...	$\vdash^6$	rewind,	...#1**#**22#...
$\vdash^4$	rewind,	...##0**#**2#...	$\vdash^{40}$	rewind,	...#1**#**20211#...
$\vdash^4$	find_2,	...#0#1**1**#...	$\vdash^8$	find_2,	...#1#11100**1**#...
$\vdash^3$	check_halt,	...#**0**#11#...	$\vdash^8$	rewind,	...#2**#**111001#...
			$\vdash^{254}$	rewind,	...#2**#**1101011202221#...

Theorem 2. There is an explicit 5-state 4-symbol Turing machine that halts if and only if Erdős' conjecture is false.

Proof. The machine is called $M_{5,4}$ and is given in Fig. 3. We index tape positions by integers in $\mathbb{Z}$. Initially (at step 0), the tape head is at position 0 and each position of the tape contains the blank symbol #. The construction organises the tape as follows: position -1 holds a counter to keep track of the 3 known special cases of the Erdős' conjecture (1, 4 and 256 in base 10 which are 1, 11 and 100111 in base 3), position 0 always contains a blank # to act as a separator, and positive positions will hold reverse-ternary representations of powers of two, i.e. least significant digit at position 1 and most significant at position $\lceil \log_3(2^n) \rceil$. States **mul2_F** and **mul2_G** reproduce the logic of the "**mul2**" FST (Fig. 2) where, in state mul2_G the blank symbol behave like ternary digit 0 while in **mul2_F** it triggers the end of the multiplication by 2.

Starting the Turing machine in state **mul2_G** appropriately initialises the process, the reader can verify that at step 5 the machine is in state **rewind**, the tape head is at position 0 and the tape content between positions -1 to 2 included is: 0#1# (tape head position underlined) and all other positions are blank.[10] From there we prove the following result (IH) by induction on $n \in \mathbb{N}$: at step $s_n = 5 + c_n + \sum_{k=1}^n 2 * (\lceil \log_3(2^k) \rceil + 1)$ either (a) $M_{5,4}$ has halted, the tape head is at position -1 and the reverse-ternary represented number written on the positive part of the tape (with digits in reverse order) is a counterexample to Erdős' conjecture, i.e. it has no digit equal to 2 and is of the form 2^{n_0} with $n_0 > 8$, or (b) $M_{5,4}$ is in state **rewind**, the tape head is at position 0 which contains a blank symbol and the reverse-ternary represented number by the digits between position 1 and $\lceil \log_3(2^n) \rceil$ is equal to 2^n and all positions coming after $\lceil \log_3(2^n) \rceil$ are blank. The number c_n in the expression of s_n accounts for

---

[10] State **mul2_G** was not designed to kick-start the process, but starting with it happens to give us what we need.

extra steps that are taken by $M_{5,4}$ to increment the counter at position -1 which keeps track of the three known special cases of the conjecture: $1 = 1_3$, $4 = 11_3$ and $256 = 100111_3$. In practice, c_n is defined by $c_n = 0$ for $n \leq 1$, $c_n = 2$ for $2 \leq n \leq 7$ and $c_n = 4$ for $n \geq 8$.

For $n = 0$ we have $s_0 = 5$ and case (b) of the induction hypothesis (IH) is verified as $1 = 2^0$ is represented on the positive part of the tape as seen above.

Let's assume that the result holds for $n \in \mathbb{N}$. If $M_{5,4}$ had already halted at step s_n, then case (a) of IH still holds and nothing is left to prove. Otherwise, by case (b) of IH, at step s_n then $M_{5,4}$ is in state **rewind** at position 0 which contains a # and between positions 1 and $\lceil \log_3(2^n) \rceil + 1$ the following word is written: $w = w_1 \ldots w_{\lceil \log_3(2^n) \rceil} \#$ such that $w_1 \ldots w_{\lceil \log_3(2^n) \rceil}$ is the reverse-ternary representation of 2^n. At step $s_n + 1$, tape position 0 still contains the blank symbol #, $M_{5,4}$ is in state **mul2_F** and the tape head is at position 1 where it begins simulating the **mul2** FST on w. By having the final # of w encode a 0 then Lemma 5 applies which means that, after scanning w, the positive part of the tape contains the reverse-ternary expression of $2 * 2^n = 2^{n+1}$. Either $M_{5,4}$ was in state **mul2_F** when it read the final # of w, in which case it will stop simulating the FST and jump to state **find_2** and move the tape head to the left. Otherwise it was in state **mul2_G** and read the # in which case it jumps to state **mul2_F** and moves the tape head to the right where, by IH, a # will be read bringing us back to the previous case. In both cases, $\lceil \log_3(2^{n+1}) \rceil + 1$ symbols have been read since s_n, $M_{5,4}$ is in state **find_2** and all positions after and including $\lceil \log_3(2^{n+1}) \rceil + 1$ are still blank.

State **find_2** will scan to the left over the reverse-ternary expression that was just computed to search for a digit equal to 2. If a 2 is found, then it switches to state **rewind** and will reach position 0 at step $s_n + 2 * (\lceil \log_3(2^{n+1}) \rceil + 1)$. In that case, we also have $c_{n+1} = c_n$ as by definition of c_n, one can verify that $c_{n+1} \neq c_n$ implies that 2^{n+1} has no ternary digit equal to 2. Hence, if a 2 was found the machine reaches tape position 0 at step $s_{n+1} = 5 + c_{n+1} + \sum_{k=1}^{n+1} 2 * (\lceil \log_3(2^k) \rceil + 1)$, position 0 contains the blank symbol, positions 1 to $\lceil \log_3(2^{n+1}) \rceil$ hold the reverse-ternary representation of 2^{n+1} and all positions after $\lceil \log_3(2^{n+1}) \rceil$ are still blank, which is what we wanted under case (b) of IH. If no 2s were found then $M_{5,4}$ will reach position -1 in state **check_halt** and if $n + 1 = 2$ or $n + 1 = 8$ it will respectively read a 0 or 1 which will respectively be incremented to 1 or 2, then $M_{5,4}$ goes back to position 0 in state **rewind** and case (b) of IH is verified as the value c_n accounts for the extra steps that were taken to increment the counter in those cases. However, if $n + 1 > 8$ then $M_{5,4}$ will read a 2 at position -1 and consequently halt with positions 1 to $\lceil \log_3(2^{n+1}) \rceil$ giving the base three representation of 2^{n+1} containing no digit equal to two: a counterexample to Erdős' conjecture was found and it is consequently false and case (a) of IH holds.

From this proof, if $M_{5,4}$ halts then Erdős' conjecture is false. For the reverse direction, note that $M_{5,4}$ will test every single successive power of two until it finds a potential counterexample meaning that if Erdős' conjecture is false $M_{5,4}$ will find the smallest counterexample and stop. Hence the 5-state, 4-symbol Turing machine in Fig. 3 halts if and only if Erdős' conjecture is false. □

4 Fifteen States, Two Symbols Turing Machine

In this section, we prove Theorem 1 by defining the 15-state 2-symbol Turing machine $M_{15,2}$ in Fig. 4 then proving it halts if and only if it finds a counterexample to Erdős' conjecture. We begin with some guiding remarks.

4.1 Intuition and Overview of the Construction

Turing machines with two symbols can be challenging to reason about and prove correctness of for two reasons: (1) the tape is a difficult-to-read stream of bits and (2) simple algorithmic concepts need to be distributed across multiple states because of the low number of symbols. We first designed $M_{5,4}$ (Fig. 3, relatively easy to understand), and then designing $M_{15,2}$ to simulate $M_{5,4}$. (With the caveat that both machines have a few program-size optimisations.)

	mul2_F_sim
a	a R mul2_F_a
b	b R mul2_F_b

	mul2_F_a
a	b R mul2_G_sim
b	a R mul2_F_sim

	mul2_F_b
a	a R mul2_F_sim
b	b L find_2_a

	mul2_G_sim (init)
a	a R mul2_G_a
b	a R mul2_G_b

	mul2_G_a
a	a R mul2_G_sim
b	a L mul2_G_extra

	mul2_G_b
a	b R mul2_F_sim
b	b R mul2_F_sim

	mul2_G_extra
a	b R mul2_G_a
b	b R rewind_b

	find_2_sim
a	a L find_2_a
b	b L find_2_b

	find_2_a
a	a L rewind_sim
b	b L find_2_sim

	find_2_b
a	a L find_2_sim
b	b L check_halt_sim

	rewind_sim
a	a L rewind_a
b	b L rewind_b

	rewind_a
a	a L rewind_sim
b	b L rewind_sim

	rewind_b
a	a L rewind_sim
b	b R mul2_G_b

	check_halt_sim
a	b L check_halt_a
b	a R rewind_b

	check_halt_a
a	Halt
b	a R mul2_G_extra

Encoding E	
0	ba
1	ab
2	aa
#	bb

Fig. 4. 15-state 2-symbol Turing machine $M_{15,2}$ that halts if and only if Erdős' conjecture is false. The initial state is **mul2_G_sim**, denoted '(init)'. The blank symbol is b, and the initial tape is empty: $\ldots bb\underline{\mathbf{b}}bb \ldots$ (tape head position bold and underlined). States are organised into 5 columns, one for each state of $M_{5,4}$ in Fig. 3, and inherit name prefixes from $M_{5,4}$. In Lemma 10 we prove that $M_{15,2}$ simulates $M_{5,4}$.

In order to avoid confusion, $M_{15,2}$ uses alphabet $\{a,b\}$ (with b blank) which is distinct to that of $M_{5,4}$. In Lemma 10 we prove that $M_{15,2}$ *simulates* $M_{5,4}$, via a tight notion of simulation given in Definition 8 and the Lemma statement; with only a linear time overhead. Symbols of $M_{5,4}$ are encoded by those of $M_{15,2}$ using the encoding function $E : \{\#, 0, 1, 2\} \to \{a, b\}^2$ defined by $E(\#) = bb$, $E(0) = ba$, $E(1) = ab$, $E(2) = aa$. Intuitively, the 15 states of $M_{15,2}$ are partitioned into 5 sets: the idea is that each of the five states of $M_{5,4}$ is simulated by one of the five corresponding *column of states* in Fig. 4.

The naming convention for states in $M_{15,2}$ is as follows: for each state of $M_{5,4}$, there is a state with the same name in $M_{15,2}$, followed by _sim (short for 'simulate'), that is responsible for imitating the behaviour of the corresponding state in $M_{5,4}$. Then, from each _sim state in $M_{15,2}$ control moves to one of two new states, suffixed by _a and _b, after reading symbol a or b. At the next step, two consecutive letters have been read and $M_{15,2}$ knows which of the 4 possible cases of the encoding function E it is considering and has sufficient information to simulate one step of $M_{5,4}$. However, in many cases, for program-size efficiency reasons, $M_{15,2}$ makes a decision before it has read both symbols of the encoding.

There are two exceptions to the $M_{15,2}$ state naming rule: there is no state **check_halt_b** as our choice of encoding function E meant not needing to consider that case when simulating **check_halt**, and the state **mul2_G_extra** which is involved in simulation of both mul2_G and **check_halt**.

Example 7. We have $bb\ ba\ bb\ ab\ bb = E(\#)E(0)E(\#)E(1)E(\#)$ which corresponds to the tape content of the second configuration shown in Example 6.

4.2 Proof of Correctness

We next define 'simulates' so that the dynamics two machine are coupled: the tape content of one is mapped in a straightforward way to that of the other.

Definition 8 (simulates). Let M and M' be single-tape Turing machines with alphabets Σ and Σ'. Then, M' *simulates* M if there exists $m \in \mathbb{N}$, a function $E : \Sigma \to \Sigma'^m$ and a computable *time-scaling* function $f : \mathbb{N} \to \mathbb{N}$, such that for all time steps $n \in \mathbb{N}$ of M, then for step $f(n)$ of M', either both M and M' have already halted, or else they are both still running and, if M has tape content $\ldots t^n_{-1} t^n_0 t^n_1 \ldots$ (where $t_i \in \Sigma$) then M' has tape content $\ldots E(t^n_{-1})E(t^n_0)E(t^n_1)\ldots$.

Definition 9 ($M_{15,2}$'s encoding function E). Let $E : \{\#, 0, 1, 2\} \to \{a, b\}^2$ be $M_{15,2}$'s encoding function, where $E(\#) = bb$, $E(0) = ba$, $E(1) = ab$, $E(2) = aa$.

Our main technical lemma in this section states that $M_{15,2}$ (Fig. 4) simulates $M_{15,2}$ (Fig. 3), with a linear time overhead:

Lemma 10. Machine $M_{15,2}$ simulates $M_{5,4}$, according to Definition 8, with encoding function E (Definition 9) and time-scaling function f recursively defined by: $f(0) = 0$ and $f(n+1) = f(n) + g((q_n, d_n), \sigma_n)$ with: q_n being the state of $M_{5,4}$ at step n, $d_n \in \{L, R\}$ the tape head direction at step $n-1$ (where $d_0 = R$), $\sigma_n \in \{\#, 0, 1, 2\}$ the read symbol at step n, and g the partial function in Fig. 5.

Proof. Let $S = \{\mathbf{mul2\_F}, mul2\_G, \mathbf{find\_2}, \mathbf{rewind}, \mathbf{check\_halt}\}$ be the set of 5 states of $M_{5,4}$ and S' be the set of 15 states of $M_{15,2}$ in Fig. 4. We prove the result by induction on $n \in \mathbb{N}$, the number of steps of the *simulated* machine $M_{5,4}$. Let $k_n = (q_n, d_n) \in S \times \{L, R\}$ be $M_{5,4}$'s state at step n together with the tape-head direction at the *previous* step $n-1$ (we take the convention $d_0 = R$), let $\sigma_n \in \{0, 1, 2, \#\}$ be $M_{5,4}$'s read symbol at step n, and let $i_n \in \mathbb{Z}$ be $M_{5,4}$'s tape head position at step n. We note that by inspecting Fig. 3 the set of possible values for k_n is K=$\{(\mathbf{mul2\_F},R), (mul2\_G,R), (\mathbf{find\_2},L), (\mathbf{check\_halt},L), (\mathbf{rewind},L), (\mathbf{rewind},R)\}$ observing that **rewind** is the only $M_{5,4}$ state reachable from both left and right tape head moves.

The induction hypothesis (IH) is the definition of simulation (Definition 8) instantiated with E (Definition 9), f as defined in the lemma statement, and the following: at step $f(n)$ machine $M_{15,2}$'s head is at position $2i_n + \pi(d_n)$ with $\pi(L) = 1$ and $\pi(R) = 0$ and the machine is in state $h(k_n)$ with $h : K \to S'$ defined by $h(\mathbf{mul2\_F}, R) = \mathbf{mul2\_F\_sim}$, $h(mul2\_G, R) = \mathbf{mul2\_G\_sim}$, $h(\mathbf{find\_2}, L) = \mathbf{find\_2\_sim}$, $h(\mathbf{check\_halt}, L) = \mathbf{check\_halt\_sim}$, $h(\mathbf{rewind}, L) = \mathbf{rewind\_sim}$ and $h(\mathbf{rewind}, R) = \mathbf{rewind\_b}$.

If $n = 0$, we have $f(0) = 0$ and both tapes are entirely blank: ...###... for $M_{5,4}$, and ...bbb... for $M_{15,2}$ which is consistent with $E(\#) = bb$. Additionally, $k_0 = (mul2\_G, R)$, the tape head of $M_{15,2}$ is at position $0 = 2i_0 + 0$ with $i_0 = 0$ being the tape head position of $M_{5,4}$ and finally, $M_{15,2}$ is in state $h(k_0) = \mathbf{mul2\_G\_sim}$ (which is also its initial state).

Let's assume that the induction hypothesis (IH) holds for $n \in \mathbb{N}$. If at step n for $M_{5,4}$, and step $f(n)$ for $M_{15,2}$, both machines have already halted they will still have halted at any future step and the IH still holds. Let's instead assume from now on, IH holds and we are in the second case of Definition 8 (both are still running and tape contents are preserved under the encoding function E). We now proceed by case distinction.

	(mul2_F,R)	(mul2_G,R)	(find_2,L)	(check_halt,L)	(rewind,L)	(rewind,R)
0	2	2	2	3	2	-
1	2	4	2	1	2	-
2	2	2	2	2	2	-
#	3	2	2	1	3	2

Fig. 5. The partial function g of Lemma 10 that defines how many steps are needed by $M_{15,2}$ to simulate one step of $M_{5,4}$, for each (state, move-direction-on-previous-step) and symbol, of an $M_{5,4}$ step.

If $k_n = (\mathbf{mul2\_F}, R)$ and $\sigma_n = 2$ then the configuration of $M_{5,4}$ at step n is $\mathbf{mul2\_F}$, ...**2**... with tape head at position i_n. By Fig. 3, at step $n + 1$ the configuration becomes $mul2\_G$, ...1*... and $i_{n+1} = i_n + 1$ (with * being whatever symbol is at position $i_n + 1$) and $k_{n+1} = (mul2\_G, R)$. By IH, at step $f(n)$ the tape head position of $M_{15,2}$ is $2i_n + \pi(R) = 2i_n$ and the configuration of $M_{15,2}$ is $\mathbf{mul2\_F\_sim}$, ...$\underline{a}a$..., since $E(2) = aa$ and $h(k_n) = \mathbf{mul2\_F\_sim}$. By Fig. 4, at step $f(n+1) = f(n) + g((\mathbf{mul2\_F}, R), 1) = f(n) + 2$, the configuration of $M_{15,2}$ is $\mathbf{mul2\_G\_sim}$, ...$ab\underline{*}$... with tape head at position $2i_n + 2$. We have $E(1) = ab$ (and no other tape positions than $2i$ and $2i + 1$ were modified), $2i_n + 2 = 2(i_n + 1) = 2i_{n+1} + \pi(R) = 2i_{n+1} + \pi(d_{n+1})$ and $\mathbf{mul2\_G\_sim} = h(k_{n+1})$ which is everything we needed to satisfy IH at step $n+1$. We leave verifications of cases $\sigma_n = 0$ and $\sigma_n = 1$ to the reader as they are very similar. If $\sigma_n = \#$ then $M_{5,4}$ writes # then goes left to state $\mathbf{find\_2}$, which gives $k_{n+1} = (\mathbf{find\_2}, L)$ and $i_{n+1} = i_n - 1$. By IH, at step $f(n)$ the tape head position of $M_{15,2}$ is $2i_n + \pi(R) = 2i_n$ and the configuration of $M_{15,2}$ is $\mathbf{mul2\_F\_sim}$, ...$\underline{b}b$.... After $g((\mathbf{mul2\_F}, R), \#) = 3$ steps, the configuration becomes $\mathbf{find\_2\_sim}$, ...$\underline{*}bb$.... This is consistent with having $E(\#) = bb$, moving the tape head to position

$2i_n - 1 = 2(i_n - 1) + 1 = 2i_{n+1} + \pi(d_{n+1})$ and having **find_2sim** = $h(k_{n+1})$ which is all we needed.

If $k_n = (mul2\_G, R)$ and $\sigma_n = 1$ then the configuration of $M_{5,4}$ at step n is $mul2\_G$, ...**1**... with tape head at position i_n. By Fig. 3, at step $n + 1$ the configuration becomes $mul2\_G$, ...0*... and $i_{n+1} = i_n + 1$ and $k_{n+1} = (mul2\_G, R)$. By IH, at step $f(n)$ the tape head position of $M_{15,2}$ is $2i_n + \pi(R) = 2i_n$ and the configuration of $M_{15,2}$ is **mul2_G_sim**, ...**a**b... as $E(1) = ab$ and $h(k_n) = $ **mul2_G_sim**. In that case, the machine will need $g((mul2\_G, R), 1) = 4$ steps as it will have to bo back one step to the left after having scanned the second symbol of $E(1) = ab$ in order to write the first symbol $E(0) = ba$. This is realised by the first transition of intermediate state **mul2_G_extra** as we are assured to read symbol a in that case. Altogether, we get that at step $f(n+1) = f(n) + 4$ the configuration of $M_{15,2}$ is **mul2_G_sim**, ...ba*... with **mul2_G_sim** = $h(k_{n+1})$ and tape head at position $2i_n + 2 = 2i_{n+1} + \pi(d_{n+1})$ which is everything we need. We leave cases $\sigma_n \in \{0, 2, \#\}$ to the reader.

If $k_n = ($**find_2**$, L)$, the simulation is straightforward for any $\sigma_n \in \{0, 1, 2, \#\}$ as the content of the tape is not modified and the tape head always moves to the left, hence we leave those cases to the reader.

If $k_n = ($**rewind**$, L)$ and $\sigma_n = \#$ then the configuration of $M_{5,4}$ at step n is $mul2\_G$, ...**#**... with tape head at position i_n. By Fig. 3, at step $n + 1$ the configuration becomes **mul2_F**, ...#*... and $i_{n+1} = i_n + 1$ and $k_{n+1} = ($**mul2_F**$, R)$. By IH, at step $f(n)$ the tape head position of $M_{15,2}$ is $2i_n + \pi(L) = 2i_n+1$ and the configuration of $M_{15,2}$ is **rewind_sim**, ...b**b**... as $E(\#) = bb$ and $h(k_n) = $ **rewind_sim**. By Fig. 4, at step $f(n+1) = f(n) + g(($**rewind**$, L), \#) = f(n) + 3$, the configuration of $M_{15,2}$ is **mul2_F_sim**, ...bb*... with $bb = E(\#)$, **mul2_F_sim** = $h(k_{n+1})$ and tape head at position $2i_n + 2 = 2i_{n+1} + \pi(d_{n+1})$, which is everything that we need. We leave cases $\sigma_n \in \{0, 1, 2\}$ to the reader, in those cases the tape head always moves left and the tape content is not modified.

If $k_n = ($**rewind**$, R)$, by the proof of Theorem 1, we know that necessarily $\sigma_n = \#$, since in that case, the tape head of $M_{5,4}$ is at position $i_n = 0$ which always holds a $\#$ (the machine is just after incrementing the '3 special cases of Erdős' conjecture' counter at position -1), hence the configuration of $M_{5,4}$ at step n is **rewind**, ...**#**... and at step $n + 1$ it becomes **mul2_F**, ...#*... (Fig. 3) and $i_{n+1} = i_n + 1$ and $k_{n+1} = ($**mul2_F**$, R)$. By IH, at step $f(n)$ the tape head position of $M_{15,2}$ is $2i_n + \pi(R) = 2i_n$ and the configuration of $M_{15,2}$ is **rewind_b**, ...**b**b... as $E(\#) = bb$ and $h(k_n) = $ **rewind_b**. By Fig. 4, at step $f(n+1) = f(n) + g(($**rewind**$, R), \#) = f(n) + 2$, the configuration of $M_{15,2}$ is **mul2_F_sim**, ...bb*... with $bb = E(\#)$, **mul2_F_sim** = $h(k_{n+1})$ and the tape head is at position $2i_n + 2 = 2i_{n+1} + \pi(d_{n+1})$, which is everything we need.

If $k_n = ($**check_halt**$, L)$ then note that for $\sigma_n \in \{0, 1, \#\}$ machine $M_{5,4}$ will 'increment' the value of σ_n then transition to $k_{n+1} = ($**rewind**$, R)$ moving its head to $i_{n+1} = i_n + 1$. Cases $\sigma_n \in \{\#, 1\}$ are arguably where we make the most use of the encoding function E in order to use very few states in $M_{15,2}$ to simulate the behavior of $M_{5,4}$. Indeed, if $\sigma_n = \#$ or $\sigma_n = 1$ then $M_{5,4}$ must respectively write 1 and 2. This means that encodings must go from $E(\#) = bb$ to $E(1) = ba$

and from $E(1) = ab$ to $E(2) = aa$. By IH, machine $M_{15,2}$ is currently reading the second symbol of the encoding (because the head is at position $2i_n + \pi(L) = 2i_n + 1$), which is symbol b, then it is enough for $M_{15,2}$, using only one step, just to turn that b into an a which deals with both cases and then go right to state $h(k_{n+1}) = $ **rewind_b** while moving the head to $2i_n + 2 = 2i_{n+1} + \pi(d_{n+1})$ which is what we need. This gives $g((\textbf{check\_halt}, L), \#) = g((\textbf{check\_halt}, L), 1) = 1$. We let the reader verify the case $\sigma_n = 0$ which gives $g((\textbf{check\_halt}, L), 0) = 3$.

If $k_n = (\textbf{check\_halt}, L)$ and $\sigma_n = 2$ then, at step $n+1$ machine $M_{5,4}$ will halt. By IH, at step $f(n)$ the tape head position of $M_{15,2}$ is $2i_n + \pi(L) = 2i_n + 1$ and the configuration of $M_{15,2}$ is **check_halt_sim**, $\ldots a\underline{a}\ldots$ as $E(2) = aa$ and $h(k_n) = $ **check_halt_sim**. By Fig. 4, at step $f(n+1) = f(n) + g((\textbf{check\_halt}, L), 2) = f(n) + 2$ machine $M_{15,2}$ halts. Hence if $M_{5,4}$ halts then $M_{15,2}$ halts. Machine $M_{15,2}$ has only one halting instruction and this proof shows that it is reached only in the case where $k_n = (\textbf{check\_halt}, L)$ and $\sigma_n = 2$ meaning that if $M_{15,2}$ halts then $M_{5,4}$ halts. Hence the machines are either both running at respectively step $n+1$ and $f(n+1)$ or both have halted.

In all cases, the induction hypothesis is propagated at step $n+1$ and we get the result: machine $M_{15,2}$ simulates $M_{5,4}$ according to Definition 8. □

4.3 Main Result and Corollaries

Theorem 1. *There is an explicit 15-state 2-symbol Turing machine that halts if and only if Erdős' conjecture is false.*

Proof. By Lemma 10, $M_{15,2}$ simulates $M_{5,4}$ which in turns means that $M_{15,2}$ halts if and only if $M_{5,4}$ halts. By Theorem 2 this means that $M_{15,2}$ halts if and only if Erdős' conjecture is false. □

Corollary 3. *Erdős' conjecture is equivalent to the following conjecture over a finite set: for all $8 < n \leq \min(\text{BB}(15), \text{BB}(5,4))$ there is at least one digit 2 in the base 3 representation of 2^n.*

Proof. From the proof of Theorem 2, If we run $M_{5,4}$ then, at step $\text{BB}(5,4) + 1 \in \mathbb{N}$ we know if Erdős' conjecture is true or not: it is true if and only if $M_{5,4}$ is still running. If $M_{5,4}$ is still running, and because the machine outputs fewer than one power of 2 per step, at step $\text{BB}(5,4) + 1$ the power of 2 written on the tape is at most $2^{\text{BB}(5,4)}$. Analogously, $M_{15,2}$ writes fewer than one power of 2 per step, hence, at step $\text{BB}(15) + 1 \in \mathbb{N}$ the power of 2 written on the tape is at most $2^{\text{BB}(15)}$. In either case, by then, we know whether the conjecture is true or not. Hence, it is enough to check Erdős' conjecture for all $n \leq \min(\text{BB}(15), \text{BB}(5,4))$ and we get the result. □

Corollary 4. *Let $n \in \mathbb{N}$ be the smallest counterexample to Erdős' conjecture, that is the smallest positive integer $n > 8$ such that 2^n does not contain any digit 2 in base 3, if it exists. Then we have: $\text{BB}(15) \geq n$ and $\text{BB}(5,4) \geq n$.*

Proof. By Theorems 1 and 2, if Erdős' conjecture has a counterexample n, then we have explicit busy beaver contenders for BB(15) and BB(5,4), respectively, i.e. machines $M_{15,2}$ and $M_{5,4}$. These machines both halt with $x = 2^n$, written on their tape in base 3 (by the proofs of Theorems 1 and 2). $M_{15,2}$ and $M_{5,4}$ enumerate less than one power of 2 per time step, hence their running time is $\geq n$, giving the stated lower bound on BB(15) and BB(5,4). □

References

1. The Busy Beaver Challenge. https://bbchallenge.org/. Accessed 04 Aug 2024
2. Aaronson, S.: The busy beaver frontier. SIGACT News **51**(3), 32–54 (2020)
3. Adamczewski, B., Faverjon, C.: Mahler's method in several variables II: applications to base change problems and finite automata (2018). https://arxiv.org/abs/1809.04826
4. Brady, A.H.: The busy beaver game and the meaning of life. In: A Half-Century Survey on the Universal Turing Machine, pp. 259–277. Oxford University Press Inc. (1988)
5. Chaitin, G.J.: Computing the busy beaver function. In: Cover, T.M., Gopinath, B. (eds.) Open Problems in Communication and Computation, pp. 108–112. Springer, New York (1987). https://doi.org/10.1007/978-1-4612-4808-8_28
6. Code Golf Addict: list27.txt (2016). https://gist.github.com/anonymous/a64213f391339236c2fe31f8749a0df6
7. Cook, M., Stérin, T., Woods, D.: Small tile sets that compute while solving mazes. In: Lakin, M., Sulc, P. (eds.) Proceedings of the 27th International Conference on DNA Computing and Molecular Programming (DNA 27). Leibniz International Proceedings in Informatics (LIPIcs), vol. 205, pp. 1–20. Schloss Dagstuhl-Leibniz-Zentrum für Informatik, Dagstuhl, Germany (2021). arxiv preprint: https://arxiv.org/abs/2106.12341
8. Dimitrov, V.S., Howe, E.W.: Powers of 3 with few nonzero bits and a conjecture of Erdős (2021). https://arxiv.org/abs/2105.06440
9. Dupuy, T., Weirich, D.E.: Bits of 3^n in binary, Wieferich primes and a conjecture of Erdős. J. Number Theory **158**, 268–280 (2016). https://doi.org/10.1016/j.jnt.2015.05.022
10. Erdős, P.: Some unconventional problems in number theory. Math. Mag. **52**(2), 67–70 (1979). https://doi.org/10.1080/0025570X.1979.11976756
11. Georgiev, G.: Busy Beaver prover. https://skelet.ludost.net/bb/index.html. Accessed 27 May 2024
12. Harland, J.: Generating candidate busy beaver machines (or how to build the zany zoo). Theor. Comput. Sci. **922**, 368–394 (2022). https://doi.org/10.1016/j.tcs.2022.04.040
13. Riebel, J.: The Undecidability of BB(748). Bachelor's thesis (2023). https://www.ingo-blechschmidt.eu/assets/bachelor-thesis-undecidability-bb748.pdf
14. Kropitz, P.: BB(6, 2) > $4\char`\^4\char`\^4\char`\^7$ (2022). https://groups.google.com/g/busy-beaver-discuss/c/-zjeW6y8ER4/m/ZBuLvbVOAgAJ
15. Kropitz, P.: BB(2, 6) > $10 \uparrow\uparrow 10 \uparrow\uparrow 10 \uparrow\uparrow 3$ (2023). https://groups.google.com/g/busy-beaver-discuss/c/UuC_Yjc5LPQ
16. Kropitz, P.: BB(3, 4) > Ack(14) (2024). https://groups.google.com/g/busy-beaver-discuss/c/dkecbR5b5Og/

17. Lagarias, J.C.: Ternary expansions of powers of 2. J. Lond. Math. Soc. (2) **79**(3), 562–588 (2009). https://doi.org/10.1112/jlms/jdn080
18. Ligocki, S.: BB(3, 4) > Ack(14) (2024). https://www.sligocki.com/2024/05/22/bb-3-4-a14.html
19. Marxen, H., Buntrock, J.: Attacking the busy beaver 5. Bull. EATCS **40**, 247–251 (1990)
20. Michel, P.: The Busy Beaver Competitions. https://bbchallenge.org/~pascal.michel/bbc.html. Accessed 04 Aug 2024
21. Michel, P.: Simulation of the Collatz 3x+1 function by Turing machines. Technical report (2014). arXiv preprint https://arxiv.org/abs/1409.7322
22. Michel, P.: The busy beaver competition: a historical survey. Technical report (2019). https://arxiv.org/abs/0906.3749v6
23. Neary, T., Woods, D.: Four small universal Turing machines. Fund. Inform. **91**(1), 123–144 (2009)
24. Radó, T.: On non-computable functions. Bell Syst. Tech. J. **41**(3), 877–884 (1962). https://archive.org/details/bstj41-3-877/mode/2up
25. Ridenour, R.: (2024). https://github.com/CatsAreFluffy/metamath-turing-machines
26. Rogozhin, Y.: Small universal Turing machines. Theor. Comput. Sci. **168**(2), 215–240 (1996)
27. Stérin, T.: Six tiles: from Collatz sequences to algorithmic DNA origami. Ph.D. thesis, Maynooth University (2023)
28. Stérin, T., Woods, D.: The Collatz process embeds a base conversion algorithm. In: Schmitz, S., Potapov, I. (eds.) RP 2020. LNCS, vol. 12448, pp. 131–147. Springer, Cham (2020). https://doi.org/10.1007/978-3-030-61739-4_9
29. Tao, T.: The Collatz conjecture, Littlewood-Offord theory, and powers of 2 and 3. https://terrytao.wordpress.com/2011/08/25/the-collatz-conjecture-littlewood-offord-theory-and-powers-of-2-and-3/. Accessed 30 June 2021
30. Woods, D., Neary, T.: On the time complexity of 2-tag systems and small universal Turing machines. In: In 47th Annual IEEE Symposium on Foundations of Computer Science (FOCS), Berkeley, California, pp. 132–143. IEEE (2006)
31. Woods, D., Neary, T.: The complexity of small universal Turing machines: a survey. Theor. Comput. Sci. **410**(4–5), 443–450 (2009)
32. Yedidia, A., Aaronson, S.: A relatively small Turing machine whose behavior is independent of set theory. Complex Syst. **25**(4), 297–328 (2016). https://doi.org/10.25088/complexsystems.25.4.297

Linear Systems and Recurrences

On the Complexity of Reachability and Mortality for Bounded Piecewise Affine Maps

Olga Tveretina[✉]

University of Hertfordshire, Hatfield, UK
o.tveretina@herts.ac.uk

Abstract. Reachability is a fundamental decision problem that arises across various domains, including program analysis, computational models like cellular automata, and finite- and infinite-state concurrent systems. Mortality, closely related to reachability, is another critical decision problem.

This study focuses on the computational complexity of the reachability and mortality problems for two-dimensional hierarchical piecewise constant derivative systems (2-HPCD) and one-dimensional piecewise affine maps (1-PAM). Specifically, we consider the bounded variants of 2-HPCD and 1-PAM, as they are proven to be equivalent regarding their reachability and mortality properties [3].

The proofs leverage the encoding of the simultaneous incongruences problem, a known NP-complete problem, into the reachability (alternatively, mortality) problem for 2-HPCD. The simultaneous incongruences problem has a solution if and only if the corresponding reachability (alternatively, mortality) problem for 2-HPCD does not. This establishes that the reachability and mortality problems are co-NP-hard for both bounded 2-HPCD and bounded 1-PAM.

Keywords: Reachability problem · Mortality problem · Complexity

1 Introduction

Reachability is a fundamental problem that arises in various contexts, including program analysis, computational models like cellular automata, and finite- and infinite-state concurrent systems. It seeks to answer the question: given a computational system with a set of rules, can a certain state (or set of states) be reached from a given initial state (or set of states)? [1].

Another fundamental problem in the analysis of hybrid dynamical systems is the mortality problem: given a computational system, decide whether the system halts when starting from any state [6]. The mortality problem relates closely to stability properties and the long-term behaviour of trajectories within the system dynamics.

Neither the reachability problem nor the mortality problem is strictly more general than the other. Both problems are generally undecidable and are only known to be decidable for specific classes [9, 10].

Studying the computational complexity of both the reachability and mortality problems is crucial for establishing theoretical boundaries on the efficiency of underlying algorithms. To date, this aspect has received less attention compared to the fundamental question of decidability, although some research has been conducted in this direction [5,6,16].

Asarin et al. explored systems that straddle the boundary between decidability and undecidability [3]. They investigated variants of two-dimensional piecewise constant derivative systems (2-PCD) and demonstrated that certain variants are equivalent, in terms of reachability, to subclasses of one-dimensional piecewise affine maps (1-PAM).

In particular, Asarin et al. considered an extension of the 2-PCD model, known as a two-dimensional hierarchical PCD (2-HPCD), where discrete locations are organised hierarchically, with each location defined by a 2-PCD. Affine reset rules govern transitions between these locations. When all locations are bounded, the 2-HPCD is referred to as a bounded 2-HPCD (2-BHPCD). Similarly, if all intervals are bounded, 1-PAM is called bounded (1-BPAM).

This study explores the computational complexity of the reachability and mortality problems for 2-HPCD and 1-PAM, aiming to elucidate the theoretical boundaries of their computational hardness. It also addresses in a broader context the open question posed in [6] regarding the complexity of mortality in dimension two, specifically focusing on restricted 2-HPCD.

Our proofs are based on encoding the NP-complete simultaneous incongruences problem [7,14] into the reachability and mortality problems for 2-BHPCD. We then extend these results to 1-BPAM. Note that, in contrast to [6], one of the challenges in this study is simulating the simultaneous incongruences problem using a 2-BHPCD, that is, using a model of a lower dimension. In [6], verifying whether the current k is a solution to the given instance of the simultaneous incongruences problem is achieved through the z-coordinate, which represents the current value of k in a restricted 3-HPCD.

It is notable that piecewise-affine models, together with reachability analysis, are applied in various domains, including gene-regulatory networks [15], biochemical kinetics [8], and qualitative biological models that depict interactions involving protein promotion or inhibition [4].

The rest of the paper is organised as follows. In Sect. 2 we introduce preliminaries. Section 3 demonstrates how the simultaneous incongruences problem can be simulated by 2-BHPCD. In Sect. 4 we prove that the reachability problem for 2-BHPCD and 1-BPAM is co-NP-hard. Section 5 establishes that the mortality problem for 2-BHPCD and 1-BPAM is also co-NP-hard. Finally, Sect. 6 contains our concluding remarks.

2 Preliminaries

In this section, we define the notions of one-dimensional piecewise affine maps (1-PAM) and two-dimensional hierarchical piecewise constant derivative systems (2-HPCD), closely following the notations used in [2,3,11,12].

2.1 Piecewise Affine Maps

A rational interval in $\mathbb{R}$ is defined as one of the following forms: $[x, y]$, $[x, y)$, $(x, y]$, (x, y), $(-\infty, y]$, $(-\infty, y)$, $[x, \infty)$, (x, ∞), where $x, y \in \mathbb{Q}$ with $x \leqslant y$.

Definition 1 (1-PAM, [3]). *Let I_i be a finite set of disjoint rational intervals. A function $f : \mathbb{R} \to \mathbb{R}$ is a one-dimensional piecewise affine map (1-PAM) if it can be expressed as $f(x) = a_i x + b_i$ for $x \in I_i$.*

Definition 2 (Trajectory). *A trajectory of a 1-PAM f is a sequence $x_1, x_2, \ldots$ such that $x_{i+1} = f(x_i)$ for all i. We say that y is reachable from x if there exists a finite trajectory starting at x and ending at y.*

Definition 3 (1-BPAM, [3]). *A 1-PAM f is called bounded (1-BPAM) if none of its intervals is infinite.*

The class of the bounded 1-PAM is particularly noteworthy because it represents a subset of piecewise affine functions that lie on the boundary between decidability and undecidability when it comes to analysing their trajectories and reachability properties. They pose interesting challenges in terms of determining the reachability of points under repeated applications of f.

2.2 Hierarchical Piecewise Constant Derivative Systems

A two-dimensional piecewise constant derivative system (2-PCD) is defined as a finite set of regions, where each region is associated with a constant vector field. In this context, the vector field within each region is characterised by a single vector in $\mathbb{R}^2$. Intuitively, the vector assigned to each region determines the direction a particle would follow upon entering the region from any of its boundaries.

Definition 4 (2-PCD, [11]). *A two-dimensional piecewise constant derivative system (2-PCD) is defined by a pair $\mathcal{H} = (\mathcal{P}, \varphi)$, where:*

1. *$\mathcal{P} = \{p_1, \ldots, p_k\}$ is a finite set of non-overlapping polygons with nonempty interiors, referred to as regions throughout this paper.*
2. *$\varphi : \mathcal{P} \to \mathbb{R}^2$ is a function that assigns a vector $\varphi(p) \in \mathbb{R}^2$ to each region $p \in \mathcal{P}$, defining the dynamics within p.*

The set of all border points of $\mathcal{P}$, denoted $\mathcal{B}(\mathcal{P}) = \bigcup_{p \in \mathcal{P}} \mathcal{B}(p)$, consists of the union of the boundaries of all regions p, where $\mathcal{B}(p)$ denotes the boundary of region p. Formal definitions can be found in [11] or [12].

Since such a system exhibits deterministic behaviour within each polygonal region, the reachability analysis primarily focuses on computing the discrete successors of points located on the boundaries of these regions.

Definition 5 (Step, [11]). *Let $\mathcal{H} = (\mathcal{P}, \varphi)$ be a 2-PCD, and let $\boldsymbol{x}$ and $\boldsymbol{x}'$ be two distinct points in $\mathbb{R}^2$. We say that the pair $(\boldsymbol{x}, \boldsymbol{x}')$ is a step if there exists a region $p \in \mathcal{P}$ and a $t > 0$ such that the following conditions hold:*

1. $x' = x + t \cdot \varphi(p)$,
2. $x, x' \in \mathcal{B}(p)$, that is, both points lie on the boundary of region p,
3. $x'' = x + t' \cdot \varphi(p) \in p$ for every $0 < t' < t$.

Intuitively, the pair (x, x') is considered a step if both points x and x' lie on the boundary of a region p, and the straight line segment connecting x and x' is entirely contained within p.

Definition 6 (Trajectory). *Let $\mathcal{H} = (\mathcal{P}, \varphi)$ be a 2-PCD, and let x_0 be a point in $\mathcal{B}(\mathcal{P})$. A trajectory rooted at x_0 is a sequence $\tau_{x_0}^\ell = x_0, x_1, \ldots, x_\ell$, where each pair (x_i, x_{i+1}) for $0 \leq i \leq \ell - 1$ is a step. We say that ℓ is the length of $\tau_{x_0}^\ell$.*

Note that for each point x_0 and each ℓ, such a trajectory is unique. Furthermore, we say that a point x_f is reachable from x_0 if x_f belongs to the trajectory $\tau_{x_0}^\ell$ for some finite ℓ.

Definition 7 (2-HPCD, [3]). *A two-dimensional hierarchical piecewise constant derivative system (2-HPCD) consists of a collection of locations, where each location is a 2-PCD system. Additionally:*

1. *Transition guards determine transitions between locations based on specified conditions.*
2. *Affine reset rules govern the behaviour of variables when transitioning between locations.*

A formal definition of a 2-HPCD system can be found in [3] and [6]. The definition of a 2-HPCD system emphasises that its trajectories largely resemble those of a 2-PCD system, but with occasional jumps induced by transition guards. We note that the notion of a trajectory in a 2-PCD can be straightforwardly extended to a 2-HPCD.

In this study we consider deterministic 2-HPCD systems: the transition guards for each location are mutually exclusive.

Definition 8 (2-BHPCD, [3]). *A 2-HPCD $\mathcal{H}$ is called bounded (abbreviated as 2-BHPCD) if none of its regions is infinite.*

2.3 Reachability and Mortality Problems

Reachability typically manifests in two forms: point-to-point reachability and interval-to-interval reachability.

Definition 9 (Reachability problem). *Given a 2-HPCD $\mathcal{H}$, the point-to-point and interval-to-interval reachability problems are defined as follows:*

- ***Point-to-point reachability:*** *given an initial point x_0 and a final point x_f, determine whether x_f is reachable from x_0.*
- ***Interval-to-interval reachability:*** *given an initial interval I_0 and a final interval I_f, determine whether some point $x_f \in I_f$ is reachable from some point $x_0 \in I_0$.*

Definition 10 (Mortality problem). *Given a 2-HPCD $\mathcal{H}$, the mortality problem asks whether there exists an initial point x_0 such that the trajectory τ_{x_0} starting from x_0 is infinite. If such a point exists, $\mathcal{H}$ is called immortal. If for every point x_0, the trajectory τ_{x_0}, starting at x_0, eventually halts, then $\mathcal{H}$ is called mortal.*

The reachability problem, including interval-to-interval reachability, and the mortality problem for 1-PAM are defined analogously to those for 2-HPCD.

2.4 Notion of Simulation

The concept of simulation typically involves a relationship between two computational systems, A and B, where one system (the simulating system, in this case A) can reproduce the behaviour of the other system (the simulated system, in this case B). This implies that every computational step or action that B can perform can also be executed by A.

Simulation is particularly relevant when comparing different models of computation or analysing their complexity. It is formally defined below in terms of state transition systems:

Definition 11 (Simulation). *We assume deterministic transition systems $\Gamma = (S, \delta, s_0, s_f)$ and $\Gamma' = (S', \delta', s'_0, s'_f)$, where:*

- *S, S' are sets of states,*
- *$\delta : S \to S$, $\delta' : S' \to S'$ are transition functions,*
- *s_0, s'_0 are initial states,*
- *s_f, s'_f are final states.*

We say that Γ can be simulated by Γ' with respect to the reachability and mortality problems if there exists a function $f : S \to S'$ such that the following conditions hold:

1. *$s'_0 = f(s_0)$,*
2. *$s'_f = f(s_f)$,*
3. *$f(\delta(s)) = \delta'(f(s))$ for any $s \in S$.*

This definition specifies that simulation involves finding a function f that maps states of system Γ to states of system Γ', preserving initial and final states, and ensuring that the transitions of Γ are mirrored by Γ''s transitions under f.

We note that it follows from Lemma 3.3 and Lemma 3.4 in [3] that bounded 1-PAM and bounded 2-HPCD can simulate each other, as summarised in the following theorem:

Theorem 1 [3]. *Every bounded 1-PAM can be simulated by a bounded 2-HPCD. Conversely, every bounded 1-HPCD can be simulated by a bounded 1-PAM.*

Furthermore, the complexity of simulating bounded 2-HPCD by bounded 1-PAM (and vice versa) is polynomial in the size of the simulated instance, as demonstrated in the proofs provided in [3].

3 Simulation of Simultaneous Incongruences Problem by Bounded 2-HPCD

Our proof of the co-NP-hardness of reachability for bounded 2-HPCD, and subsequently for bounded 1-PAM, is based on the simulation of the simultaneous incongruences problem by 2-HPCD as outlined below:

1. In Sect. 3.1, we define the simultaneous incongruences problem and discuss its feature used in our proofs.
2. In Sect. 3.2, we illustrate how to simulate an instance of the simultaneous incongruences problem using the reachability problem for 2-BHPCD. By construction, the instance of the simultaneous incongruences problem has a solution if and only if the corresponding reachability problem does not.
3. In Sect. 3.3, we discuss the complexity of the provided simulation.

3.1 Simultaneous Incongruences Problem

In this section, we define the simultaneous incongruences problem, which is known to be NP-complete [7]. We will use this problem to show that the reachability and mortality problems for 2-HPCD and 1-PAM are co-NP-hard.

Definition 12 (Simultaneous incongruences problem). *Given a set $S = \{(\alpha_1, \beta_1), \ldots, (\alpha_k, \beta_k)\}$, where $k \geqslant 1$, of ordered pairs of positive numbers such that $\alpha_i \leqslant \beta_i$ for every $1 \leqslant i \leqslant k$, the simultaneous incongruences problem asks: does there exist an integer x such that $x \not\equiv \alpha_i \pmod{\beta_i}$ for every $1 \leqslant i \leqslant k$?*

In the following, we use the notation $\mathsf{LCM}(\beta_1, \ldots, \beta_k)$ to denote the least common multiple of the numbers $\beta_1, \ldots, \beta_n$. This is the smallest positive integer that is evenly divisible by $\beta_1, \ldots, \beta_k$. We will also use the technical lemma provided below.

Lemma 1 [6]. *There exists a solution for the simultaneous incongruences problem for $S = \{(\alpha_1, \beta_1), \ldots, (\alpha_k, \beta_k)\}$ if and only if there exists a solution x such that $1 \leqslant x \leqslant \Lambda$, where $\Lambda = \mathsf{LCM}(\beta_1, \ldots, \beta_k)$, the least common multiple of the numbers $\beta_1, \ldots, \beta_k$.*

3.2 Simulation

In this section, we demonstrate the construction of a 2-HPCD that simulates the simultaneous incongruences problem. Unlike the approach taken by Bell et al. in [6], our study has a challenge of simulating this problem using a lower-dimensional system. In contrast, Bell et al. determine whether the current k is a solution to the given simultaneous incongruences problem using the z-coordinate, which tracks the current value of k in the restricted 3-HPCD.

Let $S = \{(\alpha_1, \beta_1), \ldots, (\alpha_k, \beta_k)\}$ be a set of ordered pairs of positive integers such that $\alpha_i \leqslant \beta_i$ for $1 \leqslant i \leqslant k$. In the following, we will use $\mathcal{H}^{\mathsf{sip}}(S)$ to denote the 2-HPCD that simulates the simultaneous incongruences problem for S.

The reachability problem for 2-HPCD simulates the simultaneous incongruences problem for S, incorporating the insights from Lemma 1.

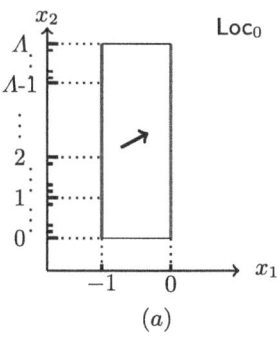 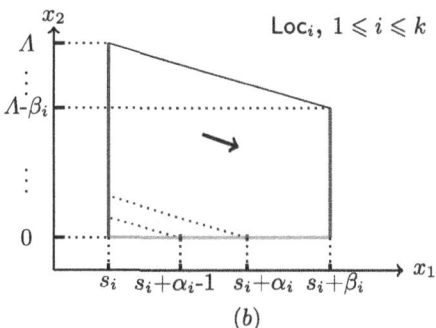

Fig. 1. Encoding of the simultaneous incongruences problem by the reachability problem for 2-HPCD, where: (a) location Loc_0 with the flow given by the vector $\boldsymbol{x} = (1, 1 - 1/(1 + \Lambda))$; (b) location Loc_i with the flow given by the vector $\boldsymbol{x} = (1, -1)$

1. For the current value of x, where $1 \leqslant x \leqslant \Lambda$ and $\Lambda = \mathsf{LCM}(\beta_1, \ldots, \beta_k)$, we check whether $x \not\equiv \alpha_i \pmod{\beta_i}$ holds for each pair (α_i, β_i), where $1 \leqslant i \leqslant k$. If for every i, x is a solution to the given simultaneous incongruences problem, then the reachability problem for the 2-HPCD $\mathcal{H}^{\mathsf{sip}}(\mathcal{S})$ will not have a solution. That is, the trajectory will not reach the final state.
2. If for any x, where $1 \leqslant x \leqslant \Lambda$, there is an i, where $1 \leqslant i \leqslant k$, such that $x \equiv \alpha_i \pmod{\beta_i}$, then x is not a solution to the simultaneous incongruences problem for $\mathcal{S}$ and the trajectory will reach the final state.

The locations of $\mathcal{H}^{\mathsf{sip}}(\mathcal{S})$ are utilised in two primary ways: firstly, to perform modulo operations, and secondly, to increment the value being tested as needed. These functionalities are schematically depicted in Fig. 1.

We assume that $s_0 = -1$ and $s_i = i + \sum_{j=1}^{i-1} \beta_j$, where $1 \leqslant i \leqslant k$. Now, we specify details of the construction of each Loc_i, where $0 \leqslant i \leqslant k$:

1. Location Loc_0:
   (a) Loc_0 is the convex polygon bounded by the straight lines $x_1 = s_0$, $x_1 = 0$, $x_2 = 0$ and $x_2 = \Lambda$.
   (b) The flow in Loc_0 is given by the vector $\boldsymbol{x} = (1, 1 - 1/(1 + \Lambda))$.
2. Location Loc_i, $1 \leqslant i \leqslant k$:
   (a) Each Loc_i is the convex polygon bounded by the straight lines $x_1 = s_i$, $x_1 = s_i + \beta_i$, $x_2 = 0$ and $x_2 = -x_1 + (\Lambda + s_i)$.
   (b) The flow in Loc_i is given by the vector $\boldsymbol{x} = (1, -1)$.

To determine if the current value of x, $1 \leqslant x \leqslant \Lambda$, is a solution to the given instance of the simultaneous incongruences problem, we simulate this check using the trajectory starting at the point $(1, x - x/(1+\Lambda))$ within location Loc_1. Thus, x is encoded as $(1, x - x/(1+\Lambda))$.

In the following, $I \times \{y\}$ will be used to denote the set $\{(x, y) \mid x \in I\}$, where I is an open or half-open or closed bounded rational interval. We will use $\{x\} \times I$ likewise. Now, each location Loc_i and its flow are used to perform the modulo operation for the pair (α_i, β_i) as follows:

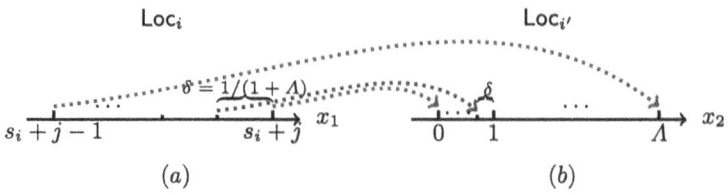

Fig. 2. Mapping the line interval $(s_i + j - 1, s_i + j) \times \{0\} \subset \text{Loc}_i$, where $1 \leq i \leq k$ and $1 \leq j \leq \beta_i$, onto the line interval $\{s_{i'}\} \times (0, \Lambda) \subset \text{Loc}_{i'}$, where either $0 \leq i + 1 = i' \leq k$ or $i' = 0$: (a) a schematic representation of the line segment $(s_i + j - 1, s_i + j) \times \{0\}$; (b) a schematic representation of the line segment $\{s_{i'}\} \times (0, \Lambda)$

- Whenever the trajectory transitions from a point in the line interval $\{s_i\} \times (0, \Lambda)$ to a point in the line interval $\{s_i + \beta_i\} \times (0, \Lambda - \beta_i)$, the value of the variable x_2 decreases by β_i.
- If the trajectory reaches a point $(x', 0) \in (s_i + j - 1, s_i + j) \times \{0\}$, where $1 \leq j \leq \beta_i$, it implies that $x \equiv j \pmod{\beta_i}$, where $x = [(s_i + j) - x'](1 + \Lambda)$.

A summary of the details of the 2-HPCD $\mathcal{H}^{\text{sip}}(\mathcal{S})$ such as its transition guards and reset relations are provided in Table 1.

Table 1. 2-HPCD $\mathcal{H}^{\text{sip}}(\mathcal{S})$: a summary of the guards, transitions and reset relations, where $1 \leq i \leq k$, and $1 \leq j \leq \alpha_i - 1$ or $\alpha_i + 1 \leq j \leq \beta_i$

Guard	Transition to	Reset relation
$G_0 = \{0\} \times (0, \Lambda)$	$\{1\} \times (0, \Lambda) \subset \text{Loc}_1$	$x_1 \to 1 \quad x_2 \to x_2$
$G_i^1 = \{s_i + \beta_i\} \times (0, \Lambda - b_i)$	$\{s_i\} \times (0, \Lambda - \beta_i) \subset \text{Loc}_i$	$x_1 \to s_i \quad x_2 \to x_2$
$G_i^2 = (s_i + \alpha_i - 1, s_i + \alpha_i) \times \{0\}$	$\{-1\} \times (0, \Lambda) \subset \text{Loc}_0$	$x_1 \to -1 \quad x_2 \to -\Lambda x_1 + \Lambda(s_i + a_i)$
$G_{i,j}^3 = (s_i + j - 1, s_i + j) \times \{0\}$	$\{s_{i+1}\} \times (0, \Lambda) \subset \text{Loc}_{i+1}$	$x_1 \to s_{i+1} \quad x_2 \to -\Lambda x_1 + \Lambda(s_i + j)$

The reset relation for variable x_1 determines the transition to the next location, and the reset relation for variable x_2 is defined as follows:

1. If $(x_1, x_2) \in G_0 \cup G_i^1$, then $x_2 \to x_2$. In this case, the reset relation for x_2 is the identity relation.
2. If $(x_1, x_2) \in G_i^2 \cup G_{i,j}^3$, then $x_2 \to -\Lambda x_1 + \Lambda(s_i + j)$ (see Fig. 2). In this case, the reset relation for x_2 is computed by solving the system of equations $a(s_i + j - 1) + b = \Lambda$ and $a(s_i + j) + b = 0$, where a and b are unknowns. This yields $a = -\Lambda$ and $b = \Lambda(s_i + j)$.

The reachability problem for $\mathcal{H}^{\text{sip}}(\mathcal{S})$ is defined as follows: is there a trajectory starting at the initial point $\boldsymbol{x}_0$ and reaching the final point $\boldsymbol{x}_f$, where

- $x_0 = (1, 1 - 1/(1+\Lambda)) \in \text{Loc}_1$, and
- $x_f = (s_0, \Lambda) \in \text{Loc}_0$.

Now, we verify if the current x, where $0 \leqslant x \leqslant \Lambda$, is a solution to the simultaneous incongruences problem for the given set $\mathcal{S}$. We assume that the trajectory has reached a point $(x', 0) \in (s_i, s_i + \beta_i) \times \{0\}$ in Loc_i, where $1 \leqslant i \leqslant k$.

1. If $x \equiv \alpha_i \pmod{\beta_i}$, then the current value of x is not a solution to the simultaneous incongruences problem for $\mathcal{S}$. In this scenario, G_i^2 forces the transition from Loc_i to Loc_0 in order to increase the value of x by one.
2. If $x \not\equiv \alpha_i \pmod{\beta_i}$, then the current value of x is a potential solution to the simultaneous incongruences problem for $\mathcal{S}$. Here, $\mathsf{G}_{i,j}^3$, where $1 \leqslant j \leqslant \alpha_i - 1$ or $\alpha_i + 1 \leqslant j \leqslant \beta_i$, forces a transition from Loc_i to Loc_{i+1} to increment the value of i by one.

We assume that all $\mathsf{G}_{k,j}^3$, where $1 \leqslant j \leqslant \alpha_k - 1$ or $\alpha_k + 1 \leqslant j \leqslant \beta_k$, have no outgoing transitions. If there exists a solution to the simultaneous incongruences problem for $\mathcal{S}$, then starting at the point $x_0 = (1, 1 - 1/(1+\Lambda))$ in Loc_1, the trajectory will eventually reach a point in $\mathsf{G}_{k,j}^3$, meaning the trajectory will halt without reaching the final state.

If there is no solution to the simultaneous incongruences problem for $\mathcal{S}$, then the trajectory will eventually reach the point $x_f = (s_0, \Lambda)$ in Loc_0, which represents the final state.

3.3 Complexity of Simulation

We adhere to the standard complexity theory convention, where both a set $\mathcal{S} = \{(\alpha_1, \beta_1), \ldots, (\alpha_k, \beta_k)\}$ and the corresponding 2-HPCD $\mathcal{H}^{\text{sip}}(\mathcal{S})$ are encoded in a "reasonable" manner, typically in binary. It is straightforward that the representation of $\mathcal{S}$ requires at least k bits. We also assume that the values of β_i's in $\mathcal{S}$ are polynomial in k.[1]

From a computational perspective, $\mathcal{H}^{\text{sip}}(\mathcal{S})$ can be specified by lists of locations (vertices), guards, and flows. By construction, the number of locations in $\mathcal{H}^{\text{sip}}(\mathcal{S})$ is $k + 1$. The number of guards in $\mathcal{H}^{\text{sip}}(\mathcal{S})$ totals $g = 1 + \sum_{i=1}^{k}(\beta_i + 1)$, which is polynomial in k provided that all β_i are polynomial in k.

Moreover, for any vertex $v = (x, y)$ in $\mathcal{H}^{\text{sip}}(\mathcal{S})$, by construction, we have $-1 \leqslant x \leqslant k + \sum_{i=1}^{k} \beta_k$ and $0 \leqslant y \leqslant \Lambda$, where Λ is the least common multiplier of $\beta_1, \ldots, \beta_k$. Since $\Lambda \leqslant \beta_1 \times \beta_2 \times \cdots \times \beta_k$, its binary representation requires $O(\sum_{i=1}^{k} \log \beta_i)$ bits. This implies that representing all vertices in $\mathcal{H}^{\text{sip}}(\mathcal{S})$ is also polynomial in k.

The complexity of representing the flow in location Loc_0, given by the vector $x = (1, 1 - 1/(1+\Lambda))$, is also polynomial in k assuming Λ is represented in

---

[1] A 3-SAT problem φ with n variables can be transformed into the simultaneous incongruences problem for some set $\mathcal{S}$, where the number of pairs in $\mathcal{S}$ is polynomial in the size of φ, and all β_i are integers polynomial in n [7,14]. Therefore, in this context, it suffices to consider a set $\mathcal{S}$ containing k pairs, where all β_i are integers polynomial in k.

binary. Therefore, we conclude that the complexity of representing $\mathcal{H}^{\text{sip}}(\mathcal{S})$, in terms of the number of bits required, is polynomial in k.

4 Complexity of the Reachability Problem

In this section, we establish our main result, demonstrating the co-NP-hardness of the reachability problem for 2-HPCD and 1-PAM. Our approach is outlined as follows:

1. We use the encoding provided in Sect. 3, which has the property that the given instance of the simultaneous incongruences problem has a solution if and only if the reachability problem does not.
2. The complexity of our simulation is polynomial in the description size of the given simultaneous incongruences problem, as discussed in Sect. 3.3. Consequently, we demonstrate that the reachability problem for 2-BHPCD is co-NP-hard in Sect. 4.1.
3. In Sect. 4.2, we extend our results to show that the reachability problem for 1-BPAM is also co-NP-hard.

4.1 Complexity of Reachability for Bounded 2-HPCD

In this section, we prove that the complexity of the reachability problem for 2-HPCD is co-NP-hard.

Theorem 2. *The reachability problem for 2-BHPCD is co-NP-hard.*

Proof. Assume a set of pairs $\mathcal{S} = \{(\alpha_1, \beta_1), \ldots, (\alpha_k, \beta_k)\}$, $k \geqslant 1$. As discussed in Sect. 3.3, we can assume without loss of generality that all β_i are of size polynomial in k.

We encode the simultaneous incongruences problem for $\mathcal{S}$ into the bounded 2-HPCD $\mathcal{H}^{\text{sip}}(\mathcal{S})$ as described in Sect. 3.2. The complexity of the construction is polynomial in k, as outlined in Sect. 3.3.

By construction, the reachability problem for the 2-HPCD $\mathcal{H}^{\text{sip}}(\mathcal{S})$ has a solution if and only if there is no solution to the simultaneous incongruences problem for $\mathcal{S}$. Since the simultaneous incongruences problem is NP-complete, this implies that the reachability problem for 2-HPCD is co-NP-hard. □

4.2 Complexity of Reachability for Bounded 1-PAM

Now, we show that the complexity of the reachability problem for a bounded 1-PAM is also co-NP-hard.

Corollary 1. *The reachability problem for 1-BPAM is co-NP-hard.*

Proof. The complexity of simulating 2-BHPCD by 1-BPAM (or vice versa) is polynomial in the instance size, as demonstrated in the proofs provided in [3]. Now, it follows from Theorem 1 and Theorem 2 that the reachability problem for 1-BPAM is co-NP-hard. □

5 Complexity of the Mortality Problem

Next, we establish that the mortality problem for bounded 2-HPCD and 1-PAM is also co-NP-hard. Unlike the reachability problem, which focuses on determining whether a trajectory reaches a final state, the mortality problem examines whether all trajectories halt. More specifically:

1. In Sect. 5.1, we demonstrate how to encode an instance of the simultaneous incongruences problem using the mortality problem for 2-BHPCD. Specifically, the system is mortal if and only if there is no solution for the corresponding simultaneous incongruences problem, otherwise the system is immortal.
2. The complexity of our simulation is polynomial in the description size of the given simultaneous incongruences problem. Consequently, we show that the mortality problem for 2-BHPCD is co-NP-hard in Sect. 5.1.
3. In Sect. 5.2 we extend our results to show that the mortality problem for 1-BPAM is also co-NP-hard.

5.1 Complexity of Mortality for Bounded 2-HPCD

Now we can state that the complexity of the mortality problem for 2-HPCD is co-NP-hard.

Theorem 3. *The mortality problem for 2-BHPCD is co-NP-hard.*

Proof. Assume a set of pairs $\mathcal{S} = \{(\alpha_1, \beta_1), \ldots, (\alpha_k, \beta_k)\}$, where $k \geq 1$. As discussed in Sect. 3.3, we can assume without loss of generality that all β_i are of size polynomial in k.

We construct the 2-HPCD $\mathcal{H}_m^{\mathsf{sip}}(\mathcal{S})$ such that the system is mortal if and only if there is no solution to the simultaneous incongruences problem for $\mathcal{S}$.

The construction follows the approach detailed in Sect. 3.2. Additionally, we assume that $G_{k,j}^3$, where $1 \leqslant j \leqslant \alpha_k - 1$ or $\alpha_k + 1 \leqslant j \leqslant \beta_k$, is reset to the interval $[x_0, x_0]$ (or, alternatively, to a sufficiently small interval around x_0), where $x_0 = (1, 1 - 1/(1 + \Lambda))$, which is the initial point in location Loc_1, as defined in Sect. 3.2. Further specifics are as follows:

1. Assuming that there is a solution to the simultaneous incongruences problem for $\mathcal{S}$:
   If there exists a solution to the simultaneous incongruences problem for $\mathcal{S}$, then starting from $x_0 = (1, 1-1/(1+\Lambda))$ in Loc_1, the trajectory will eventually reach some point $x \in G_{k,j}^3$, where $1 \leqslant j \leqslant \alpha_k - 1$ or $\alpha_k + 1 \leqslant j \leqslant \beta_k$, and it will be reset to the initial point $x_0 = (1, 1 - 1/(1 + \Lambda))$ in Loc_1. That is, there is at least one infinite trajectory. Therefore, $\mathcal{H}_m^{\mathsf{sip}}(\mathcal{S})$ is immortal.
2. Assuming that there is no solution to the simultaneous incongruences problem for $\mathcal{S}$:
   If the simultaneous incongruences problem for $\mathcal{S}$ has no solution, then starting from any point, the trajectory will eventually reach the final point $x_f = (s_0, \Lambda) \in \mathsf{Loc}_0$. We assume that there is no outgoing transition from x_f.

That is, if there is no solution to the simultaneous incongruences problem for $\mathcal{S}$, then regardless of where the trajectory starts, it will eventually halt. Therefore, $\mathcal{H}_m^{\mathsf{sip}}(\mathcal{S})$ is mortal.

Since the complexity of this simulation is polynomial in the description size of an instance of the simultaneous incongruences problem, following the reasoning provided in Sect. 3.3, we conclude that the mortality problem for 2-BHPCD is co-NP-hard. □

5.2 Complexity of Mortality for Bounded 1-PAM

Extending the results provided in Sect. 5.1, we can conclude that the complexity of the mortality problem for 1-PAM is also co-NP-hard.

Corollary 2. *The mortality problem for 1-BPAM is co-NP-hard.*

Proof. The complexity of simulating 2-BHPCD by 1-BPAM (or vice versa) is polynomial in the instance size, as demonstrated in the proofs provided in [3]. Now, it follows from Theorem 1 and Theorem 3 that the mortality problem for 1-BPAM is co-NP-hard. □

6 Conclusion

In this work, we have demonstrated that the reachability and mortality problems for bounded 2-HPCD and 1-PAM are co-NP-hard. Our proofs are based on encoding the NP-complete simultaneous incongruences problem into the reachability and mortality problems for bounded 2-HPCD, and consequently, for bounded 1-PAM.

The interval-to-interval version of the reachability problem in our proofs can be formulated by considering an ε_0-interval around the initial point to represent the initial interval, and an ε_f-interval around the final point to represent the final interval for sufficiently small ε_0 and ε_f. Furthermore, we anticipate that interval-to-interval reachability and mortality are decidable for bounded 1-PAM and 2-HPCD, which can be demonstrated by extending the results found in [12,13].

This work addresses, in a broader context, the open question regarding the complexity of mortality in dimension two posed for restricted HPCD in [6]. It also leaves several immediate questions for future research, including providing an upper bound on the computational complexity of the reachability and mortality problems in the context of bounded 2-HPCD and 1-PAM.

References

1. Alur, R., Courcoubetis, C., Henzinger, T.A., Ho, P.-H.: Hybrid automata: an algorithmic approach to the specification and verification of hybrid systems. In: Grossman, R.L., Nerode, A., Ravn, A.P., Rischel, H. (eds.) HS 1991-1992. LNCS, vol. 736, pp. 209–229. Springer, Heidelberg (1993). https://doi.org/10.1007/3-540-57318-6_30

2. Asarin, E., Maler, O., Pnueli, A.: Reachability analysis of dynamical systems having piecewise-constant derivatives. Theor. Comput. Sci. **138**(1), 35–65 (1995)
3. Asarin, E., Mysore, V., Pnueli, A., Schneider, G.: Low dimensional hybrid systems - decidable, undecidable, don't know. Inf. Comput. **211**, 138–159 (2012)
4. Aswani, A., Tomlin, C.: Reachability algorithm for biological piecewise-affine hybrid systems. In: Bemporad, A., Bicchi, A., Buttazzo, G. (eds.) HSCC 2007. LNCS, vol. 4416, pp. 633–636. Springer, Heidelberg (2007). https://doi.org/10.1007/978-3-540-71493-4_49
5. Bazille, H., Bournez, O., Gomaa, W., Pouly, A.: On the complexity of bounded time reachability for piecewise affine systems. In: Ouaknine, J., Potapov, I., Worrell, J. (eds.) RP 2014. LNCS, vol. 8762, pp. 20–31. Springer, Cham (2014). https://doi.org/10.1007/978-3-319-11439-2_2
6. Bell, P.C., Chen, S., Jackson, L.: On the decidability and complexity of problems for restricted hierarchical hybrid systems. Theor. Comput. Sci. **652**, 47–63 (2016)
7. Garey, M.R., Johnson, D.S.: Computers and Intractability: A Guide to the Theory of NP-Completeness. W. H. Freeman and Co., New York (1979)
8. Gashler, M., Martinez, T.: Geometric properties of a class of piecewise affine biological network models. J. Math. Biol. **52**, 373–418 (2006)
9. Henzinger, T.A., Kopke, P.W., Puri, A., Varaiya, P.: What's decidable about hybrid automata? J. Comput. Syst. Sci. **57**(1), 94–124 (1998)
10. Lafferriere, G., Pappas, G.J., Yovine, S.: A new class of decidable hybrid systems. In: Vaandrager, F.W., van Schuppen, J.H. (eds.) HSCC 1999. LNCS, vol. 1569, pp. 137–151. Springer, Heidelberg (1999). https://doi.org/10.1007/3-540-48983-5_15
11. Maler, O., Pnueli, A.: Reachability analysis of planar multi-linear systems. In: Courcoubetis, C. (ed.) CAV 1993. LNCS, vol. 697, pp. 194–209. Springer, Heidelberg (1993). https://doi.org/10.1007/3-540-56922-7_17
12. de Oliveira Oliveira, M., Tveretina, O.: Mortality and edge-to-edge reachability are decidable on surfaces. In: 25th ACM International Conference on Hybrid Systems: Computation and Control (2022)
13. Sandler, A., Tveretina, O.: Deciding reachability for piecewise constant derivative systems on orientable manifolds. In: Filiot, E., Jungers, R., Potapov, I. (eds.) RP 2019. LNCS, vol. 11674, pp. 178–192. Springer, Cham (2019). https://doi.org/10.1007/978-3-030-30806-3_14
14. Stockmeyer, L.J., Meyer, A.R.: Word problems requiring exponential time: preliminary report. In: Proceedings of the 5th Annual ACM Symposium on Theory of Computing, pp. 1–9. ACM (1973)
15. Sun, H., Folschette, M., Magnin, M.: Reachability analysis of a class of hybrid gene regulatory networks. In: Bournez, O., Formenti, E., Potapov, I. (eds.) RP 2023. LNCS, vol. 14235, pp. 56–69. Springer, Cham (2023). https://doi.org/10.1007/978-3-031-45286-4_5
16. Sutner, K.: On the computational complexity of finite cellular automata. J. Comput. Syst. Sci. **50**(1), 87 (1995)

Semi-linear VASR for Over-Approximate Semi-linear Transition System Reachability

Nikhil Pimpalkhare[(✉)] and Zachary Kincaid

Princeton University, Princeton, NJ 08544, USA
{np6641,zkincaid}@cs.princeton.edu

Abstract. This paper introduces Semi-Linear Integer Vector Addition Systems with Resets (SVASR). A SVASR is a labeled transition system in which the states are finite-dimensional integer-valued vectors and which transitions from one state to another by applying an orthogonal projection followed by a translation drawn from a semi-linear set. We give a polynomial-time reduction of SVASR reachability to that of Integer Vector Addition Systems with Resets.

We then consider the use of SVASRs for over-approximating the reachability relation of transition systems in which the transition relation is a semi-linear set. We show that any semi-linear transition system has a "best" SVASR that simulates its behavior, called its SVASR-reflection. The dimension of the SVASR-reflection of a semi-linear transition system T with states is exponential in the number of states; however, we show that the over-approximate reachability induced by T's SVASR-reflection can be computed in polynomial time.

Keywords: Vector Addition Systems · Linear Integer Arithmetic

1 Introduction

Vector Addition Systems (VAS) are a widely-studied class of infinite-state transition systems. Classically, states of these systems are finite vectors over the naturals and transitions increment the state by a translation vector over the integers drawn from a finite set. Haase and Halfon initiated the study of *integer* VAS, in which states are integer-valued vectors, and showed that the reachability relation for integer VAS is definable in linear integer arithmetic even in the presence of states and resets [3]. In a separate line of work, Piskac and Kunčak showed that linear integer arithmetic is effectively closed under star—i.e., if $F(x)$ is a linear integer arithmetic formula, then the set $\{\sum_{i=1}^{n} m_i : n \in \mathbb{N}, \forall i.m_i \models F\}$ is LIA-definable [9].

In this paper, we study a common generalization of these two lines of work: integer vector addition systems with resets in which the set of translation vectors is infinite but LIA-definable, or equivalently a semi-linear set. We refer to such transition systems as Semi-Linear Integer VAS with Resets (SVASR). We show that that reachability of SVASR reduces to that of integer VAS with resets (VASR).

Next, we consider the application of SVASRs to over-approximate the reachability relation of semi-linear transition systems—transition systems for which the transition relation is a semi-linear set (or equivalently LIA-definable). The reachability problem for this class of transition system is undecidable, since it generalizes counter machines. Following the strategy of [8,10], we can over-approximate the reachability relation of a semi-linear transition system by (1) computing a SVASR that simulates it and (2) computing the inverse image of the reachability relation of this SVASR under the simulation. We show that every semi-linear transition system has a best abstraction as a SVASR called its *SVASR reflection*. It is best in the sense that the reflection's transition relation over-approximates the semi-linear transition system's at least as precisely as any other SVASR.

The dimension of the SVASR-reflection of a semi-linear transition system is exponential in the size of its alphabet, so a direct attempt to compute over-approximate reachability via the SVASR-reflection results in an exponential-space algorithm. We show that we can compute an equivalent formula in polynomial time by avoiding explicit computation of the SVASR reflection.

In summary, we introduce a new extension of vector addition systems, the SVASR, and show that its reachability reduces to that of VASR. We then propose a practical technique to computing over-approximate reachability of semi-linear transition systems using its SVASR reflection; the salient feature of this technique is that the computed over-approximation is guaranteed to be at least as precise as that induced by any other SVASR abstraction.

2 Background

A semi-linear set [7] in a $\mathbb{Z}$-module V is the finite union of linear sets in V. A linear set S in V is generated by a *base point* b in V and a sequence of *periods* $p_1, \ldots, p_n$ in V:

$$S \triangleq \{b + \lambda_1 p_1 + \cdots + \lambda_n p_n : \lambda_1, \ldots, \lambda_n \in \mathbb{N}\}$$

The generator representation of a semi-linear set is not unique. For convenience, we assume that every semi-linear set has a canonical generator which we refer to as a **basis**. A basis over vector space V is an element of $(V \times (V)^*)^*$. Each element of this representation is the generator representation of a linear set; that is, a base pointer followed by a sequence of periods. Let $B(S)$ refer to the canonical generator representation of semilinear set S and let $S(B)$ be the semilinear set defined by basis B.

A **labeled** transition system[1] T over a set of variables X and alphabet Σ is a pair $\langle \mathbb{Z}^X, \to_T \rangle$ where $\mathbb{Z}^X$ is a state space and $\to_T \subseteq \mathbb{Z}^X \times \Sigma \times \mathbb{Z}^X$ is a labeled transition relation. We use the following notation:

---
[1] We restrict our attention to transition systems in which the state space is a finite-dimensional module over the integers.

- For a character $s \in \Sigma$, write $\rho \xrightarrow{s}_T \rho'$ if $\langle \rho, s, \rho' \rangle$ belongs to $\rightarrow_T$
- For a word $s_1 \ldots s_n \in \Sigma^*$, write $\rho \xrightarrow{s_1 \ldots s_n}_T \rho'$ if there exists a sequence of states $\rho_0 \ldots \rho_n$ such that $\rho = \rho_0 \xrightarrow{s_1}_T \ldots \xrightarrow{s_n}_T \rho_n = \rho'$
- For a language $L \subseteq \Sigma^*$, write $\rho \xrightarrow{L}_T \rho'$ if $\rho \xrightarrow{w}_T \rho'$ for some word $w \in L$

Various control features can be encoded using reachability constrained to a given language of paths. For instance, the reachability relation of a vector addition system with states (or with a pushdown stack) corresponds to the reachability relation of a vector addition system constrained to a regular language (or context-free language).

A linear simulation between labeled transition systems $T = \langle \mathbb{Z}^X, \rightarrow_T \rangle$ and $U = \langle \mathbb{Z}^Y, \rightarrow_U \rangle$ over the same alphabet Σ is a linear function $f : \mathbb{Z}^X \rightarrow \mathbb{Z}^Y$ such that for any $\rho, \rho' \in \mathbb{Z}^X$, if $\rho \xrightarrow{s}_T \rho'$ then $f(\rho) \xrightarrow{s}_U f(\rho')$. For any language L, if we can compute the L-reachability relation $\xrightarrow{L}_U$ of U, then we can over-approximate the L-reachability relation of T as $\left\{ \langle \rho, \rho' \rangle : f(\rho) \xrightarrow{L}_U f(\rho') \right\}$ (which must contain $\xrightarrow{L}_T$).

A labeled integer vector addition system with resets (**VASR**) over variables X and alphabet Σ is a labeled transition system $\mathcal{V} = \langle \mathbb{Z}^X, \rightarrow_\mathcal{V} \rangle$ such that for each character $s \in \Sigma$ there is an offset vector $o_s \in \mathbb{Z}^X$ and a reset vector $r_s \in \{0,1\}^X$ such that $\rho \xrightarrow{s}_\mathcal{V} \rho'$ if and only if $\bigwedge_{x \in X} \rho'(x) = r_s(x)\rho(x) + o_s(x)$. Let $RV(\mathcal{V}, s)$ and $OV(\mathcal{V}, s)$ denote r_s and o_s respectively. Note that $\rightarrow_\mathcal{V}$ is uniquely determined by $RV(\mathcal{V}, s)$ and $OV(\mathcal{V}, s)$ for all s.

A labeled semi-linear integer vector addition system with resets (**SVASR**) over variables X and alphabet Σ is a labeled transition system $\mathcal{SV} = \langle \mathbb{Z}^X, \rightarrow_{\mathcal{SV}} \rangle$. For each symbol $s \in \Sigma$ there is an offset semi-linear set $S_s \subseteq \mathbb{Z}^X$ and a reset vector $r_s \in \{0,1\}^X$ such that $\rho \xrightarrow{s}_{\mathcal{SV}} \rho'$ if and only if $\bigwedge_{x \in X} \rho'(x) = r_s(x)\rho(x) + v(x)$ for some $v \in S_s$. Let $RV(\mathcal{SV}, s)$ and $OS(\mathcal{SV}, s)$ denote r_s and S_s respectively. Note that $\rightarrow_{\mathcal{SV}}$ is uniquely determined by $RV(\mathcal{SV}, s)$ and $OS(\mathcal{SV}, s)$ for all s.

A semi-linear transition system over variables X is a labeled transition system $T = \langle \mathbb{Z}^X, \rightarrow_T \rangle$ such that $\xrightarrow{s}_T \subseteq \mathbb{Z}^X \times \mathbb{Z}^X$ is a semi-linear set. This class can also be thought of as the set of transition systems for which transitions are definable in LIA. Since counter machines are semi-linear transition systems, the reachability problem of semi-linear transition systems is undecidable.

A transition formula F over variables X is a linear integer arithmetic formula over free variables X and primed copies X'. For two states $\rho, \rho' \in \mathbb{Z}^X$, we write $[\rho, \rho'] \models F$ if F holds when all $x \in X$ are replaced with $\rho(x)$ and all $x' \in X'$ are replaced with $\rho'(x)$. $TF(X)$ denotes the set of all transition formulas over X.

3 SVASR Reachability Relations in Polynomial Time

The reachability problem for VASRs has been widely studied in the literature. Specifically, it has been shown that the reachability relation $\xrightarrow{L}_\mathcal{V}$ is LIA-definable when $L \subseteq \Sigma^*$ is a regular language [3], a communication-free Petri net language

[2], and a context-free language [8]. Given a VASR $\mathcal{V}$ over variables X and alphabet Σ and a language L in the above classes, these works compute a transition formula $F \in \mathit{TF}(X)$ such that $[\rho, \rho'] \models F$ if and only if $\rho \xrightarrow{L}_{\mathcal{V}} \rho'$ in polynomial time. Thus, these algorithms amount to polynomial-time reductions from VASR reachability to satisfiability of existential LIA formulas.

We show that regular reachability of SVASR can be reduced to regular reachability of VASR. Our reduction creates VASR transitions representing the generator representations of the semi-linear sets of the SVASR transitions and encodes the structure of these sets as a regular language over these transitions. A SVASR transition resets some part of the state then adds a vector from a semi-linear set; our key insight is that this is equivalent to a VASR transition applying the same reset and adding one of the base points of the semi-linear set followed by an arbitrary number of VASR transitions adding one of the associated periods.

Consider a SVASR $\mathcal{SV} = \langle \mathbb{Z}^X, \rightarrow_{\mathcal{SV}} \rangle$ over alphabet Σ and a language $L \subseteq \Sigma^*$. Define a new alphabet $\Sigma_{\mathcal{SV}} \subseteq \{0,1\}^X \times \mathbb{Z}^X$ to be the least set such that:

- For all $s \in \Sigma$, for all $\langle b; P \rangle \in B(OS(\mathcal{SV}, s))$, we have $\langle RV(\mathcal{SV}, s), b \rangle \in \Sigma_{\mathcal{SV}}$
- For all $s \in \Sigma$, for all $\langle b; p_1 \ldots p_n \rangle \in B(OS(\mathcal{SV}, s))$, we have $\langle \lambda x.1, p_i \rangle \in \Sigma_{\mathcal{SV}}$ for all $i \in [1, n]$

Define a VASR $\mathcal{V}(\mathcal{SV}) \triangleq \langle \mathbb{Z}^X, \rightarrow_{\mathcal{V}(\mathcal{SV})} \rangle$ over variables X alphabet $\Sigma_{\mathcal{SV}}$ where

$$RV(\mathcal{V}(\mathcal{SV}), \langle r, v \rangle) = r \qquad OV(\mathcal{V}(\mathcal{SV}), \langle r, v \rangle) = v$$

Finally, for each $s \in S$, define the following regular language $R_s \subseteq \Sigma_{\mathcal{SV}}^*$:

$$R_s \triangleq \bigcup_{\langle b; p_1, \ldots, p_n \rangle \in B(OS(\mathcal{SV}, s))} \langle RV(\mathcal{SV}, s), b \rangle \langle \lambda x.1, p_1 \rangle^* \ldots \langle \lambda x.1, p_n \rangle^*$$

Lemma 1. *Consider an SVASR $\mathcal{SV}$ over alphabet Σ. For all $s \in \Sigma$, we have:*

$$\left(\rho \xrightarrow{s}_{\mathcal{SV}} \rho' \right) \iff \left(\rho \xrightarrow{R_s}_{\mathcal{V}(\mathcal{SV})} \rho' \right)$$

Proof. ($\Longrightarrow$) Let X denote the variables of $\mathcal{SV}$. Consider any states ρ, ρ' such that $\rho \xrightarrow{s}_{\mathcal{SV}} \rho'$. We have that $\bigwedge_{x \in X} \rho'(x) = RV(\mathcal{SV}, s)(x)\rho(x) + v(x)$ for some $v \in OS(\mathcal{SV}, s)$. Then, there must be some $\langle b; p_1, \ldots, p_n \rangle \in B(OS(\mathcal{SV}, s))$ such that $v = b + \sum_{i=1}^{n} \lambda_i p_i$ for some $\lambda_1, \ldots, \lambda_n \in \mathbb{N}$. Then, observe that the word

$$w \triangleq \langle RV(\mathcal{SV}, s), b \rangle \langle \lambda x.1, p_1 \rangle^{\lambda_1} \ldots \langle \lambda x.1, p_n \rangle^{\lambda_n}$$

belongs to R_s and that $\rho \xrightarrow{w}_{\mathcal{V}(\mathcal{SV})} \rho'$, and so $\rho \xrightarrow{R_s}_{\mathcal{V}(\mathcal{SV})} \rho'$.

($\Longleftarrow$) Consider any states ρ, ρ' such that $\rho \xrightarrow{R_s}_{\mathcal{V}(\mathcal{SV})} \rho'$. Then there must be some $w \in R_s$ such that $\rho \xrightarrow{w}_{\mathcal{V}(\mathcal{SV})} \rho'$. By the definition of R_s, for some $\langle b; p_1, \ldots, p_n \rangle \in B(OS(\mathcal{SV}, s))$ we have that

$$w = \langle RV(\mathcal{SV}, s), b \rangle \langle \lambda x.1, p_1 \rangle^{\lambda_1} \ldots \langle \lambda x.1, p_n \rangle^{\lambda_n}$$

for some $\lambda_1, \ldots, \lambda_n \in \mathbb{N}$. By the definition of $\mathcal{V}(\mathcal{SV})$, this implies that $\bigwedge_{x \in X} \rho'(x) = RV(\mathcal{SV}, s)(x)\rho(x) + (b + \sum_{p \in P} \lambda_p p)$. Then, since $b + \sum_{i=1}^{n} \lambda_i p_i \in OS(\mathcal{SV}, s)$, we have that $\rho \xrightarrow{s}_{\mathcal{SV}} \rho'$.

Then, let $R(L)$ be the language replacing all characters in Σ with their corresponding regular languages: $R(L) \triangleq \{w_0 \ldots w_n : \exists s_0 \ldots s_n \in L, w_i \in R_{s_i}\}$.

Theorem 1. *For language $L \subseteq \Sigma^*$ and semi-linear VASR $\mathcal{SV}$ we have*

$$\left(\rho \xrightarrow{L}_{\mathcal{SV}} \rho'\right) \text{ if and only if } \left(\rho \xrightarrow{R(L)}_{\mathcal{V}(\mathcal{SV})} \rho'\right)$$

Proof. Follows from Lemma 1.

Therefore, L-reachability of SVASR reduces to $R(L)$-reachability of VASR. The extended language $R(L)$ is linearly larger than L with respect to the size of the bases of the semi-linear sets of the SVASR. If L is regular (resp. context-free) (resp. communication-free petri-net language), then $R(L)$ is regular (resp. context-free) (resp. communication-free petri-net language).

4 Best SVASR Abstractions of Semi-linear Transition Systems

We now shift focus to computing over-approximate L-reachability for semi-linear transition systems. This problem has a wide variety of applications, including proving safety properties of computer programs. The following definitions formalize our approach to computing the best SVASR abstraction of a semi-linear transition system for computing over-approximate reachability.

A SVASR-abstraction of labeled transition system T is a pair $\langle f, \mathcal{SV} \rangle$ composed of SVASR $\mathcal{SV}$ and linear simulation f from T to $\mathcal{SV}$. We say that a SVASR-abstraction $\langle f, \mathcal{SV} \rangle$ is a *SVASR-reflection* of T if for any other SVASR-abstraction $\langle f', \mathcal{SV}' \rangle$ of T there is a linear simulation f^* from $\mathcal{SV}$ to $\mathcal{SV}'$ such that $f^* \circ f = f'$.

L-reachability of a SVASR abstraction of T can be used to over-approximate L-reachability of T. A SVASR-reflection $\langle f, \mathcal{SV} \rangle$ is best because its implied over-approximate reachability is at least as precise as any other SVASR abstraction $\langle f', \mathcal{SV}' \rangle$, as $\left\{\langle \rho, \rho' \rangle : f(\rho) \xrightarrow{L}_{\mathcal{SV}} f(\rho')\right\} \subseteq \left\{\langle \rho, \rho' \rangle : f^*(f(\rho)) \xrightarrow{L}_{\mathcal{SV}'} f^*(f(\rho'))\right\}$.

This section shows how to compute the SVASR reflection $\langle f, \mathcal{SV} \rangle$ of a semi-linear transition system. We then show how it allows use to define over-approximate L-reachability of semi-linear transition systems in LIA. However, this formula is exponentially sized with respect to the semi-linear transition system, making it impractical for direct usage; we show how to compute a smaller equivalent formula in Sect. 5.

4.1 SVASR-Reflections of Semi-linear Transition Systems

The SVASR reflection $\langle f, \mathcal{SV} \rangle$ of a semi-linear transition system T over variables X and alphabet Σ can be computed as follows. The state space of $\mathcal{SV}$ is $\mathbb{Z}^{X \times \{0,1\}^\Sigma}$. Intuitively, for each variable x and character $s \in \Sigma$, we must make a choice of whether to treat s as a reset of x. Since the choice is arbitrary, we introduce $2^{|\Sigma|}$ copies of each variable x to encode all possible choices. Thus, a "variable" of the SVASR reflection is a pair $\langle x, C \rangle \in X \times \{0,1\}^\Sigma$ where x is variable of T, and $C(s)$ indicates whether s is to be treated as a increment ($C(s) = 1$) or reset ($C(s) = 0$). The simulation $f : \mathbb{Z}^X \to \mathbb{Z}^{X \times \{0,1\}^\Sigma}$ is defined as:

$$f(\rho) = \lambda \langle x, C \rangle . \rho(x)$$

The relation $\to_{\mathcal{SV}}$ is defined in terms of $RV(\mathcal{SV}, s)$ and $OS(\mathcal{SV}, s)$ as:

$$RV(\mathcal{SV}, s) = \lambda \langle x, C \rangle . C(s)$$
$$OS(\mathcal{SV}, s) = \left\{ \lambda \langle x, C \rangle. \text{ if } C(s) \text{ then } \rho'(x) - \rho(x) \text{ else } \rho'(x) : \rho \xrightarrow{s}_T \rho' \right\}$$

A basis for $OS(\mathcal{SV}, s)$ can be computed from a basis for $\xrightarrow{s}_T$ as follows. For any $t = \langle \rho, \rho' \rangle \in \mathbb{Z}^X \times \mathbb{Z}^X$, define

$$\hat{t} \triangleq \lambda \langle x, C \rangle. \text{ if } C(s) \text{ then } \rho'(x) - \rho(x) \text{ else } \rho'(x) \ .$$

Then we define

$$B(OS(\mathcal{SV}, s)) \triangleq \left\{ \langle \hat{b}; \hat{p}_1, \ldots, \hat{p}_n \rangle : \langle b; p_1, \ldots, p_n \rangle \in B(\xrightarrow{s}_T) \right\} \ .$$

Lemma 2. *The pair $\langle f, \mathcal{SV} \rangle$ is an abstraction of T.*

Proof. Consider any states ρ, ρ' such that $\rho \xrightarrow{s}_T \rho'$. By the definition of $OS(\mathcal{SV}, s)$, we have that $v = (\lambda \langle x, C \rangle. \text{ if } C(s) \text{ then } \rho'(x) - \rho(x) \text{ else } \rho'(x)) \in OS(\mathcal{SV}, s)$. We can conclude that $f(\rho) \xrightarrow{s}_{\mathcal{SV}} f(\rho')$, since:

$$\bigwedge_{\langle x, C \rangle \in X \times \{0,1\}^\Sigma} f(\rho')(\langle x, C \rangle) = RV(\mathcal{SV}, s)(x, C) f(\rho)(x, C) + v(\langle x, C \rangle)$$

$$\iff \bigwedge_{\langle x, C \rangle \in X \times \{0,1\}^\Sigma, C(s)=1} \rho'(x) = 1\rho(x) + (\rho'(x) - \rho(x))$$

$$\land \bigwedge_{\langle x, C \rangle \in X \times \{0,1\}^\Sigma, C(s)=0} \rho'(x) = 0\rho(x) + (\rho'(x))$$

Theorem 2. *The SVASR-abstraction $\langle f, \mathcal{SV} \rangle$ is a SVASR reflection of T.*

Proof. Consider another SVASR abstraction $\langle f', \mathcal{SV}' \rangle$ over variables Y. We will show that the following function $f^* : \mathbb{Z}^{X \times \{0,1\}^\Sigma} \to \mathbb{Z}^Y$ is a simulation from $\mathcal{SV}$ to $\mathcal{SV}'$.

$$f^*(\sigma) = \lambda y. f'(\lambda x. \sigma(\langle x, C_y \rangle))(y) \text{ with } C_y = \lambda s. RV(\mathcal{SV}', s)(y)$$

A piece of intuition for this definition is that $\langle x, C_y \rangle$ is the variable of $\mathcal{SV}$ that abstracts the variable x of T and that experiences the same resets per-character as y in $\mathcal{SV}'$.

First, observe that $f^* \circ f = f'$:

$$f^*(f(\rho)) = \lambda y.f'(\lambda x.f(\rho)(x, C_y))(y)$$
$$= \lambda y.f'(\lambda x.\rho(x))(y) = f'(\rho)$$

To show f^* is a simulation from $\mathcal{SV}$ to $\mathcal{SV}'$, consider states $\sigma, \sigma' \in \mathbb{Z}^{X \times \{0,1\}^\Sigma}$ such that $\sigma \xrightarrow{s}_{\mathcal{SV}} \sigma'$. By the definition of $OS(\mathcal{SV}, s)$, there must be ρ, ρ' such that $\rho \xrightarrow{s}_T \rho'$ and:

$$\bigwedge_{\langle x,C \rangle \in X \times \{0,1\}^\Sigma, C(s)=1} \sigma'(\langle x, C \rangle) = 1\sigma(\langle x, C \rangle) + (\rho'(x) - \rho(x))$$

$$\wedge \bigwedge_{\langle x,C \rangle \in X \times \{0,1\}^\Sigma, C(s)=0} \sigma'(\langle x, C \rangle) = 0\sigma(\langle x, C \rangle) + (\rho'(x))$$

Thus, for all $\langle x, C \rangle \in X \times \{0,1\}^\Sigma$, if $C(s) = 0$ then $\sigma'(\langle x, C \rangle) = \rho'(x)$ and if $C(s) = 1$ then $\sigma(\langle x, C \rangle) - \sigma'(\langle x, C \rangle) = \rho'(x) - \rho(x)$.

Since $\rho \xrightarrow{s}_T \rho'$, we have that $f'(\rho) \xrightarrow{s}_{\mathcal{SV}'} f'(\rho')$. Then, there is some $v \in OS(\mathcal{SV}', s)$ such that:

$$\bigwedge_{y \in Y} f'(\rho')(y) = RV(\mathcal{SV}', s)(y) f'(\rho)(y) + v(y) \qquad (1)$$

Note that $RV(\mathcal{SV}', s)(y) = C_y(s)$ for all s. Then, for all variables $y \in Y$ of SVASR $\mathcal{SV}'$, if $RV(\mathcal{SV}', s)(y) = 0$ then $\sigma'(x, C_y) = \rho'(x)$ by previous reasoning. In such cases:

$$f^*(\sigma')(y) = f'(\lambda x.\sigma'(x, C_y))(y) = f'(\lambda x.\rho'(x))(y) = f'(\rho')(y)$$

Substituting $f'(\rho')(y)$ with $f^*(\sigma')(y)$ in Eq. 1 and using $RV(\mathcal{SV}', s)(y) = 0$, we have that $f^*(\sigma')(y) = RV(\mathcal{SV}', s)(y) f^*(\sigma)(y) + v(y)$.

In the other case, if $RV(\mathcal{SV}, s)(y) = 1$ then $C_y(s) = 1$ and so $\sigma'(x, C_y) - \sigma(x, C_y) = \rho'(x) - \rho(x)$. Then, using the linearity of f', we have:

$$f^*(\sigma')(y) - f^*(\sigma)(y) = f'(\lambda x.\sigma'(x, C_y))(y) - f'(\lambda x.\sigma(x, C_y))(y)$$
$$= f'(\lambda x.\sigma'(x, C_y) - \sigma(x, C_y))(y)$$
$$= f'(\lambda x.\rho'(x) - \rho(x))(y) = f'(\rho')(y) - f(\rho)(y)$$

Then, substituting $f'(\rho')(y) - f'(\rho)(y)$ with $f^*(\sigma')(y) - f^*(\sigma)(y)$ in (1), we have that $f^*(\sigma')(y) = RV(\mathcal{SV}', s)(y) f^*(\sigma)(y) + v(y)$.

Therefore, by cases, we have that:

$$\bigwedge_{y \in Y} f^*(\sigma')(y) = RV(\mathcal{SV}', s)(y) f^*(\sigma')(y) + v(y)$$

We can conclude that $f^*(\sigma) \xrightarrow{s}_{\mathcal{SV}'} f^*(\sigma')$, that f^* is a simulation from $\mathcal{SV}$ to $\mathcal{SV}'$, and that $\langle f, \mathcal{SV} \rangle$ is a reflection of T.

4.2 Over-Approximate Semi-linear Transition System Reachability

Using the contents of Sect. 3 and Subsect. 4.1, we have a procedure to compute over-approximate L-reachability of semi-linear transition systems. Given a semi-linear transition system T over variables X, we first compute its SVASR reflection $\langle f, \mathcal{SV} \rangle$ over variables $X \times \{0,1\}^\Sigma$. We then compute an over-approximation of the L-reachability relation of T via Sect. 3 and the following lemma:

Lemma 3. *Consider a language $L \subseteq \Sigma^*$, a semi-linear transition system T and its SVASR reflection $\langle f, \mathcal{SV} \rangle$ as defined in Sect. 4. Let $F \in TF(X \times \{0,1\}^\Sigma)$ be a formula such that $[\sigma, \sigma'] \models F$ if and only if $\sigma \xrightarrow{L}_{\mathcal{SV}} \sigma'$. Define $G \triangleq F[\langle x, C \rangle \mapsto x]$. Then we have*

$$[\rho, \rho'] \models G \text{ if and only if } f(\rho) \xrightarrow{L}_{\mathcal{SV}} f(\rho')$$

Proof. ($\implies$) If $[\rho, \rho'] \models G$ then by the definition of f we have that $[f(\rho), f(\rho')] \models F$ and so $f(\rho) \xrightarrow{L}_{\mathcal{SV}} f(\rho')$. ($\impliedby$) If $f(\rho) \xrightarrow{L}_{\mathcal{SV}} f(\rho')$ then $[f(\rho), f(\rho')] \models F$ and so by the definition of f we have that $[\rho, \rho'] \models G$.

The formula G is an over-approximation of the L-reachability relation of T, as if $\rho \xrightarrow{L}_T \rho'$ then $f(\rho) \xrightarrow{L}_{\mathcal{SV}} f(\rho')$ and so $[\rho, \rho'] \models G$, so it can be used to prove safety properties about T. However, it is too large to be practically useful. The SVASR reflection of a semi-linear transition system has an exponentially larger state space and defining SVASR reachability in LIA takes polynomial space, so the size of G is exponential with respect to the semi-linear transition system.

5 Over-Approximate Semi-linear Transition System Reachability in Polynomial Time

As in Subsect. 4.2, given a semi-linear transition system T, we can compute a formula G such that $[\rho, \rho'] \models G$ if and only if $f(\rho) \xrightarrow{L}_{\mathcal{SV}} f(\rho')$ where $\langle f, \mathcal{SV} \rangle$ is the SVASR-reflection of T. This formula is an over-approximation of the L-reachability relation of T, but requires exponential space w.r.t. T because the dimension of $\mathcal{SV}$ is exponential in the size of the alphabet. We show here that we can compute a formula equivalent to G in polynomial time w.r.t. T.

The key is to never explicitly compute the SVASR-reflection $\langle f, \mathcal{SV} \rangle$. Given a semi-linear transition system T, let $M_T : \Sigma \to (\mathbb{Z}^{X \times \{0,1\}} \times (\mathbb{Z}^{X \times \{0,1\}})^*)^*$ be the function mapping each $s \in \Sigma$ to a basis of the following semilinear set:

$$\left\{ \lambda \langle x, r \rangle.\ \text{if } r = 1 \text{ then } \rho'(x) - \rho(x) \text{ else } \rho'(x) : \rho \xrightarrow{s}_T \rho' \right\}$$

A basis of this set can be computed from a basis of $\xrightarrow{s}_T$ in polynomial time, as in Sect. 4.1. Given a semi-linear transition system T over variables X, this subsection computes $G' \in TF(X)$ such that $[\rho, \rho'] \models G'$ if and only if $f(\rho) \xrightarrow{L}_{\mathcal{SV}} f(\rho')$ in polynomial time w.r.t. M_T.

Haase and Halfon [3] recognized that the L-reachability relation of a VASR can be computed via a counting abstraction of L; for each word in L, one must compute the final time that each dimension of the state is reset from left to right and compute the character counts of the subwords in between these final resets. This information is sufficient to compute the composition of VASR transitions along the word because the final reset nullifies the effects of all characters before it, and the effects of all characters after it commute with respect to the counter because they do not reset it; then, their net effect is computable from the character count after the final reset. We consider *abstract trajectories*, a counting abstraction which we will use to identify final resets.

Definition 1. *An d-marked **abstract trajectory** is a function $n : (\Sigma \times [1, 2d+1]) \to \mathbb{N}$ such that for all even i we have that $\sum_{s \in \Sigma} n(s, i) \leq 1$.*

For a trajectory $w \in \Sigma^*$ and an abstract trajectory n, write $w \Vdash n$ if and only if there exists a decomposition $w = w_1 w_2 \ldots w_{2d+1}$ such that for all $i \in [1, 2d+1]$ and $s \in \Sigma$, the value of $n(s, i)$ is the number of times character s appears in subword w_i. The definition ensures that n uniquely identifies w_i for all even i. In this sense, abstract trajectories are a counting abstraction which identifies up to d characters in order from a word and captures the character counts of the subwords in between. We restrict our attention to languages for which the set of abstract trajectories is LIA-definable, and use additional constraints to identify the abstract trajectories that mark the final reset of each dimension.

Formally, we restrict our attention to languages $L \subseteq \Sigma^*$ for which we can compute a formula $AT(L, |\Sigma|)$ over free variables $c_{s,i}$ for all $s \in \Sigma$ and all $i \in \{1 \ldots 2|\Sigma| + 1\}$; for any $|\Sigma|$-marked abstract trajectory n, $AT(L, |\Sigma|)$ holds when each $c_{s,i}$ is replaced with $n(s, i)$ if and only if there exists some $w \in L$ such that $w \Vdash n$. Pimpalkhare and Kincaid [8] gave an explicit definition for $AT(L, |\Sigma|)$ in the case that L is context-free; one can adapt techniques from the literature to compute $AT(L, |\Sigma|)$ in the cases that L is regular [3] or a communication-free Petri-net language [2].

Our approach to computing G' is to compute a formula representing the abstract-trajectories of L, constrain the free variables to ensure the final reset of each dimension of the SVASR reflection occurs at an even index, and to encode the resulting SVASR transition for all variables $\langle x, C \rangle$ of the SVASR reflection on the corresponding variable x of T. Directly considering every variable of the SVASR produces an exponentially sized formula. However, the final reset of every SVASR variable will occur at the final occurrence of some character, and at least one SVASR variable for each variable x of the semi-linear transition system will have its final reset at the final occurrence of every character. Our approach is then to mark the final occurrence of each symbol and to conjoin a formula per final occurrence representing the transition of all variables which experience their final reset there.

Leveraging techniques from the literature, we first compute $AT(L, |\Sigma|)$, a formula defining the $|\Sigma|$-marked abstract trajectories of L. We then define a formula $WF(\Sigma)$ ensuring that our symbolic $|\Sigma|$-marked abstract trajectory marks the final occurrence of every $s \in \Sigma$:

$$WF(\Sigma) = \bigwedge_{s \in \Sigma} \left(\bigwedge_{i=1}^{2|\Sigma|+1} \left(c_{s,i} > 0 \implies \bigvee_{\text{even } k \geq i} c_{s,k} > 0 \right) \wedge \left(\sum_{j=1}^{|\Sigma|} c_{s,2j} \leq 1 \right) \right)$$

Then, we define the formula $Transition(M_T, \Sigma)$ which describes the transition corresponding to the symbolic abstract trajectory value. This formula uses the following sets of variables to symbolically pick the SVASR translation vector corresponding to each occurrence of $c_{s,i}$:

$$D \triangleq \{d_{s,b,i} : s \in \Sigma, \langle b, P \rangle \in M_T(s), i \in [1, 2|\Sigma|+1]\}$$
$$E \triangleq \{e_{s,b,p,i} : s \in \Sigma, \langle b, P \rangle \in M_T(s), p \in P, i \in [1, 2|\Sigma|+1]\}$$

The following formula $Corr(\Sigma, M_T)$ corresponds variables $c_{s,i}$ to the relevant variables $d_{s,b,i}$ and $e_{s,b,p,i}$. The variable $c_{s,i}$ captures how many times s appears in subword i - for each such appearance, we must pick a single base vector and any number of periods.

$$Corr(\Sigma, M_T) \triangleq \bigwedge_{s \in \Sigma} \bigwedge_{i=1}^{2|\Sigma|+1} \left(\sum_{\langle b, P \rangle \in M_T(s)} d_{s,b,i} = c_{s,i} \right) \wedge$$
$$\bigwedge_{\langle b, P \rangle \in M_T(s)} \bigwedge_{p \in P} (e_{s,b,p,i} > 0 \implies d_{s,b,i} > 0)$$

We define $ResetAt(i, x, M_T, \Sigma)$ to compute the value that x would be reset to by the ith even subword and $AddsAfter(i, x, M_T, \Sigma)$ to compute the value of the increments to x after the ith subword.

$$ResetAt(i, x, M_T, \Sigma) = \sum_{s \in \Sigma} \sum_{\langle b, P \rangle \in M_T(s)} \left(d_{s,b,i} b(\langle x, 0 \rangle) + \sum_{p \in P} e_{s,b,p,i} p(\langle x, 0 \rangle) \right)$$

$$AddsAfter(i, x, M_T, \Sigma) = \sum_{j=i+1}^{2|\Sigma|+1} \sum_{s \in \Sigma} \sum_{\langle b, P \rangle \in M_T(s)} \left(\begin{array}{c} d_{s,b,j} b(\langle x, 1 \rangle) + \\ \sum_{p \in P} e_{s,b,p,j} p(\langle x, 1 \rangle) \end{array} \right)$$

And subsequently define:

$$Transition(X, M_T, \Sigma) \triangleq \bigwedge_{x \in X} \left(\begin{array}{c} x' = x + AddsAfter(0, x, M_T, \Sigma) \wedge \\ \bigwedge_{k=1}^{|\Sigma|} \left(\begin{array}{c} \sum_{s \in \Sigma} c_{s,2k} > 0 \implies \\ x' = ResetAt(2k, x, M_T, \Sigma) + \\ AddsAfter(2k, x, M_T, \Sigma) \end{array} \right) \end{array} \right)$$

Finally, we conjoin these formulae to produce our procedure summary.

$$G'(X, M_T, L, \Sigma) = \begin{array}{c} \exists \{c_{s,i} \geq 0 : s \in \Sigma, i \in [1, 2|\Sigma|+1]\} \\ \exists \{d_{s,b,i} \geq 0 : d_{s,b,i} \in D\} \exists \{e_{s,b,p,i} \geq 0 : e_{s,b,p,i} \in E\} \\ \left(\begin{array}{c} Transition(X, M_T, \Sigma) \wedge AT(L, |\Sigma|) \\ \wedge WF(\Sigma) \wedge Corr(\Sigma, M_T) \end{array} \right) \end{array}$$

Theorem 3. *Consider a semi-linear transition system T and a language $L \subseteq \Sigma^*$. Let $\langle f, \mathcal{SV} \rangle$ be the SVASR-reflection of T defined in Sect. 4. Then,*

$$[\rho, \rho'] \models G'(X, M_T, L, \Sigma) \iff f(\rho) \xrightarrow{L}_{\mathcal{SV}} f(\rho')$$

Proof. Firstly, note that there is a one-to-one correspondence between the elements of $S(M_T(s))$ and $OS(\mathcal{SV}, s)$, as both are defined by ρ, ρ' such that $\rho \xrightarrow{s}_T \rho'$. For each $s \in \Sigma$, define $\psi_s : \mathbb{Z}^{X \times \{0,1\}} \to \mathbb{Z}^{X \times \{0,1\}^\Sigma}$ by $\psi_s(v)(x, C) \triangleq v(x, C(s))$ to translate between $S(M_T(s))$ and $OS(\mathcal{SV}, s)$. Let ψ_s^{-1} be a left inverse.

($\implies$) Consider ρ, ρ' such that $[\rho, \rho'] \models G'(X, M_T, L, \Sigma)$. By its definition, there exists a valuation $A : \{c_{s,i} : s \in \Sigma, i \in [1, 2|\Sigma|+1]\} \cup D \cup E \to \mathbb{N}$ such that $(Transition(X, M_T, \Sigma) \land AT(L, |\Sigma|) \land WF(\Sigma) \land Corr(\Sigma, M_T))$ holds when each $c_{s,i}$ is replaced with $A(c_{s,i})$, each $d_{s,b,i}$ is replaced with $A(d_{s,b,i})$, and each $e_{s,b,p,i}$ is replaced with $A(e_{s,b,p,i})$.

Let $n : (\Sigma \times [1, 2|\Sigma|+1]) \to \mathbb{N}$ be the function mapping each $\langle s, i \rangle$ to $A(c_{s,i})$. Since this valuation satisfies $AT(L, |\Sigma|)$, we have that n is a $|\Sigma|$-marked abstract trajectory over Σ such that there exists a word $w = s_1 \ldots s_{|w|} \in L$ such that $w \Vdash n$. We will show that $f(\rho) \xrightarrow{w}_{\mathcal{SV}} f(\rho')$, therefore showing $f(\rho) \xrightarrow{L}_{\mathcal{SV}} f(\rho')$.

Consider any $\langle x, C \rangle \in X \times \{0, 1\}^\Sigma$ such that $RV(\mathcal{SV}, s_i)(\langle x, C \rangle) = 1$ for all s_i in w. Observe that the first conjunct of $Transition(X, M_T, \Sigma)$ ensures that $\rho'(x) = \rho(x) + \sum_{i=1}^{|w|} v_i(\langle x, 1 \rangle)$ where each $v_i \in S(M_T(s_i))$. Since it is the case that $v_i(x, 1) = \psi_{s_i}(v_i)(\langle x, C \rangle)$ for all i in $[\|w\|]$ since $C(s_i) = 1$, we have that $f(\rho')(\langle x, C \rangle) = f(\rho)(\langle x, C \rangle) + \sum_{i=1}^{|w|} \psi_{s_i}(v_i)(\langle x, C \rangle)$.

Consider any $\langle x, C \rangle \in X \times \{0, 1\}^\Sigma$ such that $RV(\mathcal{SV}, s_i)(\langle x, C \rangle) = 0$ for some s_i in w. Let j be the highest index such that $RV(\mathcal{SV}, s_j)(\langle x, C \rangle) = 0$. $WF(\Sigma)$ ensures that $A(c_{s_j, 2k}) = 1$ for some k. The corresponding conjunct of $Transition(X, M_T, \Sigma)$ ensures that $\rho'(x) = v_j(\langle x, 0 \rangle) + \sum_{i=j+1}^{|w|} v_i(\langle x, 1 \rangle)$ where each $v_i \in S(M_T(s_i))$. Since it is the case that $v_j(\langle x, 0 \rangle) = \psi_{s_j}(v_j)(\langle x, C \rangle)$ since $C(s_j) = 0$ and $v_i(\langle x, 1 \rangle) = \psi_{s_i}(v_i)(\langle x, C \rangle)$ for all $i \in [j+1, |w|]$ since $C(s_i) = 1$, we have that $f(\rho')(\langle x, C \rangle) = \psi_{s_j}(v_j)(\langle x, C \rangle) + \sum_{i=j+1}^{|w|} \psi_{s_i}(v_i)(\langle x, C \rangle)$.

Observe that for all $i \in [1, |w|]$, we have $\psi_{s_i}(v_i) \in OS(\mathcal{SV}, s_i)$. Then, by the above casework over all $\langle x, C \rangle$, we have that $f(\rho) \xrightarrow{L}_{\mathcal{SV}} f(\rho')$.

($\impliedby$) Consider ρ, ρ' such that $f(\rho) \xrightarrow{L}_{\mathcal{SV}} f(\rho')$. There exists $w = s_1 \ldots s_n \in L$ such that $f(\rho) \xrightarrow{w}_{\mathcal{SV}} f(\rho')$. Let d be the number of unique characters in w and let $i_1 \ldots i_d$ be the indexes of the final occurrence of each letter; that is, character s_{i_j} does not appear in subword $s_{i_j+1} \ldots s_n$. For all $j \in [1, d]$, let word w_{2j} be the character s_{i_j} and let word w_{2j-1} be the subword $s_{i_{j-1}+1} \ldots s_{i_j-1}$; let w_{2d+1} through $w_{2|\Sigma|+1}$ be empty. Let $n : (\Sigma \times [1, 2|\Sigma|+1]) \to \mathbb{N}$ be the function such that $n(s, i)$ is the number of occurrences of s in w_i. Observe that n is a $|\Sigma|$-marked abstract trajectory and $w \Vdash n$.

By assumption, we have $f(\rho) = \sigma_1 \xrightarrow{s_1}_{\mathcal{SV}} \ldots \xrightarrow{s_n} \sigma_{n+1} = f(\rho')$. For all $i \in [1, n]$, let $o_i \in OS(\mathcal{SV}, s_i)$ be the offset vector used in the transition $\sigma_i \xrightarrow{s_i} \sigma_{i+1}$. There exists some $\langle b, P \rangle \in M_T(s_i)$ such that $\psi(o_i) = b + \sum_{p \in P} \lambda_p p$. Fix such a representation for each o_i. Let $\phi : (D \cup E) \to \mathbb{N}$ be the function mapping each $d_{s,b,i}$ to the number of times b occurs in the representations of the o_j corresponding to all s_j in w_i and mapping each $e_{s,b,p,i}$ to the sum of λ_p in the representations of the o_j corresponding to all s_j in w_i.

Finally, we can observe that $G'(X, M_T, L, \Sigma)$ holds when all $c_{s,i}$ are set to $n(s,i)$, all $d \in D$ are set to $\phi(d)$, and all $e \in E$ are set to $\phi(e)$. The subformulas $AT(L, |\Sigma|) \wedge WF(\Sigma)$ and $Corr(\Sigma, M_T)$ hold by construction of n and ϕ respectively. The first conjunct of $Transition(X, M_T, \Sigma)$ holds because the transition $f(\rho) \xrightarrow{w}_{SV} f(\rho')$ implies that $f(\rho')(x, \lambda s.1) = f(\rho)(x, \lambda s.1) + \sum_{i=1}^{n} o_i(x, \lambda s.1)$ or equivalently $\rho'(x) = \rho(x) + \sum_{i=1}^{n} \psi(o_i)(x, 1)$. For the remainder of the conjuncts, with $k \in [1, d]$ the transition implies that $f(\rho')(x, \lambda s.s = s_{i_k}) = o_{i_k}(x, \lambda s.s = s_{i_k}) + \sum_{j=i_k+1}^{n} o_j(x, \lambda s.s = s_{i_k})$ or equivalently $\rho'(x) = \psi(o_{i_k})(x, 0) + \sum_{j=i_k+1}^{n} \psi(o_j)(x, 1)$. Therefore, $Transition(X, M_T, \Sigma)$ holds, and we therefore have that $G'(X, M_T, L, \Sigma)$ holds.

Observe that $G'(X, M_T, L, \Sigma)$ is polynomially-sized with respect to M_T. We can therefore over-approximate L-reachability of a semi-linear transition system via its exponentially sized SVASR reflection in polynomial time.

6 Related Work

Reachability for Vector Addition Systems over the naturals is decidable [6] but non-elementary [4], prompting the study of integer VAS [3] which operate over integral state vectors. The reachability of integer VAS with resets has been studied widely [2,3,8]; this paper extends such work by considering transition systems in which the set of translation vectors is a potentially infinite semi-linear set. Blondin et al. investigated the extension of integer VAS to affine transformations beyond resets in [1]. Another line [5,9] has investigated extending linear integer arithmetic to include a star operator, effectively computing reachability for Integer Semi-Linear Vector Addition Systems, but not considering resets as we do in this paper. A recent line of work [8,10] has used vector addition systems to compute logical summaries of loops and procedures in computer programs; this work applies a similar recipe with the strictly more powerful domain of SVASRs.

References

1. Blondin, M., Haase, C., Mazowiecki, F., Raskin, M.: Affine extensions of integer vector addition systems with states. Logical Methods Comput. Sci. **17**(3) (2021). https://doi.org/10.46298/lmcs-17(3:1)2021
2. Chistikov, D., Haase, C., Halfon, S.: Context-free commutative grammars with integer counters and resets. Theor. Comput. Sci. **735**, 147–161 (2018). https://doi.org/10.1016/j.tcs.2016.06.017. https://www.sciencedirect.com/science/article/pii/S0304397516302535. Reachability Problems 2014: Special Issue
3. Haase, C., Halfon, S.: Integer vector addition systems with states. In: Ouaknine, J., Potapov, I., Worrell, J. (eds.) RP 2014. LNCS, vol. 8762, pp. 112–124. Springer, Cham (2014). https://doi.org/10.1007/978-3-319-11439-2_9

4. Lazic, R.: The reachability problem for vector addition systems with a stack is not elementary (2013). https://arxiv.org/abs/1310.1767
5. Levatich, M., Bjørner, N., Piskac, R., Shoham, S.: Solving LIA* using approximations. In: Beyer, D., Zufferey, D. (eds.) VMCAI 2020. LNCS, vol. 11990, pp. 360–378. Springer, Cham (2020). https://doi.org/10.1007/978-3-030-39322-9_17
6. Mayr, E.W.: An algorithm for the general Petri net reachability problem. In: Proceedings of the Thirteenth Annual ACM Symposium on Theory of Computing, STOC 1981, pp. 238–246. Association for Computing Machinery, New York (1981). https://doi.org/10.1145/800076.802477
7. Parikh, R.J.: On context-free languages. J. ACM **13**(4), 570–581 (1966). https://doi.org/10.1145/321356.321364
8. Pimpalkhare, N., Kincaid, Z.: Monotone procedure summarization via vector addition systems and inductive potentials. https://nikhilpim.github.io/images/oopsla24.pdf
9. Piskac, R., Kuncak, V.: Linear arithmetic with stars. In: Gupta, A., Malik, S. (eds.) CAV 2008. LNCS, vol. 5123, pp. 268–280. Springer, Heidelberg (2008). https://doi.org/10.1007/978-3-540-70545-1_25
10. Silverman, J., Kincaid, Z.: Loop summarization with rational vector addition systems. In: Dillig, I., Tasiran, S. (eds.) CAV 2019. LNCS, vol. 11562, pp. 97–115. Springer, Cham (2019). https://doi.org/10.1007/978-3-030-25543-5_7

Reachability in Linear Recurrence Automata

Mika Hirvensalo[1], Akitoshi Kawamura[2], Igor Potapov[3(✉)], and Takao Yuyama[2]

[1] Turku University, Turku, Finland
mikhirve@utu.fi
[2] Kyoto University, Kyoto, Japan
{kawamura,yuyama}@kurims.kyoto-u.ac.jp
[3] University of Liverpool, Liverpool, UK
potapov@liverpool.ac.uk

Abstract. This paper studies the linear recurrence automata model in which multiple linear recurrences can be used to generate sequences of numbers. We parameterised the model by varying several structural properties to identify under which constraints or extensions the reachability problems become computationally hard and undecidable. We show first the reduction of classical matrix semigroup problems to reachability in linear recurrence automata and then analyse several variants of the model restricting the state structures, the number and the depth of recurrences.

Keywords: Linear recurrences · Matrix Semigroups · Reachability

1 Introduction

Reachability problems in linear recurrence sequences represent a fundamental area of research with wide-ranging applications in theoretical computer science, mathematics, population dynamics, etc. [3,5,6,11]. Reachability in linear recurrence sequences is related to Diophantine approximation of the irrational numbers, matrix equations, regular grammars and finite state machines. Although significant progress has been made in understanding the dynamics of linear recurrence sequences, there exist many challenges and open problems in this area [4,13,15]. This paper studies nondeterministic automata model when more than one linear recurrence can be used to generate sequences and we aim to understand under which constraints the reachability problems for such systems become computationally hard, undecidable or equivalent to other problems.

One notable example of reachability problems in linear recurrence sequences is the renowned Skolem-Pisot problem [5]. This is the problem of determining whether a given linear recurrent sequence includes the number zero. In the biological context the linear recurrence describes the population dynamics of species when the age-specific survival rate is equal to 1. In more general form, the sub-diagonal of 1's in the companion matrix of a linear recurrence can be replaced by the age-specific survival rates $s_1, s_2, \ldots$. then such matrix M is known in biological literature as a Leslie Matrix [6]. For example the matrix product $M_1^p M_2^q$ may correspond to a case where a population which is governed by two different sets of fertility and survival parameters represented by M_1 for one "season" within p discrete time steps and M_2, e.g. for one in the next "season" (or a change in environmental conditions) for q steps [7]. It is natural to consider a more complex dynamics where the order of changes in biological or technological processes can be described by automata structure and formulated as reachability questions.

The reachability problems with several linear recurrences has not been studied in the literature and in this paper we introduce and study the model of automata with linear recurrence updates and generalised reachability subsequence problem[1]. The simplest form of such automata is an automata with one state and one linear recurrence. The reachability of subsequence of length 1 which is equal to zero with single linear recurrence corresponds to the Skolem-Pisot problem which decidability status in general is still open. The generalisation of linear recurrence system can be achieved by increasing the number, depth and order of applications of linear recurrences. In this paper we show that in the contrast to stateless systems like Skolem problem, enriching the structure with states and additional recurrences leads to undecidability of the reachability problem. In Sect. 2 the reductions from several matrix semigroup problems can be used to identify under which parameters the reachability problems become undecidabile. In Sect. 3 is focused on the limitations to the state structure where it was shown that the single loop system with multiple linear recurrences is equivalent to the system with a single linear recurrence and the sequence linearly connected loops still lead to undecidability. In Sect. 4 provides motivation to study non-deterministic recurrence systems of the order two, by showing the PSPACE hardness in this case as well as decidability for the order one systems and undecidability for the order three. Finally in Sect. 5 we provide a new model of linear recurrences over matrices generalising the original concept.

2 Linear Recurrence Sequences and Automata

Possibly the simplest, and certainly the most well known linear recurrence sequence is the *Fibonacci sequence*. It is the sequence $(x_t)_{t \in \mathbb{N}} \in \mathbb{Z}^\mathbb{N}$ defined by the initial conditions $x_1 = 1$ and $x_0 = 0$ and the linear recurrence $x_t = x_{t-1} + x_{t-2}$, or equivalently,

$$\begin{pmatrix} x_t \\ x_{t-1} \end{pmatrix} = \begin{pmatrix} 1 & 1 \\ 1 & 0 \end{pmatrix} \cdot \begin{pmatrix} x_{t-1} \\ x_{t-2} \end{pmatrix}, \qquad (1)$$

for $t \in \mathbb{N} \setminus \{0, 1\}$. In general, a *linear recurrence* of *depth* (or *order*) $k \in \mathbb{N}$ over a field $\mathbb{F}$ is given by a k-tuple $a = (a_1, \ldots, a_k) \in \mathbb{F}^k$, which we regard as a recursive definition

$$x_t = a_1 \cdot x_{t-1} + \cdots + a_k \cdot x_{t-k}, \qquad (2)$$

---

[1] The similar model name with a focus on a language recognition was discussed in [10].

for $t \in \mathbb{N} \setminus \{0, \ldots, k-1\}$, of a sequence $(x_t)_{t \in \mathbb{N}} \in \mathbb{F}^{\mathbb{N}}$ starting with an initial condition $(x_{k-1}, \ldots, x_0)$. This recurrence (1) can be written as

$$\begin{pmatrix} x_t \\ \vdots \\ x_{t-k+1} \end{pmatrix} = C(a) \cdot \begin{pmatrix} x_{t-1} \\ \vdots \\ x_{t-k} \end{pmatrix} \qquad (3)$$

using the *companion matrix* of a:

$$C(a) = \begin{pmatrix} a_1 & a_2 & \cdots & a_{k-1} & a_k \\ 1 & 0 & \cdots & 0 & 0 \\ 0 & 1 & & 0 & 0 \\ \vdots & & \ddots & \vdots & \vdots \\ 0 & 0 & \cdots & 1 & 0 \end{pmatrix}. \qquad (4)$$

We may generalize this further and consider (nondeterministic) automata with transitions labelled by linear recurrences. A *linear recurrence automaton* $\mathcal{A} = (G, a, q_0, Q_{\mathrm{acc}})$ of *depth* $k \in \mathbb{N}$ over a field $\mathbb{F}$ consists of

- a directed graph $G = (Q, E, \delta^-, \delta^+)$ (possibly with self-loops and parallel edges), i.e., finite sets Q and E whose elements are called *states* and *transitions*, respectively, together with a pair of functions $\delta^-, \delta^+ : E \to Q$ specifying that a transition $e \in E$ is from state $\delta^-(e)$ to state $\delta^+(e)$;
- for each edge $e \in E$, a linear recurrence $a^{(e)} \in \mathbb{F}^k$ of depth k;
- a vertex $q_0 \in Q$ called the *initial state*; and
- a set $Q_{\mathrm{acc}} \subseteq Q$ of *final states*.

A *walk* in this graph G is a sequence $p = (e_0, e_1, \ldots, e_{|p|-1}) \in E^*$ of transitions such that $\delta^-(e_t) = \delta^+(e_{t-1})$ for each $t = 1, \ldots, |p|-1$. We say that p is *from* the state $\delta^-(e_0)$ and *to* the state $\delta^+(e_{|p|-1})$. From an initial k-tuple $u \in \mathbb{F}^k$ (called an *initial condition*), a walk $p = (e_k, e_{k+1}, \ldots, e_{k+|p|-1}) \in E^{|p|}$ generates a unique sequence $(x_0, \ldots, x_{k+|p|-1}) \in \mathbb{F}^{k+|p|}$ such that

- $(x_0, \ldots, x_{k-1}) = u$, and
- for each $t = k, \ldots, |p|-1$, we have (3) with $a = a^{(e_t)}$.

Repeated application of (3) shows that x_t, for each $t \geq k$, is the first component of the vector

$$C(a^{(e_t)}) \cdot C(a^{(e_{t-1})}) \cdots C(a^{(e_k)}) \cdot \begin{pmatrix} x_{k-1} \\ \vdots \\ x_0 \end{pmatrix}. \qquad (5)$$

The automaton $\mathcal{A}$ is said to *accept* a numerical sequence $x \in \mathbb{F}^*$ of length $\geq k$ if $\mathcal{A}$ has a walk p from the initial state q_0 to a final state $\in Q_{\mathrm{acc}}$ such that x is generated by p from the initial segment $x_{<k} = (x_0, \ldots, x_{k-1})$ of x.

Definition 1. *Let $s, m, k \in \mathbb{N}$. We write $\mathrm{LRA}_{\mathbb{F}}(s, m, k)$ for the class of linear recurrence automata with less than or equal to s states and less than or equal to m transitions labeled by linear recurrences of depth k over $\mathbb{F}$.*

Definition 2. *Let $\mathscr{C}$ be the class of linear recurrence automata of depth $k \in \mathbb{N}$ over a field $\mathbb{F}$. For $l \in \mathbb{N}$ with $l \leq k$, the l-subsequence reachability problem (l-SR) for $\mathscr{C}$ is to decide, for a given $\mathcal{A} \in \mathscr{C}$, an initial sequence $u \in \mathbb{F}^k$ and a target sequence $v \in \mathbb{F}^l$, whether $\mathcal{A}$ accepts a sequence over $\mathbb{F}$ starting with u and ending with v.*[2]

If every state of an automaton $\mathcal{A} \in \mathscr{C}$ is final, SR corresponds to the problem whether v can appear in the infinite orbit generated by $\mathcal{A}$. Note that if $\mathcal{A} \in \mathrm{LRA}_{\mathbb{Z}}(1, 1, k)$ (i.e. $\mathcal{A}$ has a single state and a single linear recurrence of order k) and v is the length-1 sequence (0), then the problem corresponds the well-known Skolem-Pisot problem.

Definition 3. *Let $\mathbb{F}$ be a field and let $r, n \in \mathbb{N}$. Let us denote the semigroup generated by a set $\mathcal{M} \subseteq \mathbb{F}^{n \times n}$ of matrices over a field $\mathbb{F}$ by $\langle \mathcal{M} \rangle$.*

- *The (r, n)-scalar reachability problem over $\mathbb{F}$ is to decide, given two vectors $u, v \in \mathbb{F}^n$, a scalar $\lambda \in \mathbb{F}$ and a finite set $\mathcal{M} \subseteq \mathbb{F}^{n \times n}$ of r matrices, whether there exists $M \in \langle \mathcal{M} \rangle$ such that $u^\top \cdot M \cdot v = \lambda$.*
- *The (r, n)-vector reachability problem over $\mathbb{F}$ is to decide, given two vectors $u, v \in \mathbb{F}^n$ and a finite set $\mathcal{M} \subseteq \mathbb{F}^{n \times n}$ of r matrices, whether there exists $M \in \langle \mathcal{M} \rangle$ such that $M \cdot v = u$.*
- *The (r, n)-membership problem over $\mathbb{F}$ is to decide, given a finite set $\mathcal{M} \subseteq \mathbb{F}^{n \times n}$ of r matrices and another matrix $M \in \mathbb{F}^{n \times n}$, whether $M \in \langle \mathcal{M} \rangle$.*

The case with iteration of one linear recurrence (Skolem-Pisot problem) can be represented as the scalar reachability problem for the matrix semigroup generated by a single companion matrix. Here we show the other direction that various matrix reachability problems can be reduced to subsequence reachability problems for linear recurrence automata. By applying Theorem 1 in combination with known hardness or undecidability results on matrix problems can provide a landscape of boundaries for hard cases of linear recurrence automata model. In order to tighten the results more sophisticated constructions could be used like shown in the Sect. 4.

Theorem 1. *Let $\mathbb{F}$ be a field and $r, n \in \mathbb{N}$.*

- *The (r, n)-scalar reachability problem over $\mathbb{F}$ can be reduced to 1-SR for*

$$\mathrm{LRA}_{\mathbb{F}}(r(n-1)+2, rn+1, 2n-1).$$

- *The (r, n)-vector reachability problem over $\mathbb{F}$ can be reduced to n-SR for*

$$\mathrm{LRA}_{\mathbb{F}}(r(n-1)+1, rn, 2n-1).$$

---

[2] In contrast to standard subsequence reachability on at least one run, it is possible to define the *strong subsequence reachability* problem of deciding whether a particular subsequence can be reached on all possible non-deterministic runs on $\mathcal{A} \in \mathscr{C}$.

– The (r, n)-membership problem over $\mathbb{F}$ can be reduced to n^2-SR for
$$\text{LRA}_{\mathbb{F}}(r(n^2 - 1) + 1, rn^2, 2n^2 - 1).$$

Proof. Let us consider a reduction from the (r, n)-scalar reachability problem over $\mathbb{F}$. Suppose we are given two vectors $u, v \in \mathbb{F}^n$, a scalar $\lambda \in \mathbb{F}$ and a set $\mathcal{M} \subseteq \mathbb{F}^{n \times n}$ of r matrices. The problem is to decide whether there exists $M \in \langle \mathcal{M} \rangle$ such that $u^\top \cdot M \cdot v = \lambda$. We will build an automaton in $\text{LRA}_{\mathbb{F}}(r(n-1)+2, rn+1, 2n-1)$ by decomposing an application of each linear transformation into a sequence of linear recurrences of depth upto $2n - 1$.

In the product $u^\top \cdot M \cdot v = \lambda$ let us consider $M \cdot v$ part. The initial values of linear recurrence will be $v = (v_1, \ldots, v_n) \in \mathbb{F}^n$. We start the reduction from a set of generators that are not in the form of the companion matrices. So for each $n \times n$ matrix

$$M = \begin{pmatrix} a_{11} & a_{12} & \ldots & a_{1n} \\ a_{21} & a_{22} & \ldots & a_{2n} \\ \vdots & \vdots & \ddots & \vdots \\ a_{n1} & a_{n2} & \ldots & a_{nn} \end{pmatrix} \in \mathcal{M},$$

we consider a directed graph that consists of a path with $n + 1$ states $q_{M,0}$, $q_{M,1}, \ldots, q_{M,n}$ and, for each $i \in \{1, \ldots, n\}$, a transition from $q_{M,i-1}$ to $q_{M,i}$, associated the linear recurrences of depth $n + i - 1$ and a vector of coefficients

$$(\underbrace{0, \ldots, 0}_{i-1}, a_{in}, \ldots, a_{i1}), \text{ corresponding to } x_t = a_{in} \cdot x_{t-i} + \cdots + a_{i1} \cdot x_{t-i-n+1}$$

to compute the next n values in the sequence $v'_1, \ldots, v'_n$ s.t. $v' = (v'_1, \ldots, v'_n)^T = M \cdot v$ with the walk from $q_{M,0}$ to $q_{M,n}$ (of length n). Now we merge the $2 \cdot r$ states $q_{M,0}$ and $q_{M,n}$, for all $M \in \mathcal{M}$, into a single common initial state q_0, obtaining a petal graph with r cycles (which has altogether $r \cdot (n - 1) + 1$ states and $r \cdot n$ transitions).

Finally for scalar reachability reduction we add to this petal graph one transition from q_0 to a unique final state q_f, with the linear recurrence corresponding to $u^\top$, to compute the value $\lambda = u^\top \cdot M \cdot v$. The number of different linear recurrence functions used in the automata is bounded by the number of transitions which is $rn + 1$ and the number of states in the automaton is $r(n - 1) + 2$. If the sequence contains a value λ that is generated by the automaton in the state q_f then λ can be reached in the scalar reachability problem.

The reduction from the vector reachability (whether there exists $M \in \langle \mathcal{M} \rangle$ such that $M \cdot v = u$) is identical, with the only difference that, instead of adding this new final state, we simply make q_0 the final state reducing to the question whether n-subsequence corresponding to the values of the vector u. Finally for the membership problem, let us denote by X a matrix that can be generated in the matrix semigroup. The linear recurrence automaton should start with an initial sequence of length n^2 that correspond to values of the matrix X

$$x_{1,1}, \ldots, x_{n,1}, x_{1,2}, \ldots, x_{n,2}, \ldots, x_{1,j}, \ldots, x_{n,j}, \ldots, x_{1,n}, \ldots, x_{n,n}$$

and X will be equal to identity matrix at the initialisation step. Then each loop of an automata will be a path of n^2 linear recurrences to compute sequentially the next n^2 values in the sequence corresponding to multiplication from the left $M \cdot X$. As in the previous construction the initial and final states of the paths corresponding to each matrix in a generator need to be merged to get the petal graph. Then the membership problem for a matrix semigroup corresponds to the n^2-subsequence reachability problem in constructed automata with $r(n^2-1)+1$ states, rn^2 transitions with linear recurrences of depth $2n^2 - 1$. □

Theorem 1 can be used to derive undecidability for different classes $\text{LRA}_\mathbb{F}(s, m, k)$ utilizing matrix semigroup results. For example, since the $(5, 3)$-scalar reachability problem (i.e., the one for semigroups generated by five integer matrices of dimension 3) is undecidable [9], it follows that 1-SR is undecidable for $\text{LRA}_\mathbb{Z}(12, 16, 5)$, i.e., for automata having 12 states, 16 distinct linear recurrences of order 5.

Theorem 2. *The k-subsequence reachability problem (k-SR) for $\text{LRA}_\mathbb{Z}(s, m, k)$ with a single final state can be reduced to the $(m, k+2)$-vector reachability problem over $\mathbb{Z}$.*

Proof. Let $\mathcal{A} = (G, a, q_0, Q_{\text{acc}}) \in \text{LRA}_\mathbb{Z}(s, m, k)$, where we may assume that the set of states in $G = (Q, E, \delta^-, \delta^+)$ is $Q = \{0, \ldots, s-1\}$. For every transition $e \in E$, we build a matrix

$$M_e = \begin{pmatrix} C(a^{(e)}) & \bar{0} \\ \bar{0}^T & D_e \end{pmatrix}, \qquad (6)$$

where $C(a^{(e)})$ is the companion matrix of $a^{(e)}$ (see (4)) and

$$D_e = \begin{pmatrix} s & \delta^+(e) - s \cdot \delta^-(e) \\ 0 & 1 \end{pmatrix}. \qquad (7)$$

Then the reachability of $v \in \mathbb{Z}^k$ from $u \in \mathbb{Z}^k$ in a state $q \in Q_{\text{acc}}$ would correspond to the vector reachability of a vector $(v, q, 1)^T$ from $(u, q_0, 1)^T$ for $\langle \{M_e : e \in E\} \rangle$. To see this, let (Q, E) be a directed graph with vertices $q_0, \ldots, q_{s-1}$. Then there exists a set of affine functions F such that for an edge

$$e = (\delta^-(e), \delta^+(e)) \in E,$$

- there exists a unique $f_{\delta^-(e)\delta^+(e)} \in F$ such that $f_{\delta^-(e)\delta^+(e)}(\delta^-(e)) = \delta^+(e)$
- $f_{\delta^-(e)\delta^+(e)}(x) \notin [0, s-1]$ for all $x \neq \delta^-(e)$ and
- $f_{\delta^-(e)\delta^+(e)}(x) = s \cdot x + b = \delta^+(e)$, where $b = \delta^+(e) - s \cdot \delta^-(e) \in [-s^2+s, s-1]$.

The idea of the encoding is to represent each state by integers in $[0, s-1]$ and every edge by affine functions of the form $ax + b$, where a value of x changes according to the automata structure taking as an input the value of one state and computing the value of the resulting state. In the event of application of affine function which is not representing outgoing edge from the current state the coefficient s ensures that the resulting value of the register will either exceed $s-1$ or fall below zero. Moreover applying a function f to the value which is out side of a range $[0, s-1]$ cannot return it to the original interval range. □

Remark 1. In case of multiple final states $q \in Q_{\text{acc}}$, where $|Q_{\text{acc}}| = \tau$ the k-subsequence reachability problem can be reduces to τ vector reachability problems for the same matrix semigroup but different target vectors following the above construction.

3 Linear Recurrence Automata Without Nested Loops

In this subsection we show that reachability for an automata a single loop structure over different linear recurrences corresponds to reachability in a single linear recurrence of a higher order (depth) preserving the number of non-zero coefficients. At the same time we show that the subsequence reachability problem for a class of automata with a state structure consisting of multiple cycles connected in sequence is undecidable.

It is well known that the characteristic polynomial of the companion matrix $C(a)$ (see (4)) of a linear recurrence $a = (a_1, \ldots, a_k)$ is given by $\Phi_{C(a)}(x) = x^k - a_1 x^{k-1} - a_2 x^{k-2} - \cdots - a_k$. Conversely, for a polynomial $f(x) = x^k - a_1 x^{k-1} - a_2 x^{k-2} - \cdots - a_k$, we say that $a = (a_1, \ldots, a_k)$ is *the linear recurrence defined by* $f(x)$.

Lemma 1. *Let A be a $k \times k$ matrix and $\boldsymbol{x}$ and $\boldsymbol{y}$ be k-dimensional vectors. Let $(w_t)_{t \in \mathbb{N}}$ be the sequence defined as $w_t = \boldsymbol{x}^\top \cdot A^t \cdot \boldsymbol{y}$. Then $(w_t)_{t \in \mathbb{N}}$ satisfies the linear recurrence defined by the characteristic polynomial $\Phi_A(x)$.*

Proof. Let $\Phi_A(x) = x^k - a_1 x^{k-1} - a_2 x^{k-2} - \cdots - a_k$. By the Cayley-Hamilton theorem, we have $\Phi_A(A) = O$, hence $A^t \cdot \Phi_A(A) = A^{t+k} - a_1 A^{t+k-1} - a_2 A^{t+k-2} - \cdots - a_k A^t = O$ and thus $w_{t+k} - a_1 w_{t+k-1} - a_2 w_{t+k-2} - \cdots - a_k w_t = \boldsymbol{x}^\top \cdot (A^{t+k} - a_1 A^{t+k-1} - a_2 A^{t+k-2} - \cdots - a_k A^t) \cdot \boldsymbol{y} = 0$. □

Proposition 1. *Let $a^{(1)}, \ldots, a^{(m)}$ be m linear recurrences of depth k, and $(w_t)_{t \in \mathbb{N}}$ be a sequence generated by applying $a^{(1)}, \ldots, a^{(m)}$ successively and cyclically to the initial sequence $(w_{k-1}, \ldots, w_0)$. Then $(w_t)_{t \in \mathbb{N}}$ satisfies the linear recurrence of depth $k \cdot m$ defined by the characteristic polynomial $\Phi_{C(a^{(m)}) \cdots C(a^{(1)})}(x^m)$.*

Proof. For each $r = 0, 1, \ldots, m-1$, by using Lemma 1, we show that the subsequence $(w_t \mid t \equiv r \pmod{m})_{t \in \mathbb{N}} = (w_{mt+r})_{t \in \mathbb{N}}$ satisfies the linear recurrence defined by $\Phi_{C(a^{(m)}) \cdots C(a^{(1)})}(x)$ (we write $\Psi(x)$ for short).

Suppose $0 \leq r < m$ and then $w_{mt+r} = \begin{pmatrix} 0 & \cdots & 0 & 1 \end{pmatrix} \begin{pmatrix} w_{mt+r+k-1} \\ w_{mt+r+k-2} \\ \vdots \\ w_{mt+r} \end{pmatrix}$

$$= \begin{pmatrix} 0 & \cdots & 0 & 1 \end{pmatrix} C(a^{(r)}) \cdots C(a^{(1)}) \cdot (C(a^{(m)}) \cdots C(a^{(1)}))^t \cdot \begin{pmatrix} w_{k-1} \\ w_{k-2} \\ \vdots \\ w_0 \end{pmatrix}. \text{ Hence, by}$$

Lemma 1, the subsequence $(w_{mt+r})_{t \in \mathbb{N}}$ satisfies the linear recurrence defined by $\Psi(x)$. Suppose $\Psi(x) = x^k - a_1 x^{k-1} - a_2 x^{k-2} - \cdots - a_k$. Then we have

$$w_{mt+r} = a_1 w_{m(t-1)+r} + a_2 w_{m(t-2)+r} + \cdots + a_k w_{m(t-k)+r}$$
$$= a_1 w_{(mt+r)-m} + a_2 w_{(mt+r)-2m} + \cdots + a_k w_{(mt+r)-km}$$

for each $t \geq k$ and $r = 0, 1, \ldots, m-1$. Thus $(w_t)_{t \in \mathbb{N}}$ satisfies the recurrence relation

$$w_t = a_1 w_{t-m} + a_2 w_{t-2m} + \cdots + a_k w_{t-km} \tag{8}$$

of depth $k \cdot m$, which is defined by $\Psi(x^m) = x^{km} - a_1 x^{km-m} - a_2 x^{km-2m} - \cdots - a_k = \Phi_{C(a^{(m)}) \cdots C(a^{(1)})}(x^m)$. □

Example 1. Let $a^{(1)} = (1, 2)$ and $a^{(2)} = (3, 4)$. If we put $w_0 = w_1 = 1$, then the alternating applications of $a^{(1)}$ and $a^{(2)}$ to (w_1, w_0) generates the sequence

$$w_2 = a^{(1)} \begin{pmatrix} w_1 \\ w_0 \end{pmatrix} = 1 \cdot 1 + 2 \cdot 1 = 3,$$

$$w_3 = a^{(2)} \begin{pmatrix} w_2 \\ w_1 \end{pmatrix} = 3 \cdot 3 + 4 \cdot 1 = 13,$$

$$w_4 = a^{(1)} \begin{pmatrix} w_3 \\ w_2 \end{pmatrix} = 1 \cdot 13 + 2 \cdot 3 = 19$$

Continuing this process, we have

$$(w_t)_{t \in \mathbb{N}} = (1, 1, 3, 13, 19, 109, 147, 877, 1171, 7021, \ldots). \tag{9}$$

The companion matrices are

$$C(a^{(1)}) = \begin{pmatrix} 1 & 2 \\ 1 & 0 \end{pmatrix}, \quad C(a^{(2)}) = \begin{pmatrix} 3 & 4 \\ 1 & 0 \end{pmatrix}, \quad C(a^{(2)})C(a^{(1)}) = \begin{pmatrix} 7 & 6 \\ 1 & 2 \end{pmatrix}.$$

Thus we have $\Phi_{C(a^{(2)})C(a^{(1)})}(x^2) = x^4 - 9x^2 + 8$. In fact, the sequence (9) satisfies the linear recurrence relation $w_t = 9w_{t-2} - 8w_{t-4}$.

Remark 2. The depth of linear recurrence relation obtained in Proposition 1 is not minimal (in respect to the depth) in general. Indeed, the sequence (9) satisfies the relation $w_t = -w_{t-1} + 8w_{t-2} + 8w_{t-3}$ of depth three. Alternative construction of a single linear recurrence has been considered in [18]. However, the linear recurrence obtained in Proposition 1 does not depend on the initial sequence $w_{k-1}, \ldots, w_0$ and has at most k non-zero coefficients.

Proposition 2. *1-Subsequence Reachability Problem for an automaton $\mathcal{A} \in \mathscr{C}$ with a single loop constructed from m linear recurrences $f_1, f_2, \ldots, f_m$ of order k can be reduced to the reachability problem for a linear recurrence sequence of order $k \cdot m$ with k non-zero coefficients.*

Proof. Let an automaton $\mathcal{A} \in \mathscr{C}$ have a single loop with m states and $m-1$ edges labeled by different linear recurrences. By applying Proposition 1 the edges of the loop can be labeled with a single type linear recurrence of a larger order and then the length of this loop can be reduced to one edge. The right-hand side of the recurrence relation (8) has at most k non-zero coefficients, while its depth is $k \cdot m$. So, at most k values out of $k \cdot m$ previous values will be used in each linear recurrence to generate the next value w_t. □

Remark 3. The reduction in the other direction is not always possible. For example the linear recurrence of the order k : $u_n = u_{n-k}$ can not be decomposed to a finite sequence of linear recurrences of any smaller order than k.

Proposition 3. *A linear recurrence automata (without nested loops) can generate zero values in the positions corresponding to a non-semilinear set.*

Proof. Let us consider the linear recurrence automaton with five states $s_1, \ldots, s_5$, initial state s_1 and the following edges with correspondent linear recurrences:

(s_1, s_2): $u_n = u_{n-1} + u_{n-2}$ (s_3, s_4): $u_n = 2u_{n-2}$
(s_2, s_1): $u_n = u_{n-2}$ (s_4, s_3): $u_n = u_{n-2}$
(s_1, s_3): $u_n = u_{n-2}$ (s_3, s_5): $u_n = u_{n-1} - u_{n-2}$

The automaton can reach the state s_5 only with the value $u_n = 2^p - q$ starting from an initial value $u_0 = 1, u_2 = 1$, where $p, q \in \mathbb{N}$ and in general the sequence is of the form $1, 1, 2, 1, 3, 1, 4, 1, \ldots, q, 1, q, 2^1, q, 2^2, \ldots, q, 2^p, 2^p - q$. The loop between s_1 and s_2 computes the value q reaching the subsequence $1, q$ and the loop with states s_3 and s_4 compute the value 2^p reaching the subsequence $q, 2^p$ and the final transition to s_5 compute the difference between the last two values. □

Now we show that 1-subsequence reachability problem for a class of automata with a linear structure of connected cycles in general is undecidable.

Proposition 4. *Let $\mathcal{L}_1^y, \ldots, \mathcal{L}_z^y$ be cyclic walks of a given length $y \in \mathbb{N}$ based on a set of linear recurrences $\mathcal{L}_1, \ldots, \mathcal{L}_z$ that are sequentially connected by a set of linear recurrences $P_1, \ldots, P_{z-1}$ in a LRA $\mathcal{A}$ to form walks for any $i_j \in \mathbb{N}$:*

$$\mathcal{L}_1^{y \cdot i_1} P_1 \ldots P_{j_1} \mathcal{L}_2^{y \cdot i_2} P_{j_1+1} \ldots P_{j_2} \mathcal{L}_3^{y \cdot i_3} P_{j_2+1} \ldots P_{j_3} \ldots P_{j_{z-2}} \ldots P_{j_{z-1}} \mathcal{L}_z^{y \cdot i_z}.$$

The 1-subsequence reachability problem for $\mathcal{A}$ is undecidable.

Proof. It is known that the matrix equation (aka Knapsack problem in $\mathbb{Z}^{n \times n}$)

$$M_1^{x_1} \cdots M_r^{x_r} = O \qquad (10)$$

is undecidable for sufficiently large $r, n \in \mathbb{N}$, where $M_i \in \mathbb{Z}^{n \times n}$ and O is $n \times n$ zero matrix [1]. We will construct Linear Recurrence Automata to encode the sequential multiplications of matrices in matrix equation which is undecidable in general[3]. By applying a technique from Theorem 1 we can convert each matrix M_i into a loop of n^2 linear recurrences $f_{i_1}, \ldots f_{i_{n^2}}$ of the order $2n^2 - 1$ with n^2 states $s_0, s_1 \ldots s_{i_n^2-1}$ computing the values of a matrix product. Following the order of matrix Eq. (10) two loops corresponding to multiplication by M_i followed by M_j will be connected sequentially by an edge labeled with f_{j_1} from a state s_{i_1} to s_{j_2}. Finally by applying Proposition 1 we can replace a sequence of linear recurrences $f_{i_1}, \ldots f_{i_{n^2}}$ by a loop with a single linear recurrence $\mathcal{L}_i$ of the order $n^2(2n^2 - 1)$, but a single edge connecting i-th and j-th loops need to be replaced by recalculating next $n^2(2n^2 - 1)$ values by a sequence of linear recurrences $f_{j_1}, \ldots f_{j_{n^2(2n^2-1)}}$. Also the initial sequence of $n^2(2n^2-1)$ values need to be precomputed to start the automata. So from this directly follows that the n^2-subsequence reachability problem of reaching a sequence with n^2 zero values for the constructed automata with linear recurrences of the order $n^2(2n^2 - 1)$ and a constant $y = n^2$ is undecidable. □

4 Linear Recurrence Automata of the Low Order

Theorem 3. *1-Subsequence Reachability Problem for a Linear Recurrence Automata $\mathcal{A} \in LRA_{\mathbb{Q}(i)}(s, m, 1)$, i.e. with linear recurrences of the order one is decidable in polynomial time, where $\mathbb{Q}(i)$ is Gaussian rational numbers.*

Proof. Let us consider an automaton with s states labelled with numbers from 1 to s and transition labelled by linear recurrences of depth 1, which corresponds to a multiplication by complex numbers, e.g. with the update in the form $x_t = c \cdot x_{t-1}$, $c \in \mathbb{C}$. The configuration of an automata is a pair of a current state q and a value x_{t-1} can be represented as a vector of dimension $s+1$:

$$(\underbrace{0,\ldots,0}_{q-1}, 1, \underbrace{0,\ldots,0}_{s-q}, x_{t-1})^T$$

Every transition in the automata with an update $x_t = c \cdot x_{t-1}$ will be represented by a $(s+1) \times (s+1)$ matrices, where top $s \times s$ block is a permutation matrix to change a position of 1 representing the state transition and the lower block of a single value in the position s, s that contains c. Every matrix is a monomial matrix with elements over a commutative semigroup and the vector reachability for the monomial matrix semigroup is decidable in polynomial time [16]. Since Gaussian rational numbers $\mathbb{Q}(i)$ can be represented by commutative 2×2 matrices with rational elements this completes the proof. □

Theorem 4. *1-Subsequence Reachability Problem for a Linear Recurrence Automata $\mathcal{A} \in LRA_{\mathbb{Z}}(s, m, 2)$, i.e. with linear recurrences of the order two is PSPACE-hard.*

Proof. The reachability problem for an affine register machine over intergers is PSPACE-hard [12]. Affine register machine over integers is defined as an automaton with a finite number of states and the transition labelled by an affine updates

---
[3] A subclass of matrix equations has been shown to be decidable in [3].

of the form $x' = ax + b$, where $a, b \in \mathbb{Z}$. The reachability problem for this machine is the questions whether the value "y" can appear in the register with affine updates at the final state of a machine starting form an initial state and the initial register value x_0. The affine update $ax+b$. could be represented by the matrix $M = \begin{pmatrix} b & a \\ 1 & 0 \end{pmatrix}$ applying to the vector $\begin{pmatrix} x \\ 1 \end{pmatrix}$ and resulting in $\begin{pmatrix} ax+b \\ 1 \end{pmatrix}$. Also this can be seen as a sequential application of two linear recurrences of the order two. The first one is $\begin{pmatrix} b & a \\ 1 & 0 \end{pmatrix} \begin{pmatrix} u_{n-1} \\ u_{n-2} \end{pmatrix} = \begin{pmatrix} u_n \\ u_{n-1} \end{pmatrix}$ and the second one $\begin{pmatrix} 0 & 1 \\ 1 & 0 \end{pmatrix} \begin{pmatrix} u_{n-1} \\ u_{n-2} \end{pmatrix} = \begin{pmatrix} u_n \\ u_{n-1} \end{pmatrix}$ and starting with initial values $\begin{pmatrix} 1 \\ x_0 \end{pmatrix}$. We can construct linear recurrence automaton from the register machine with affine updates by inheriting the state structure with the same initial and the final state and by replacing each transition $ax + b$ by two transitions with sequential linear recurrences $u_n = bu_{n-1} + au_{n-2}$ and $u_n = u_{n-2}$. So by applying in alternative way the first and the second linear recurrence we generate the sequence $x_0, 1, a(x_0) + b, 1, a(a(x_0) + b) + b, 1, \ldots$. □

Remark 4. So the 1-subsequence reachability problem for a Linear Recurrence Automata is at least PSPACE-hard and upto our knowledge it is not known whether this problem with linear recurrences of the order (depth) two is decidable or not. Following Theorem 2 the 2-subsequence reachability could be reduced to the vector reachability for 4×4 matrices which is in general undecidable [8,17].

Theorem 5. *2-Subsequence Reachability Problem for a Linear Recurrence Automata $\mathcal{A} \in LRA_\mathbb{Q}(7, 11, 3)$, i.e. with linear recurrences of the order three is undecidable.*

Proof. It was shown in [2] that the point-to-point reachability is undecidable for a set of five affine transformations of dimension two over rational numbers. In fact a class of affine transformations of dimension two with undecidable reachability can be restricted to the following form:

$$f(p,q) = \begin{cases} p' = ap + bq + c \\ q' = dq \end{cases}$$

which corresponds to undecidable vector reachability for matrix semigroups generated by 3×3 matrices over rational numbers

$$M = \begin{pmatrix} a & b & c \\ 0 & d & 0 \\ 0 & 0 & 1 \end{pmatrix} \text{ applying to the vector } \begin{pmatrix} x \\ y \\ 1 \end{pmatrix}$$

with starting from a vector $(0, 1, 1)^T$ and target vector $(0, 1, 1)^T$. Following construction of the Theorem 1 we can build an automaton that can generate the sequence $p_0, q_0, 1, p_1, q_1, 1, p_2, q_2, 1, \ldots, p_i, q_i, 1$. Moreover 2-subsequence reachability problem and in particular reaching 2-subsequence $0, 1$ at any position is undecidable since it can only appear when 0 is in the position $i = 0 \pmod 3$ and 1 in the position $i = 1 \pmod 3$ based on the matrix form. At the same time undecidability holds for reaching 1-subsequence 0 at the position $i = 0 \pmod 3$ which will correspond to reaching 0 when automaton in the final state which is also an initial state of the constructed automaton. Note that the linear recurrences for one iteration of matrix multiplications will be of order 3 and the last transformation to copy value 1 in the position $i = 2 \pmod 3$ is the same for any matrix M. So we have three sequential transformations

$$x_j = ax_{j-3} + bx_{j-2} + cx_{j-1} \to x_j = dx_{j-3} \to x_j = x_{j-3}.$$

In this case only $m+2$ states and $2m+1$ recurrences of order 3 are needed, where m is the number of affine transformations and therefore we have undecidability for automata $\mathcal{A} \in LRA_\mathbb{Q}(7, 11, 3)$. □

Theorem 6. *The reachability for stateless polynomial recurrent system in dimension two is undecidable.*

Proof. It is know that the reachability problem is undecidable for stateless affine automata in dimension three and for two dimensional affine automata with states where updates are of the form: $x' = ax+b$ and $y' = cy+d$ [14]. Two dimensional affine transformations can be encoded by one dimensional recurrence automata with states and the sequence $x_1, y_1, 1, x_2, y_2, 1, \ldots$ by a sequence of recurrences

$$u_n = a \cdot u_{n-3} + b \cdot u_{n-1} \to u_n = c \cdot u_{n-3} + d \cdot u_{n-2} \to u_n = u_{j-n}.$$

In the second dimension it is possible to encode transitions between states with a polynomial updates where state i can be encoded as 10^i and then transition from state i to state j is encoded by a function $v_n = \frac{v_{n-1}^s}{10^{((s \cdot i - j))}}$. □

5 Multidimensional Linear Recurrences

Another extension of the automata model that can be considered is a case of multidimensional linear recurrences which can be seen as a generalisation of standard linear recurrences. For example, in the biological context the model can be used to describe the populations of a group of highly interconnected species or organisms (e.g. symbiotic bacteria). Let us define a sequence X_0, X_1, X_2, ... of $r \times s$-matrices over $\mathbb{F}$ is called *linearly recursive of depth k* or *recursion* or *recurrence* if there are $r \times r$ matrices C_1, ..., C_k so that $X_n = C_1 X_{n-1} + C_2 X_{n-2} + \ldots + C_k X_{n-k}$ for $n > k$. Let us then consider a number $i \in \{1, \ldots, m\}$ of linear recurrences

$$X_n = C_1^{(i)} X_{n-1} + C_2^{(i)} X_{n-2} + \ldots + C_k^{(i)} X_{n-k}. \quad (11)$$

The reachability problems for automata with multidimensional linear recurrences of the order two are undecidable for dimension four when C_i are diagonal by simulation of reachability in independent affine updates [14] alternating affine updates with copying constant value like in Theorem 4. Also for dimension three and for more general case of C_j^i values the stateless automata has undecidable reachability problem based on [2,9]. However in case of limiting the matrix structure or parameters the undecidability results would not hold and more general tools for analysis of this model may be required.

Proposition 5. *A sequence $X_0, X_1, X_2, \ldots$ of $r \times s$ matrices is linearly recursive if and only if there are matrices A ($r \times d$), M ($d \times d$), and B ($d \times s$) so that for $n \geq 0$*

$$X_n = AM^n B. \tag{12}$$

Proof. Assume first (11) with only one recurrence $C_j = C_j^{(i)}$, let $d = kr$ and create a $kr \times kr$ companion matrix in block form

$$M = \begin{pmatrix} C_1 & C_2 & \ldots & C_{k-1} & C_k \\ I & 0 & \ldots & 0 & 0 \\ 0 & I & \ldots & 0 & 0 \\ \vdots & \vdots & \ddots & \vdots & \vdots \\ 0 & 0 & \ldots & I & 0 \end{pmatrix}.$$

It follows immediately that

$$M \begin{pmatrix} X_{n-1} \\ X_{n-2} \\ \vdots \\ X_{n-d} \end{pmatrix} = \begin{pmatrix} X_n \\ X_{n-1} \\ \vdots \\ X_{n-d+1} \end{pmatrix},$$

and by induction

$$M^n \begin{pmatrix} X_{d-1} \\ X_{d-2} \\ \vdots \\ X_0 \end{pmatrix} = \begin{pmatrix} X_{d+n} \\ X_{d+n-1} \\ \vdots \\ X_n \end{pmatrix}.$$

Letting $A = (0\ 0 \ldots I)$ ($r \times kr$ matrix) and $B = \begin{pmatrix} X_{d-1} \\ X_{d-2} \\ \vdots \\ X_0 \end{pmatrix}$ ($kr \times s$ matrix) we see that

$$X_n = AM^n B.$$

Assume then (12) holds for some $d \times d$-matrix M. By Cayley-Hamilton theorem,

$$M^d - c_1 M^{d-1} - c_2 M^{d-2} - \ldots - c_{d-1}M - c_d I = 0, \tag{13}$$

where $p(x) = x^d - c_1 x^{d-1} - c_2 c^{d-2} - \ldots - c_{d-1} x - c_d$ is the characteristic polynomial of matrix M.

Multiplying (13) by M^{n-k} we get

$$M^n = c_1 M^{n-1} + c_2 M^{n-2} + \ldots + c_d M^{n-d}, \tag{14}$$

and left and right multiplication of (14) by A and B respectively gives

$$X_n = c_1 X_{n-1} + c_2 X_{n-2} + \ldots + c_d X_{n-d}.$$

Claim follows now by choosing $C_i = c_i I$, where I is $r \times r$ identity matrix. □

Definition 4. *Sequence X_0, X_1, X_2, $\ldots$ matrices is said to be nondeterministically recurrent or recurrent through choices i_0, i_1, i_2, $\ldots$ if for $n > k$ X_n is obtained by recurrence (11) by selecting the indices i_k nondeterministically in set $\{1, \ldots, m\}$. If $m = 1$, the sequence is said to be deterministically recurrent.*

Definition 5. *Let $w = a_1 a_2 \ldots a_l$ be a word over alphabet $\Sigma = \{1, \ldots, m\}$ and M_1, $\ldots$, M_m a set of matrices indexed by the elements of Σ. Then we define*

$$M_w = M_{a_1} M_{a_2} \ldots M_{a_l},$$

which is just another way of expressing the fact that $w \to M_w$ is a morphism from Σ^* to the semigroup generated by matrices M_a, $a \in \Sigma$.

Proposition 6. *Sequence X_n of $r \times s$ matrices is linearly recurrent through choices i_0, i_1, i_2, $\ldots$ if and only if*

$$X_n = A M_{w^R} B, \tag{15}$$

where $w = i_0 i_1 \ldots i_l \in \{1, \ldots, m\}^*$ and $w^R = i_l i_{l-1} \ldots i_1 i_0$ stands for the mirror image of w, A is a $r \times d$-matrix, each M_i is a $d \times d$-matrix, and B is a $d \times s$-matrix.

Proof. Assume first that there are m linear recurrences as in (11). By convention, we suppose that all those have their k initial values $X_0^{(i)}$, $\ldots$, $X_{k-1}^{(i)}$. Matrix B in the claim is constructed as a block matrix $B = (X_{k-1}^{i_{k-1}} \ldots X_0^{i_0})^T$, where i_0, i_1, $\ldots$, i_{k-1} are selected from the set $\{1, 2, \ldots, m\}$. Each matrix M_i and A are constructed as a block-formed companion matrix of i:th recurrence just like in Proposition 5 and the claim follows straightforwardly.

For the opposite direction, we need to show that each finite set of matrices $\{M_1, \ldots, M_n\}$ gives raise to a set of linear recurrences via Eq. (15).

Multiple choices for recurrence means that we do not have only one sequence $B \to MB \to M^2 B$, $\ldots$ as in Proposition 5, but rather a tree with B as a root, and all $M_i B$ as nodes with distance 1 from the root, all $M_j M_i B$ as nodes with distance 2 from the root, all nodes $M_k M_j M_i B$ with distance 3 from the root, etc., see Fig. 1.

To generalize, we define $D_i = \{M_w B \mid |w| = i\}$ as the set of all nodes with distance i from the root.

If $D = D_1 \cup D_2 \cup \ldots$ is finite, only finitely many matrices of form $M_w B$ exist. Then for each w long enough and each i, we have $M_i M_w B \in D$, and hence the set of recursions is trivial.

We can therefore assume that set D is infinite. Since set $D \subseteq \mathbb{F}^{d \times s}$ (a vector space of dimension $d \cdot s$, D generates a subspace $V \subseteq \mathbb{F}^{d \times s}$, and hence there exists a basis $\mathcal{B} \subset D$ of V of cardinality at most ds. The basis is by no means unique, but we choose a basis $\mathcal{B}$ so that $\max\{|w| \mid M_w B \in D\}$ is minimal. Now, the set $\{AM_w B \mid M_w B \in \mathcal{B}\}$ is considered as the set of initial values of recurrence, and for longer words, the recurrence is obtained in the following way:

For each matrix M_i and each $M_w B \in \mathcal{B}$, matrix $M_i M_w B$ belongs to the subspace V, and hence

$$M_i M_w B = \sum_{M_v B \in \mathcal{B}} c^{(i)}_{M_v} M_v B,$$

where coefficient $c_{M_v} \in \mathbb{F}$, and hence value $AM_i M_w B$ can be obtained by using initial values:

$$AM_i M_w B = \sum_{M_v B \in \mathcal{B}} c^{(i)}_{M_v} AM_v B. \qquad \square$$

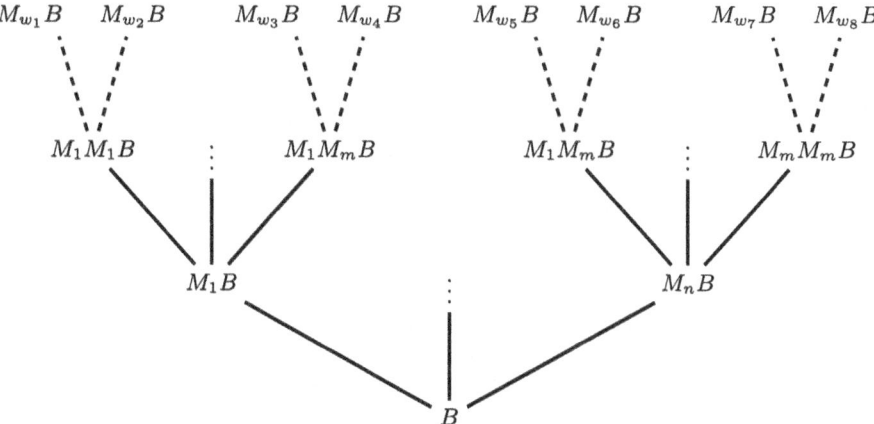

Fig. 1. A tree of possibilities with multiple choices for recurrences.

Corollary 1. *If X_n and Y_n are recurrent sequences (deterministically or non-deterministically), so are*

1. $X_n \otimes Y_n$ *(tensor product).*
2. $X_n + Y_n$, *if the dimensions of X_n and Y_n match.*
3. $CX_n D$, *if the dimensions of C and D are compatible.*

The next statement follows from Propositions 5 and 6 since $X_n = AM^n_{w_1^R} M_{w_0^R} B = AM^n_{w_1^R} B_1$, where $B_1 = M_{w_0^R} B$.

Corollary 2. *Let X_n be a sequence obtained by nondeterministic linear recurrences in set $\{1, 2, \ldots, m\}$ appearing eventually in cyclic order: $w = w_0 w^n$, where $w_0 \in \{1, 2, \ldots, m\}$ is arbitrary. Then there is a deterministic linear recurrence generating sequence X_n.*

Acknowledgements. This work was supported by Royal Society International Exchanges IES\R1\191184 and by JSPS KAKENHI Grant Number JP20H05961.

References

1. Bell, P., Halava, V., Harju, T., Karhumäki, J., Potapov, I.: Matrix equations and Hilbert's tenth problem. Int. J. Algebra Comput. **18**(8), 1231–1241 (2008)
2. Bell, P., Potapov, I.: On undecidability bounds for matrix decision problems. Theor. Comput. Sci. **391**(1), 3–13 (2008). Combinatorics, Automata and Number Theory
3. Bell, P.C., Potapov, I., Semukhin, P.: On the mortality problem: from multiplicative matrix equations to linear recurrence sequences and beyond. Inf. Comput. **281**, 104736 (2021)
4. Bilu, Y., Luca, F., Nieuwveld, J., Ouaknine, J., Purser, D., Worrell, J.: Skolem meets Schanuel. In: Szeider, S., Ganian, R., Silva, A. (eds.) 47th International Symposium on Mathematical Foundations of Computer Science (MFCS 2022). Leibniz International Proceedings in Informatics (LIPIcs), Dagstuhl, Germany, vol. 241, pp. 20:1–20:15. Schloss Dagstuhl – Leibniz-Zentrum für Informatik (2022)
5. Blondel, V.D., Portier, N.: The presence of a zero in an integer linear recurrent sequence is NP-hard to decide. Linear Algebra Appl. **351–352**, 91–98 (2002). Fourth Special Issue on Linear Systems and Control
6. Caswell, H.: Matrix Population Models, 2nd edn. Sinauer (2001)
7. Caswell, H., Trevisan, M.C.: Sensitivity analysis of periodic matrix models. Ecology **75**(5), 1299–1303 (1994)
8. Halava, V., Harju, T., Hirvensalo, M.: Undecidability bounds for integer matrices using Claus instances. Int. J. Found. Comput. Sci. **18**(05), 931–948 (2007)
9. Halava, V., Hirvensalo, M.: Improved matrix pair undecidability results. Acta Informatica **44**(3), 191–205 (2007)
10. Hirvensalo, M., Yakaryilmaz, A.: Decision problems on unary probabilistic and quantum automata. Balt. J. Mod. Comput. **4**(4) (2016)
11. Hitarth, S., Kenison, G., Kovács, L., Varonka, A.: Linear loop synthesis for quadratic invariants. In: Beyersdorff, O., Kanté, M.M., Kupferman, O., Lokshtanov, D. (eds.) 41st International Symposium on Theoretical Aspects of Computer Science (STACS 2024). Leibniz International Proceedings in Informatics (LIPIcs), Dagstuhl, Germany, vol. 289, pp. 41:1–41:18. Schloss Dagstuhl – Leibniz-Zentrum für Informatik (2024)
12. Jaax, S., Kiefer, S.: On affine reachability problems. In: Esparza, J., Král', D. (eds.) 45th International Symposium on Mathematical Foundations of Computer Science (MFCS 2020). LIPIcs, vol. 170, pp. 48:1–48:14 (2020)
13. Kenison, G.: On the Skolem problem for reversible sequences. In: Szeider, S., Ganian, R., Silva, A. (eds.) 47th International Symposium on Mathematical Foundations of Computer Science (MFCS 2022). Leibniz International Proceedings in Informatics (LIPIcs), Dagstuhl, Germany, vol. 241, pp. 61:1–61:15. Schloss Dagstuhl – Leibniz-Zentrum für Informatik (2022)
14. Ko, S., Niskanen, R., Potapov, I.: Reachability problems in low-dimensional nondeterministic polynomial maps over integers. Inf. Comput. **281**, 104785 (2021)
15. Lipton, R., Luca, F., Nieuwveld, J., Ouaknine, J., Purser, D., Worrell, J.: On the Skolem problem and the Skolem conjecture. In: Baier, C., Fisman, D. (eds.) LICS 2022: 37th Annual ACM/IEEE Symposium on Logic in Computer Science, Haifa, Israel, 2–5 August 2022, pp. 5:1–5:9. ACM (2022)
16. Lisitsa, A., Potapov, I.: Membership and reachability problems for row-monomial transformations. In: Fiala, J., Koubek, V., Kratochvíl, J. (eds.) MFCS 2004. LNCS, vol. 3153, pp. 623–634. Springer, Heidelberg (2004). https://doi.org/10.1007/978-3-540-28629-5_48

17. Potapov, I.: From post systems to the reachability problems for matrix semigroups and multicounter automata. In: Calude, C.S., Calude, E., Dinneen, M.J. (eds.) DLT 2004. LNCS, vol. 3340, pp. 345–356. Springer, Heidelberg (2004). https://doi.org/10.1007/978-3-540-30550-7_29
18. Shallit, J.: Numeration systems, linear recurrences, and regular sets. Inf. Comput. **113**(2), 331–347 (1994)

Games and Abstractions

Robust Deterministic Abstractions for Supervising Discrete-Time Continuous Systems

Gwendal Priser, Elena Vanneaux, and Goran Frehse(✉)

U2IS, ENSTA Paris, Institut Polytechnique de Paris, Paris, France
{gwendal.priser,elena.vanneaux,goran.frehse}@ensta-paris.fr

Abstract. We present a method for constructing discrete abstractions for discrete-time, continuous-state systems. Related approaches construct a discrete bisimulation, which leaves little room for nondeterminism in the outputs and quickly leads to highly complex models since all concrete behavior is covered. Our approach is to relax these requirements and build a satisficing solution: a discrete abstraction that is deterministic, robust, and as complete as possible under the given parameters. This allows us to balance granularity and computational feasibility. We leverage linearization and linear feedback control to extend the approach from globally contractive systems to systems with contractive cycles. The resulting abstraction directly induces a supervisor policy. The approach is illustrated with numerical experiments and has potential applications in various domains where system safety and reversibility are essential.

Keywords: continuous dynamical systems · discrete abstraction · robustness · supervision

1 Introduction

We are interested in supervising the behavior of dynamical systems that are subject to a sequence of control inputs, also called actions. The system's reaction to an input can be nondeterministic, i.e., the same input can lead to different successor states. The supervisor's job is to ensure that the control actions always lead to runs that satisfy a given specification. Ideally, we want to identify the maximally permissive supervisor that achieves this. However, this is known to be a hard problem, so the approach we pursue in this paper is to trade off some of the maximality to reduce the computational cost.

The design of controllers and supervisors often involves the construction of an abstraction, i.e., a substitute dynamical system that is simple enough to carry out the actual synthesis process [4,23]. This is usually a two-step process: First, designing an abstraction that captures the system response to all possible control actions in all states. Second, using the abstraction to identify the control actions that ensure a given specification. One problem with this two-step approach is that a complete abstraction may be extremely complex. It stands to reason that a simpler abstraction may be sufficient to find an acceptable supervisor, so we forego the objective of finding the maximal one.

Discrete abstractions of continuous systems are usually associated with a notion of contractiveness, like global asymptotic stability [11, 12, 20, 24]. We propose to construct *deterministic* abstractions that use feedback control policies to induce contractiveness locally. Deterministic policy abstractions provide an extraordinarily strong link to the concrete system. One can pick any succession of policies from the abstraction, apply it to the concrete system and is guaranteed to obtain the same outputs – without having to adapt to the concrete state. If even the policies are independent of the concrete state, which we call a reach set abstraction, *any trace from a reach set abstraction can be enforced completely in an open loop*. This can be advantageous in critical situations, under degraded operating conditions, or when feedback can be compromised due to damage, malfunction, or communication problems. For example, open-loop control is preferred for the control surfaces of certain missiles [6].

Given the power of a deterministic abstraction, the question is under what conditions and with which tools they can be constructed. We show that deterministic abstractions exist under conditions similar to standard bisimulations. While finding the maximal policy involves robust backward reachability, the forward reachability operator suffices to obtain a solution, and we show maximality under certain conditions. To synthesize the local feedback policies, we use linearization. The linearized system allows us to describe the sufficient conditions for the abstraction with a single linear constraint system, which can then be fed to an off-the-shelf solver. If the linear constraint system is infeasible, we can use feedback from the LP solver to select a subset of transitions that is satisfiable yet preserves properties like connectedness.

Related Work. The formal basis of our abstractions is provided by the well-established notions of simulation [17] and alternating simulation [2], which were applied to continuous control systems in [20]. Inspired by the approach in [9], we admit nondeterministic outputs and consider specifications that are upwards-closed with respect to sets of outputs. While simulation preserves safety properties, bisimulation preserves much stronger properties like those captured with temporal logics like CTL* [8]. However, bisimulation with both state and input labels requires an extremely close correspondence between the two systems so that abstractions that bisimulate the original system may be found only in a restricted class of systems; see, for example, [22].

We consider nondeterministic dynamical systems, which may end up in different states for the same input. For this situation, *alternating simulation* [2] provides a stronger relationship, which is particularly suited to control systems [20], and we build on this concept. An even stronger notion, called *feedback refinement relations* [21], requires that inputs in the concrete and abstract system are matched exactly; our policy abstractions require this implicitly in the sense that applying the policy to the concrete state must in all cases lead to states that are covered by the relation. Controllers synthesized with the approach in [21] can immediately be mapped to concrete control actions; in our case, this is achieved through policies, which, however, require continuous state information. In [20], the antagonistic choice of successor states is explicitly represented by disturbance labels. In our approach, disturbances are implicitly represented by successor sets, which encode a nondeterministic choice. We formalize a novel abstraction-based approach in which the transitions are not labeled by discretized control inputs, but by local state-feedback controllers that can ensure the determinism of the symbolic system, even when the concrete system is non-deterministic and not incrementally stable [13]. This stateful policy differentiates our approach from much of the literature, where the supervisor is often based only on the state of the concrete system.

To make our abstraction deterministic, we use closed-loop strategies to ensure contractions. To compute a closed-loop strategy, we first locally linearize the system and then design a linear feedback control to force the system to be locally contractive. Our approach is similar to [10,25] and can be seen as an instance of an ε-close bisimulation as in [11]. However, in contrast to [25], the states of our abstraction can overlap, which is crucial for building non-trivial smart abstractions [10]. As opposed to [10], we solve a linear programming problem to ensure contraction in terms of infinite norms, not quadratic norms.

Structure. The remainder of the paper is structured as follows. In Sect. 2, we describe the fundamental building blocks of the abstraction, notably LTS semantics and alternating simulation. In Sect. 3, we define a particular type of alternating simulation that we call policy abstraction, which maps abstract transitions to feedback policies in the concrete system. We discuss the particular power of deterministic abstractions and show their existence for systems that are globally Lipschitz and contractive. Section 4 extends the approach to more general classes of systems by applying continuous feedback control to induce contractiveness locally. Numerical experiments illustrate the approach in Sect. 5.

2 Discrete Abstractions of Continuous Systems

In this section, we present the fundamental notions that describe the relationship between the continuous dynamical system that we want to supervise and the discrete abstractions that we use to build the supervisor: LTS semantics, simulation, and alternating simulation.

2.1 Discrete-Time Dynamical Control Systems

We consider discrete-time dynamical systems with control inputs, with a set of states $\mathcal{X} \subseteq \mathbb{R}^n$ and a closed and bounded set of control actions $\mathcal{U} \subset \mathbb{R}^m$. The dynamics are described by a set-valued map $F \colon \mathcal{X} \times \mathcal{U} \rightrightarrows \mathcal{X}$ and $F(x,u) \neq \emptyset$ for any $x \in \mathcal{X}, u \in \mathcal{U}$. For a given initial state x_0, a trajectory consists of a sequence

$$x'_{k+1} \in F(x_k, u_k), \qquad x_k \in \mathcal{X}, \; u_k \in \mathcal{U}. \tag{1}$$

We extend $F(x,u)$ to sets $X \in \mathcal{X}$, $U \subseteq \mathcal{U}$ as $F(X,U) = \bigcup_{x \in X, u \in U} F(x,u)$. Based on the one-step forward reachset F, the robust one-step backward reachset $B(\mathcal{X}', \mathcal{U}')$ is defined as the states where choosing the right action from $\mathcal{U}' \subseteq \mathcal{U}$ leads always inside the target set $\mathcal{X}' \subseteq \mathcal{X}$:

$$B(\mathcal{X}', \mathcal{U}') = \{x \mid \exists u \in \mathcal{U}' : F(x,u) \subseteq \mathcal{X}'\}. \tag{2}$$

A *(feedback) policy* is a set-valued function $\pi : \mathcal{X} \rightrightarrows \mathcal{U}$ associating each state with a set of available inputs. Connecting a system with dynamics F with a feedback policy π leads to the closed-loop dynamics F_π defined by $F_\pi(x) = \bigcup_{u \in \pi(x)} F(x,u)$. Let $\Pi(\mathcal{X}, \mathcal{U})$ be the set of all policies over $\mathcal{X}$ and $\mathcal{U}$. A policy is called *non-blocking* if for all $x \in \mathcal{X}$, we have that $\pi(x) \neq \emptyset$.

2.2 LTS Semantics

Two main types of transition systems that are used as abstractions are Kripke structures (KS), whose states are labeled with atomic propositions and whose transitions are unlabeled, and Labeled Transition Systems (LTS), whose states are unlabeled and whose transitions are labeled. Both formalisms are essentially equivalent [7], but KS-type abstractions seem to have been favored for continuous systems, e.g., in [20]. We are concerned with both state information, wishing to direct the system to certain states, and the transitions associated with input actions. Following the example of [9], we include both kinds of labels. This has been called a *doubly labeled transition system* [8], but for the sake of simplicity, we will stick with LTS, defined as follows.

Definition 1. *A labeled transition system (LTS) is $L = (S, s_0, \Sigma, \rightarrow, P, O)$ with*

- *a set of states S, including a state s_0 called the initial state,*
- *a set of action labels Σ,*
- *a transition relation $\rightarrow \subseteq S \times \Sigma \times S$, where $s \xrightarrow{\alpha} s'$ denotes that the system can transition from state s to s' if the action α is applied,*
- *a set of atomic observations P,*
- *an observation function $O : S \rightrightarrows P$ that attributes to each state s all observations that hold in s.*

An LTS is called nonblocking *if every state has at least one outgoing transition. It is called* complete *over a set of labels Σ' if every state has at least one outgoing transition for every label in Σ'.*

The semantics of the LTS are defined over runs and traces of observations, which then lead us to specifications, following the approach proposed in [9]:

Definition 2 (Run, trace, specification). *A run $\rho = s_0 \xrightarrow{\pi_0} s_1 \xrightarrow{\pi_1} s_2 \ldots$ is a (finite or infinite) sequence of alternating states and labels, starting in the initial state and connected by the transition relation. Applying the observation function to each state in a run maps it to a trace, i.e., the sequence of sets of* observations $\tau = O(\rho) = o_0, o_1, o_2 \ldots$ with $o_k = O(s_k)$ for all k. We denote with $\tau \leq \tau'$ for traces $\tau = o_0, o_1, o_2 \ldots$, $\tau' = o'_0, o'_1, o'_2 \ldots$ if $o_k \subseteq o'_k$ for all k, i.e., all observations in τ also hold in τ'. A specification Spec is a set of traces that is upwards closed, i.e., for any $\tau \in$ Spec and τ' with $\tau \leq \tau'$, we have $\tau' \in$ Spec. We say that a run ρ satisfies a specification Spec if $O(\rho) \in$ Spec.

The upwards-closedness of specification follows intuition: When the specification requires that observations $\{a, b\}$ hold, it should also be satisfied if $\{a, b, c\}$ hold since $a \wedge b \wedge c$ implies $a \wedge b$. Our dynamical systems are cast as LTS as follows:

Definition 3. *The semantics of the dynamical system* (1) *is a labeled transition system* $L = (\mathcal{X}, x_0, \mathcal{U}, \rightarrow_{\mathcal{X}}, P, O)$, *with state domain* $\mathcal{X}$, *the label set* $\mathcal{U}$ *and transition relation* $\rightarrow_{\mathcal{X}}$ *defined by* $x \xrightarrow{u}_{\mathcal{X}} x'$ *if and only if* $x' \in F(x, u)$.

The set of atomic observations P and the output function O depend on the type of observation we wish to consider. Assuming perfect state information, we have $P = \mathcal{X}$ and $O(x) = \{x\}$. Alternatively, O can represent a quantizer that indicates states up to a neighborhood; P consists then of the different possible neighborhoods:

Definition 4. *Let* $P = \{\mathcal{S}_1, \mathcal{S}_2, \ldots\}$ *be a collection of sets that cover* $\mathcal{X}$. *We call state quantizer the output function* $O_P(x) = \{\mathcal{S} \in P \mid x \in \mathcal{S}\}$.

If P is a partition of $\mathcal{X}$, the state quantizer is deterministic. If the sets in P overlap, e.g., to model noisy measurements, then O is nondeterministic.

2.3 Abstractions on LTS

We follow the classical route by defining abstractions using simulation relations. From KS, we adopt that outputs match, and from LTS, that labels match.

Definition 5 (Simulation). *Given a pair of LTS* (L_1, L_2) *and* $P_1 = P_2$, *a relation* $R \subseteq S_1 \times S_2$ *is a KS-simulation relation if* $(s_{0,1}, s_{0,2}) \in R$ *and for all* $(s_1, s_2) \in R$

(i) $O_2(s_2) \subseteq O_1(s_1)$ *and*[1]
(ii) $\forall s_1 \xrightarrow{u_1}_1 s'_1$ *there exists* u_2, s'_2 *such that* $s_2 \xrightarrow{u_2}_2 s'_2$ *and* $(s'_1, s'_2) \in R$.

R is a LTS-simulation relation if in (ii), $u_1 = u_2$. R is a bisimulation relation if it is a simulation relation of (L_1, L_2) and its converse R^T is a simulation relation of (L_2, L_1). L_2 is said to simulate L_1 if there is a simulation relation R for (L_1, L_2).

---

[1] [9] uses the converse condition, i.e., $O_1(s_1) \subseteq O_2(s_2)$. We consider our version consistent with the upward-closure requirement for specifications; other work also uses this direction [16].

In typical use, L_1 would be the concrete system and L_2 the abstraction, which may admit more transitions than the concrete system. Simulation preserves safety properties: If the abstract system L_2 remains inside a set of safe states, then so does L_1. For example, assume safe $\in O(s_2)$ if and only if s_2 is considered safe in L_2, and let only safe states be reachable from the initial state $s_{0,2}$. If S_2 simulates S_1, then safe $\in O(s_1)$ for all reachable s_1 in L_1.

We now turn to the question of when a transition in the abstraction can be enforced in the concrete system. We consider nondeterministic dynamical systems, which may end up in different states for the same input. This can be modeled as a two-player game in which an agent plays the input action, and the opponent gets to pick the successor state. For this situation, *alternating simulation* [2] provides a relationship that is particularly suited to control systems [20]:

Definition 6 (Alternating Simulation, adapted from [20]). *Given a pair of LTS (L_1, L_2) with $P_1 = P_2$, a relation $R \subseteq S_1 \times S_2$ is an* alternating simulation relation *if $(s_{0,1}, s_{0,2}) \in R$ and for all $(s_1, s_2) \in R$*

- $O_1(s_1) \subseteq O_2(s_2)$,
- $\forall s_1 \xrightarrow{u_1}_1 s_1'$ *there exist* u_2, s_2' *such that* $s_2 \xrightarrow{u_2}_2 s_2'$ *with* $(s_1', s_2') \in R$ *and* $\forall s_2 \xrightarrow{u_2}_2 s_2''$ *there exist* s_1'' *such that* $s_1 \xrightarrow{u_1}_1 s_1''$ *with* $(s_1'', s_2'') \in R$.

In our case, L_2 is the concrete system and L_1 the abstraction: If the abstraction L_1 proposes a move, then L_2 must be able to realize that move with an action, and all other possible successor states for that action (s_2'') must also be in the relation. Intuitively, an alternating simulation over (L_1, L_2) guarantees that any action in L_1 can be implemented in L_2. Similarly, any sequence of observations that can be realized in L_1 can also be realized in L_2 without any risk of nondeterminism leading to different traces [26, Thm. 1].

3 Supervision with Deterministic Policy Abstractions

Our goal is to represent and synthesize supervisors efficiently. We employ discrete—ideally, finite—LTS as models for this. In what we call a *policy abstraction*, every transition $s \xrightarrow{\pi}_A s'$ attributes a policy π to the change from s to s'. In the literature, abstractions are frequently based on the forward reach set $F(\mathcal{X}', \mathcal{U}')$, where $\mathcal{X}'$ are the concrete states associated with s, so the input is the same for all states in $\mathcal{X}'$. This case, which we call *reach set abstraction*, is covered by letting $\pi(x) = \mathcal{U}'$ for all $x \in \mathcal{X}$.

3.1 Policy Abstractions

To relate the abstraction to the control system, we use a special case of alternating simulation, where the relationship between the abstract labels (policies) and concrete labels (input actions) is not entirely arbitrary: the abstract labels are policies, i.e., they map concrete states to a set of concrete labels.

Definition 7 (Reach-set and Policy Abstraction). *Given a dynamical system* (1), *a set of observations P and an observation function $O : \mathcal{X} \rightrightarrows P$, a policy abstraction is an LTS $A = (S, s_0, \Pi, \to_A, P, O_A)$, where Π is a set of nonblocking policies (mapping $\mathcal{X}$ to nonempty subsets of $\mathcal{U}$), such that there exists a relation $R \subseteq S \times \mathcal{X}$ (an alternating simulation relation), with $(s_0, x_0) \in R$ and for all $(s, x) \in R$*

(i) $O_A(s) \subseteq O(x)$,
(ii) $\forall s \xrightarrow{\pi}_A s', \forall x'' \in F(x, \pi(x))$ there exists $s'' : s \xrightarrow{\pi}_A s''$ with $(s'', x'') \in R$.

If all policies in $\xrightarrow{\pi}_A$ are independent of the continuous state, i.e., $\pi = \mathcal{X} \times \mathcal{U}'$ for some $\mathcal{U}' \subseteq \mathcal{U}$, we call A a reach set abstraction.

Simply put, the abstraction must cover the forward reach set in the concrete system with abstract transitions that have the same label. The relation R above is an alternating simulation relation over A and the LTS semantics of F, with an additional constraint on the correspondence between labels.

We can operate a policy abstraction as a supervisor on the system. This leads to a hybrid system, i.e., discrete states representing the policy abstraction, similar to the approach in [10].

Definition 8 (Supervision by a PA). *The semantics of the dynamical system* (1) *supervised by a policy abstraction A is the LTS $L||A = (\mathcal{X}_A, x_{A,0}, \mathcal{U}, \to_{X||A}, P, O)$ with $\mathcal{X}_A = \mathcal{X} \times S$, $x_{A,0} = (x_0, s_0)$, and $(x, s) \xrightarrow{u}_{X||A} (x', s')$ if and only if there exists $s \xrightarrow{\pi}_A s'$ such that $u \in \pi(x)$ and $x' \in F(x, u)$. These dynamics correspond to the one-step forward reachset*

$$F_A\big((x,s)\big) = \Big\{(x',s') \;\Big|\; \exists \pi \in \Pi, s' \in S : s \xrightarrow{\pi}_A s', x' \in F(x, \pi(x))\Big\}.$$

The supervised control system matches the policy abstraction if the output function matches exactly. We formalize this relationship as a bisimulation.

Proposition 1. *If for all (s, x) in the relation R in Definition 7, $O_A(s) = O(x)$, then $L||A$ is a KS-bisimulation of A.*

We are particularly interested in the special case where the policy abstraction is *deterministic*, i.e., in Definition 7, $s' = s''$. A deterministic policy abstraction A encodes traces that can be realized by applying the corresponding sequence of policies from A to the concrete system L. We formalize this by specializing a result from [9] to deterministic policy abstractions:

Theorem 1 (Deterministic PA). *Let $s_0 \xrightarrow{\pi_0} s_1 \xrightarrow{\pi_1} s_2 \ldots$ be a run of a deterministic policy abstraction A that satisfies a given specification* Spec. *Then any run $x_0 \xrightarrow{u_0} x_1 \xrightarrow{u_1} x_2 \ldots$ of L with $u_k \in \pi_k(x_k)$ for all k also satisfies* Spec.

Proof. Let $o_0, o_1, \ldots$ be the trace of $s_0 \xrightarrow{\pi_0}_A s_1 \xrightarrow{\pi_1}_A s_2 \ldots$ and $o'_0, o'_1, \ldots$ the trace of $x_0 \xrightarrow{u_0} x_1 \xrightarrow{u_1} x_2 \ldots$ We first show by induction that $(s_k, x_k) \in R$ for all $k \geq 0$. Induction start: For $x = 0$, we have $(s_0, x_0) \in R$ from the definition of R in Definition 7. Induction step: Let $(s_k, x_k) \in R$. With $s_k \xrightarrow{\pi_k}_A s_{k+1}$, $\forall x'' \in F(x_k, \pi_k(x_k))$ we get $(s_{k+1}, x'') \in R$. Since $x_{k+1} \in F(x_k, \pi_k(x_k))$, we get $(s_{k+1}, x_{k+1}) \in R$. Since $(s_k, x_k) \in R$, $O_A(s_k) \subseteq O(x_k)$ and therefore $t \leq t'$. Since $t \in \mathsf{Spec}$ and Spec is upwards closed, we get $t' \in \mathsf{Spec}$, which concludes the proof.

In the sequel, we use abstractions based on a discrete cover of the states and inputs; we give a generic definition below and will the following sections present ways to compute the corresponding parameters and identify suitable policies. We extensively use vector norms to define sets of states since this reduces set containment relationships to linear inequalities. The results in the remainder of the paper hold for arbitrary vector and matrix norms provided that they are consistent. Let $\mathbb{B}(\varepsilon)$ denote the n-dimensional ball with vector norm ε (centered around the origin), and $\Delta(X)$ the diameter of the set X, i.e., the diameter of the smallest ball containing X.

Definition 9 (Neighborhood-based Abstraction). *Consider a dynamical system (1) equipped with quantizer output, i.e., with LTS semantics $L = (\mathcal{X}, x_0, \mathcal{U}, \to_\mathcal{X}, P, O_P)$, and consider covers of $\mathcal{X}$ and $\mathcal{U}$, defined with distance parameters α and β as follows. Let $\hat{X}$ be a set of pairwise distinct points such that $\hat{X} \oplus \mathbb{B}(\alpha)$ covers $\mathcal{X}$ and let $\hat{U}$ be a set of pairwise distinct points such that $\hat{U} \oplus \mathbb{B}(\beta)$ covers $\mathcal{U}$. We associate each $\hat{x} \in \hat{X}$ with its neighborhood of radius $\varepsilon_{\hat{x}} \geq \alpha$. Let the observations be $P = \{\hat{x} \oplus \mathbb{B}(\varepsilon_{\hat{x}}) \mid \hat{x} \in \hat{X}\}$ and let O_P be the state quantizer in Definition 4. Let A be the LTS $(\hat{X}, \hat{x}_0, \hat{X} \times \hat{U}, \to_A, P, O_A)$ defined as follows. Let $\hat{x}_0$ be any of the points in $\hat{X}$ that are closest to x_0. Let $O_A(\hat{x}) = \hat{x} \oplus \mathbb{B}(\varepsilon_{\hat{x}})$. We call $\varepsilon = \sup\{\varepsilon_{\hat{x}} \mid \hat{x} \in \hat{X}\}$ the accuracy of A.*

It is straightforward to show that the above abstraction satisfies the conditions of a policy abstraction:

Proposition 2. *Any Neighborhood-based Abstraction A, as defined in Definition 9, is a deterministic policy abstraction (witnessed by an alternating simulation relation) if for all $\hat{x}, \pi, \hat{x}'$ with $\hat{x} \xrightarrow{\pi}_A \hat{x}'$ holds that*

$$x \in \hat{x} \oplus \mathbb{B}(\varepsilon_{\hat{x}}) \quad \Rightarrow \quad \|F(x, \pi(x)) - \hat{x}'\| \leq \varepsilon_{\hat{x}'}. \tag{3}$$

3.2 Existence of Deterministic Reach Set Abstractions

We consider dynamical systems from (1) satisfying the following assumption:

Assumption 1. *1. The radius of the reach sets is globally bounded by*

$$\omega = \sup_{x \in X, u \in U} \tfrac{1}{2}\Delta(F(x, u)).$$

2. There are constants $K_x, K_u \geq 0$ such that for all $x, x' \in \mathcal{X}$, $u, u' \in \mathcal{U}$:

$$F(x', u') \subseteq F(x, u) \oplus \mathbb{B}(K_x \|x - x'\| + K_u \|u - u'\|).$$

The above assumptions may, in general, not be satisfied. E.g., if X or U are unbounded, the sup might not exist. If the assumptions are satisfied, a discrete abstraction exists:

Proposition 3. *Consider a dynamical system satisfying Assumption 1 and let A be the LTS $(\hat{X}, \hat{x}_0, \Pi(\mathcal{X},\mathcal{U}), \to_A, P, O_A)$ as defined in Definition 9, i.e., a neighborhood-based abstraction. Let the accuracy ε be uniform, i.e., $\varepsilon_{\hat{x}} = \varepsilon$ for all $\hat{x}$. Let $\hat{U}$ be a set of pairwise distinct points such that $\hat{U} \oplus \mathbb{B}(\beta)$ covers $\mathcal{U}$. Let $\to_A$ be defined as $\hat{x} \xrightarrow{\pi}_A \hat{x}'$ for all combinations of $\hat{x}, \hat{x}' \in \hat{X}, \hat{u} \in \hat{U}$ that satisfy $||F(\hat{x},\hat{u}) - \hat{x}'|| \leq \omega + \alpha$, with $\pi = \mathcal{X} \times (\hat{u} \oplus \mathbb{B}(\beta)) \cap \mathcal{U}$. Let the discretization parameters α, ε, β be such that*

$$K_x < 1 \text{ and } \varepsilon \geq \frac{\omega + K_u \beta + \alpha}{1 - K_x}, \tag{4}$$

Under the above conditions, the LTS A is a deterministic reach set abstraction of the concrete system L, and consequently, there is an alternating simulation relation R that witnesses this relationship. Furthermore, R^T is also a witness that A KS-simulates L.

Proof. We first show by structural induction that $R = \{(\hat{x}, x) \mid \hat{x} \in \hat{X}, x \in \hat{x} \oplus \mathbb{B}(\varepsilon)\}$ is a PA relation. Since $K_x < 1$, we have $\varepsilon > \alpha$. Hence $x_0 \in \hat{x}_0 \oplus \mathbb{B}(\varepsilon) = O_A(\hat{x}_0)$ and $(\hat{x}_0, x_0) \in R$. Assume $(\hat{x}, x) \in R$, so that $x \in \hat{x} \oplus \mathbb{B}(\varepsilon)$. For all $x \in O_A(\hat{x})$, $O_A(\hat{x}) \in O_P(x)$ by Definition 4, which satisfies condition (i). Now assume $\hat{x} \xrightarrow{\pi}_{PA} \hat{x}'$. Under the hypothesis,

$$F(\hat{x} \oplus \mathbb{B}(\varepsilon), \hat{u} \oplus \mathbb{B}(\beta)) \subseteq F(\hat{x}, \hat{u}) \oplus \mathbb{B}(K_x \varepsilon + K_u \beta).$$

Under the hypothesis, $F(\hat{x},\hat{u}) \subseteq \hat{x}' \oplus \mathbb{B}(\omega + \alpha)$, so that with (4):

$$F(\hat{x} \oplus \mathbb{B}(\varepsilon), \hat{u} \oplus \mathbb{B}(\beta)) \subseteq \hat{x}' \oplus \mathbb{B}(\alpha + \omega + K_x \varepsilon + K_u \beta) \subseteq \hat{x}' \oplus \mathbb{B}(\varepsilon).$$

This satisfies condition (ii) and concludes the proof for R.

We now show that R^T is a simulation relation for (L, A). By definition, $x \xrightarrow{u}_X x'$ means that $x' \in F(x, u)$. Since $\mathcal{X}$ is covered by $\hat{X} \oplus \mathbb{B}(\alpha)$ and $\mathcal{U}$ is covered by $\hat{U} \oplus \mathbb{B}(\beta)$, there are $\hat{x}$, π and $\hat{u}$ such that $x \in \hat{x} \oplus \mathbb{B}(\varepsilon)$ and $u \in \pi(x) = \hat{u} \oplus \mathbb{B}(\beta)$. Since $x' \in F(x, u)$, under the hypothesis,

$$x' \in F(\hat{x} \oplus \mathbb{B}(\varepsilon), \hat{u} \oplus \mathbb{B}(\beta)) \subseteq F(\hat{x}, \hat{u}) \oplus \mathbb{B}(K_x \varepsilon + K_u \beta),$$

and there is some $\hat{x}' \in \hat{X}$ with

$$F(\hat{x},\hat{u}) \oplus \mathbb{B}(K_x \varepsilon + K_u \beta) \subseteq \hat{x}' \oplus \mathbb{B}(\alpha + \omega + K_x \varepsilon + K_u \beta) \subseteq \hat{x}' \oplus \mathbb{B}(\varepsilon).$$

Therefore, $x' \in \hat{x}' \oplus \mathbb{B}(\varepsilon)$, and $(x', \hat{x}' \oplus \mathbb{B}(\varepsilon)) \in R^T$. The transition follows from

$$F(\hat{x},\hat{u}) \oplus \mathbb{B}(K_x \varepsilon + K_u \beta) \subseteq \hat{x}' \oplus \mathbb{B}(\alpha + \omega + K_x \varepsilon + K_u \beta) \Leftrightarrow$$
$$F(\hat{x},\hat{u}) \subseteq \hat{x}' \oplus \mathbb{B}(\alpha + \omega) \Leftrightarrow$$
$$||F(\hat{x},\hat{u}) - \hat{x}'|| \leq \alpha + \omega.$$

Therefore $\hat{x} \xrightarrow{\pi}_A \hat{x}'$. The LTS A is deterministic since the $\hat{u}$, and therefore the labels π for each transition, are pairwise distinct.

Corollary 1. *If F is deterministic ($\omega = 0$) and Lipschitz over $\mathcal{X}$ with constant $K_x < 1$ and Lipschitz over $\mathcal{U}$, then a deterministic forward reach set approximation can be constructed with arbitrarily small accuracy ε by setting α and β small enough.*

The above result on the existence of a discrete abstraction is consistent with results from the literature on nondeterministic abstractions, e.g., [20], where it is associated with global asymptotic stability. According to our above result, we can obtain a deterministic abstract and, therefore, the full power of Theorem 1, without any particular downside.

4 Policy Abstraction Through Linear Feedback

In Sect. 3.2, we described a deterministic abstraction obtained using only the forward reach set operator. This works for dynamics that are globally contractive. We now extend the applicable cases to systems that are not contractive everywhere by designing policies. One way to achieve this would be to use the robust backward reach set operator: The states for which there exists a policy that drives them towards a target set is exactly the backward reach set in (2). However, the backward reach set operator is much more expensive to compute than the forward operator, even for linear systems [26]. There is a fundamental limitation to backward reachability, even without robust control: a system that is stable going forward in time is unstable going backward in time [18]. Our approach is to stick with forward reachability and use linear state feedback to achieve contractiveness. The reach set computation and feedback design is based on linear control.

4.1 Linearizing over Discrete States

We base our analysis on linearization and assume a global limit $\varepsilon_{\max}$ on the size of the neighborhoods around the states we consider. Let the input set be a ball $\mathcal{U} = \mathbb{B}(\delta)$ with $\delta > 0$. We assume the following linearization is available. Given linearization points $(\hat{x}, \hat{u}) \in \hat{X} \times \hat{U}$, let $A_{\hat{x},\hat{u}}, B_{\hat{x},\hat{u}}$ be matrices, $c_{\hat{x},\hat{u}}$ a vector and $\gamma_{\hat{x},\hat{u}}, \omega$ be scalars such that for all $\varepsilon \leq \varepsilon_{\max}$, $x \in \hat{x} \oplus \mathbb{B}(\varepsilon), u \in \mathbb{B}(\delta) \cap (\hat{u} \oplus \mathbb{B}(\beta))$,

$$F(x,u) \subseteq c_{\hat{x},\hat{u}} \oplus A_{\hat{x},\hat{u}}(x-\hat{x}) \oplus B_{\hat{x},\hat{u}}(u-\hat{u}) \oplus \mathbb{B}\Big(\gamma_{\hat{x},\hat{u}}(||x-\hat{x}|| + ||u-\hat{u}||) + \omega\Big). \quad (5)$$

In the following, we associate each transition $\hat{x} \xrightarrow{\pi_{\hat{x},\hat{u},\hat{x}'}}_A \hat{x}'$ in the abstraction with an input $\hat{u}$ and a constant feedback control policy

$$\pi_{\hat{x},\hat{u},\hat{x}'}(x) = \left\{ \hat{u} - K_{\hat{x},\hat{u},\hat{x}'} \frac{x-\hat{x}}{\varepsilon_{\hat{x}}} \oplus \mathbb{B}(\beta) \right\} \cap \mathcal{U}, \quad (6)$$

where $K_{\hat{x},\hat{u},\hat{x}'}$ is a real-valued matrix mapping concrete states to the space of control inputs. We exploit this feedback matrix to locally induce contractiveness as needed (but not necessarily everywhere). Note that scaling the feedback matrix by $1/\varepsilon_{\hat{x}}$ is a trick to obtain linear constraints in the sequel, e.g., in (8) and Proposition 4. To avoid saturating the input signal, we must ensure that $\pi_{\hat{x},\hat{u},\hat{x}'}(x) \in \mathcal{U}$ for all $x \in \hat{x} \oplus \mathbb{B}(\varepsilon_{\hat{x}})$, which is surely the case if

$$\left\|\hat{u} - K_{\hat{x},\hat{u},\hat{x}'}\mathbb{B}(1) + \beta\right\| \leq \delta \tag{7}$$

Substituting the policy (6) in (5), we obtain for all $x \in \hat{x} \oplus \mathbb{B}(\varepsilon)$:

$$F(x, \pi(x)) \subseteq c_{\hat{x},\hat{u}} \oplus (A_{\hat{x},\hat{u}} - B_{\hat{x},\hat{u}}K_{\hat{x},\hat{u},\hat{x}'}/\varepsilon_{\hat{x}})(x - \hat{x}) \oplus B_{\hat{x},\hat{u}}\mathbb{B}(\beta)$$
$$\oplus \mathbb{B}\Big((\gamma_{\hat{x},\hat{u}} + \|K_{\hat{x},\hat{u},\hat{x}'}\|/\varepsilon_{\hat{x}})\|x - \hat{x}\| + \omega\Big).$$

We apply the above to the containment relationship (3) to derive a constraint that is sufficient for a deterministic policy abstraction:

$$\left\|c_{\hat{x},\hat{u}} - \hat{x}' \oplus (A_{\hat{x},\hat{u}}\varepsilon_{\hat{x}} - B_{\hat{x},\hat{u}}K_{\hat{x},\hat{u},\hat{x}'})\mathbb{B}(1) \oplus B_{\hat{x},\hat{u}}\mathbb{B}(\beta)\right\|$$
$$+\gamma_{\hat{x},\hat{u}}\varepsilon_{\hat{x}} + \|K_{\hat{x},\hat{u},\hat{x}'}\| + \omega \leq \varepsilon_{\hat{x}'}. \tag{8}$$

Proposition 4. *Consider a dynamical system (1) with LTS L and linearization (5). Let A be the LTS $(\hat{X}, \hat{x}_0, \Pi(\mathcal{X},\mathcal{U}), \rightarrow_A, P, O_A)$ as defined in Definition 9, i.e., a neighborhood-based abstraction. Let $\hat{U}$ be a set of pairwise distinct points Let $I \subseteq \hat{X} \times \hat{U} \times \hat{X}$ be a given set of tuples $(\hat{x}, \hat{u}, \hat{x}')$ such that the following conjunction of linear constraints on the variables $\varepsilon_{\hat{x}}, K_{\hat{x},\hat{u},\hat{x}'}, \beta$ is satisfiable:*

$$\bigwedge_{(\hat{x},\hat{u},\hat{x}')\in I} \left\|c_{\hat{x},\hat{u}} - \hat{x}' \oplus (A_{\hat{x},\hat{u}}\varepsilon_{\hat{x}} - B_{\hat{x},\hat{u}}K_{\hat{x},\hat{u},\hat{x}'})\mathbb{B}(1) \oplus B_{\hat{x},\hat{u}}\mathbb{B}(\beta)\right\| + \gamma_{\hat{x},\hat{u}}\varepsilon_{\hat{x}}$$
$$+\|K_{\hat{x},\hat{u},\hat{x}'}\| + \omega \leq \varepsilon_{\hat{x}'} \wedge \quad \|\hat{u} - K_{\hat{x},\hat{u},\hat{x}'}\mathbb{B}(1) + \beta\| \leq \delta \wedge \alpha \leq \varepsilon_{\hat{x}} \leq \varepsilon_{\max}. \tag{9}$$

Let the transition relation consist of $\hat{x} \xrightarrow{\pi_{\hat{x},\hat{u},\hat{x}'}}_A \hat{x}'$ for all $(\hat{x}, \hat{u}, \hat{x}') \in I$, with $\pi_{\hat{x},\hat{u},\hat{x}'}$ as defined in (6). Then A is a deterministic policy abstraction, witnessed by an alternating simulation relation.

It is straightforward to show that any solution of Proposition 3 also satisfies Proposition 4 ($K_x = \|A_{\hat{x},\hat{u}}\|$, $K_u = \|B_{\hat{x},\hat{u}}\|$, $\gamma_{\hat{x},\hat{u}} = 0$, $\|c_{\hat{x},\hat{u}} - \hat{x}'\| \leq \alpha$ and letting $K_{\hat{x},\hat{u},\hat{x}'} = 0$), so we can rest assured that Proposition 4 is strictly more powerful. The constraint system (9) is linear in the variables $\varepsilon_{\hat{x}}, K_{\hat{x},\hat{u},\hat{x}'}, \beta$ if we use the infinity norm (the 1-norm works, too). This means that a solution can be found efficiently. In addition, we cast it as a linear optimization problem to find a solution with the highest precision (minimize a global bound on $\varepsilon_{\hat{x}}$) or permissiveness (maximize β). The constraints can be further simplified; we limit the discussion for lack of space. An important special case is $B_{\hat{x},\hat{u}}$ having full row rank, since it maximizes the capacity of the feedback controller to make the system contractive. A standard reduction to this case consists of modeling the system at every p-th step for some p: The dynamics of this sub-sampled system are given by

$$F_p(x, [u_{(1)}; \ldots; u_{(p)}]) = F\Big(\cdots F\big(F(x, u_{(1)}), u_{(2)}\big) \cdots, u_{(p)}\Big),$$

where the augmented input vector $u = [u_{(1)}; \ldots; u_{(p)}]$ consists of the concatenation of p input vectors of the original system (one for each time step). The outputs of the system must, of course, be adapted accordingly to preserve the desired properties.

4.2 Selecting Transitions

The main challenge in constructing the abstraction proposed in Proposition 4 is selecting which tuples of transitions $(\hat{x}, \hat{u}, \hat{x}')$ to include in the set I. First, we note that in cases where the linearization $A_{\hat{x},\hat{u}}, B_{\hat{x},\hat{u}}, c_{\hat{x},\hat{u}}$ is independent of $\hat{u}$, we can declare $\hat{u}$ as a variable in the constraint system (9). This reduces the search to pairs $(\hat{x}, \hat{x}')$.

Transitions that are infeasible in the concrete system can be ruled out, e.g., by forward or backward reachability analysis (which one is more precise depends on whether the system is locally stable or instable [18]). However, even if we somehow include only transitions that individually can be concretized, this does not mean that the constraint system (9) is satisfiable.

We propose an iterative approach: Starting from an initial set I_0 (possibly very conservative), we identify one or more problematic transitions, remove them to obtain I_1, and repeat the process until (9) becomes satisfiable (possibly because there are no transitions left). Each time the constraint system (9) is infeasible, the LP solver helpfully provides us with a collection of Irreducible Infeasible Subsets (IIS) of constraints. In our abstraction, each IIS corresponds to a cycle in the transition graph, and by encoding the problem accordingly, the IIS allows us to detect the transitions in the cycle.

Depending on the properties that we wish to preserve, we may be able to prioritize transitions. An approach to identify a minimal subset that preserves reachability relationships between states is described in [14]. A stronger requirement is to preserve strongly connected components (see discussion in Sect. 5); an approach to find the such minimal transition set is given in [3]. Finally, the slack variables returned by the solver can provide quantitative information on which constraints are harder to satisfy, which we can translate into priorities on transitions.

5 Experiments

We present experiments on discrete-time versions of two nonlinear systems with unstable equilibria: the inverted pendulum (IP) and the Van der Pol oscillator (VdP) [15]. Both examples have unstable equilibria, and VdP has a stable limit cycle, so constructing an abstraction is challenging. The continuous-time dynamics are given by the ODE $\dot{x} = f(x, u)$, with:

$$f_{\text{IP}}(x, u) = \begin{pmatrix} x_2 \\ -\frac{g}{l}\sin x_1 + \frac{\beta}{ml^2}x_2 + u \end{pmatrix} \quad f_{\text{VdP}}(x, u) = \begin{pmatrix} 2x_2 \\ -0.8x_1 + 2x_2 - 10x_1^2 x_2 + u \end{pmatrix}, \quad (10)$$

Table 1. Policy Abstractions obtained through Linearization

	Inv. Pendulum	Van der Pol
Number of transitions	835	452
Maximum out-degree	14	3
Average out-degree	6.9	2.0
$\min_{\hat{x}} \varepsilon_{\hat{x}}$	0.100	0.050
$\max_{\hat{x}} \varepsilon_{\hat{x}}$	0.100	0.069
$\min_{\hat{x},\hat{u}} \left\| A_{\hat{x},\hat{u}} - B_{\hat{x},\hat{u}} K_{\hat{x},\hat{u},\hat{x}'} / \varepsilon_{\hat{x}} \right\|_\infty$	0.42	0.71
$\max_{\hat{x},\hat{u}} \left\| A_{\hat{x},\hat{u}} - B_{\hat{x},\hat{u}} K_{\hat{x},\hat{u},\hat{x}'} / \varepsilon_{\hat{x}} \right\|_\infty$	0.95	2.27
avg. $\left\| A_{\hat{x},\hat{u}} - B_{\hat{x},\hat{u}} K_{\hat{x},\hat{u},\hat{x}'} / \varepsilon_{\hat{x}} \right\|_\infty$	0.77	1.21

with parameters $g = -9.81, l = 1, \beta = 1, m = 1$. We consider inputs in $\mathcal{U} = [-8, 8]$ for IP and $\mathcal{U} = [-2, 2]$ for VdP. The nonlinear dynamics are linearized, and the approximation error is over-estimated using Taylor models from interval analysis [1,5]. The parameters $\gamma_{\hat{x},\hat{u}}, \omega$ of the linearization are chosen such that the linearization is guaranteed to contain the original forward reach set by taking into account all approximation and linearization errors. The linearized dynamics are then integrated to obtain the discrete dynamics over a given time step h ($h = 1$ for IP and $h = 0.04$ for VdP), keeping the inputs constant over the time interval. To achieve full rank in B, we take two time steps at a time, i.e., the input of the abstraction is two-dimensional.

Some statistics on the abstractions are given in Table 1, for a discretization parameter $\alpha = 0.05$. For the inverted pendulum, the closed-loop dynamics of the system with feedback are strictly contractive everywhere, i.e.,

$$\max_{\hat{x},\hat{u}} \left\| A_{\hat{x},\hat{u}} - B_{\hat{x},\hat{u}} K_{\hat{x},\hat{u},\hat{x}'} / \varepsilon_{\hat{x}} \right\|_\infty = 0.95 < 1.$$

As a consequence, we can use the same precision $\varepsilon = 0.1$ everywhere. Since the closed-loop system is quite contractive on average, the average out-degree is elevated, i.e., we can deterministically choose from an average of 6.9 possible successor states. Note that the average out-degree is biased by states on the border of the domain, many of which have out-degree zero.

For the Van der Pol oscillator, the closed-loop dynamics are, on average, enlarging the set of successor states, i.e.,

$$\text{avg.} \left\| A_{\hat{x},\hat{u}} - B_{\hat{x},\hat{u}} K_{\hat{x},\hat{u},\hat{x}'} / \varepsilon_{\hat{x}} \right\|_\infty = 1.21 > 1.$$

The expansion of the sets of successor states in some transitions is balanced by a number of contractive transitions. In consequence, the precision ε varies between 0.05 and 0.069. Because the closed-loop system is less contractive compared to the inverted pendulum, the average out-degree is also lower, with only two possible successor states to choose from.

The graphs of the obtained abstractions are shown in Fig. 1 and 2. The greyed nodes represent the strongly connected components (SSCs). The fact that the

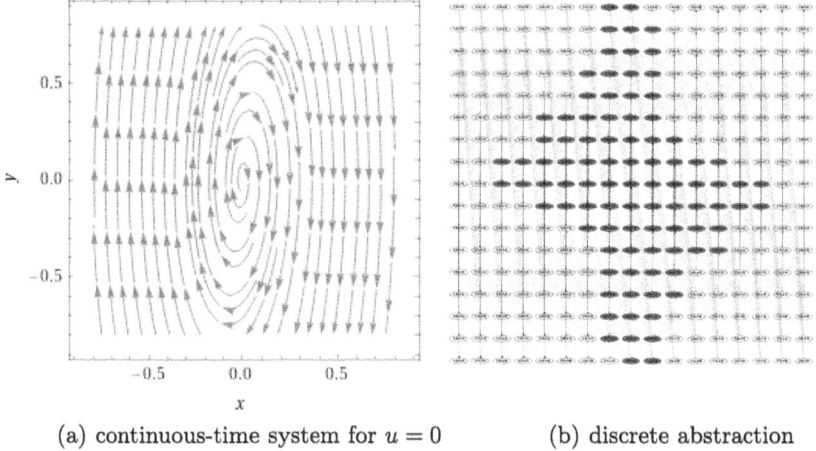

(a) continuous-time system for $u = 0$ (b) discrete abstraction

Fig. 1. Streamline plot of a continuous-time version of the concrete system and a discrete abstraction of the inverted pendulum, with strongly connected components indicated in dark grey (node size not to scale)

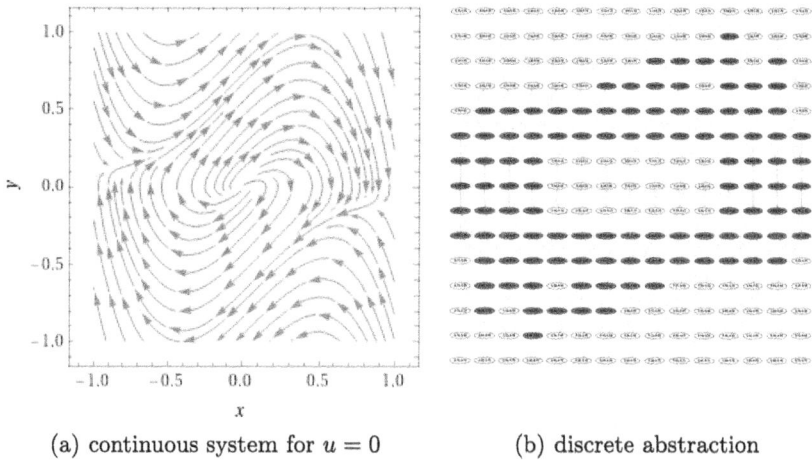

(a) continuous system for $u = 0$ (b) discrete abstraction

Fig. 2. Streamline plot of a continuous-time version of the concrete system and a discrete abstraction of the Van der Pol oscillator, with strongly connected components indicated in dark grey (node size not to scale)

SSCs include a significant portion of the abstract states demonstrates that our abstraction provides a certain degree of completeness even for the chosen coarse grid. We are interested in computing strongly connected components since, for deterministic abstractions, they represent a set of states where all actions are reversible (they can be undone), and any state may be revisited an infinite number of times. This is particularly important when designing supervisors for a reinforcement learning agent [19], which is a goal for future research.

6 Conclusions

The presented method for constructing discrete abstractions for discrete-time, continuous-state systems provides a trade-off between accuracy, completeness, and computational feasibility. By relaxing the stringent completeness requirements of traditional discrete bisimulations, this approach produces deterministic and robust solutions. By leveraging linearization and linear feedback control, we make the approach applicable to non-contractive nonlinear systems. Numerical experiments with classic nonlinear systems illustrate the practicality of this method. The proposed deterministic abstractions provide strong guarantees of system behavior, making them particularly useful for applications requiring high levels of safety and robustness. Numerical experiments with unstable nonlinear systems illustrate the practicality of this method. In future research, we will explore different types of properties that can be guaranteed by adapting these abstractions online, such as preserving safety and reversibility in the presence of dynamic obstacles.

Acknowledgments. This work was supported in part by the project TRAITS, under the French National Research Agency (ANR) grant number ANR-21-FAI1-0005.

References

1. Althoff, M., Grebenyuk, D., Kochdumper, N.: Implementation of Taylor models in CORA 2018. In: Proceedings of the 5th International Workshop on Applied Verification for Continuous and Hybrid Systems (2018)
2. Alur, R., Henzinger, T.A., Kupferman, O., Vardi, M.Y.: Alternating refinement relations. In: Sangiorgi, D., de Simone, R. (eds.) CONCUR 1998. LNCS, vol. 1466, pp. 163–178. Springer, Heidelberg (1998). https://doi.org/10.1007/BFb0055622
3. Bellitto, T., Bergougnoux, B.: On minimum connecting transition sets in graphs. In: Brandstädt, A., Köhler, E., Meer, K. (eds.) WG 2018. LNCS, vol. 11159, pp. 40–51. Springer, Cham (2018). https://doi.org/10.1007/978-3-030-00256-5_4
4. Belta, C., Yordanov, B., Gol, E.A.: Formal Methods for Discrete-Time Dynamical Systems. Springer, Cham (2017). https://doi.org/10.1007/978-3-319-50763-7
5. Berz, M., Hoffstätter, G.: Computation and application of Taylor polynomials with interval remainder bounds. Reliable Comput. **4**(1), 83–97 (1998)
6. Chomachar, S.A., Fard, A.M.: Flight control system for guided rolling-airframe missile. In: 2016 IEEE Aerospace Conference, pp. 1–9 (2016)
7. De Nicola, R., Vaandrager, F.: Action versus state based logics for transition systems. In: Guessarian, I. (ed.) LITP 1990. LNCS, vol. 469, pp. 407–419. Springer, Heidelberg (1990). https://doi.org/10.1007/3-540-53479-2_17
8. De Nicola, R., Vaandrager, F.: Three logics for branching bisimulation. J. ACM (JACM) **42**(2), 458–487 (1995)
9. Demangeon, R., Dima, C., Varacca, D.: Observational preorders for alternating transition systems. In: Malvone, V., Murano, A. (eds.) EUMAS 2023. LNCS, vol. 14282, pp. 312–327. Springer, Cham (2023). https://doi.org/10.1007/978-3-031-43264-4_20

10. Egidio, L.N., Lima, T.A., Jungers, R.M.: State-feedback abstractions for optimal control of piecewise-affine systems. In: 2022 IEEE 61st Conference on Decision and Control (CDC), pp. 7455–7460 (2022)
11. Girard, A., Pappas, G.J.: Approximation metrics for discrete and continuous systems. IEEE Trans. Autom. Control **52**(5), 782–798 (2007)
12. Girard, A., Pappas, G.J.: Approximate bisimulation: a bridge between computer science and control theory. Eur. J. Control. **17**(5–6), 568–578 (2011)
13. Girard, A., Pola, G., Tabuada, P.: Approximately bisimilar symbolic models for incrementally stable switched systems. IEEE Trans. Autom. Control **55**(1), 116–126 (2010). https://doi.org/10.1109/TAC.2009.2034922
14. Khuller, S., Raghavachari, B., Young, N.: Approximating the minimum equivalent digraph. SIAM J. Comput. **24**(4), 859–872 (1995)
15. Korda, M., Mezić, I.: Linear predictors for nonlinear dynamical systems: Koopman operator meets model predictive control. Automatica **93**, 149–160 (2018)
16. Liu, J., Ozay, N.: Finite abstractions with robustness margins for temporal logic-based control synthesis. Nonlinear Anal. Hybrid Syst. **22**, 1–15 (2016)
17. Milner, R.: An Algebraic Definition of Simulation Between Programs. Citeseer (1971)
18. Mitchell, I.M.: Comparing forward and backward reachability as tools for safety analysis. In: Bemporad, A., Bicchi, A., Buttazzo, G. (eds.) HSCC 2007. LNCS, vol. 4416, pp. 428–443. Springer, Heidelberg (2007). https://doi.org/10.1007/978-3-540-71493-4_34
19. Moldovan, T.M., Abbeel, P.: Safe exploration in Markov decision processes. In: Proceedings of the Conference on Machine Learning, ICML 2012, Madison, WI, USA, pp. 1451–1458. Omnipress (2012)
20. Pola, G., Tabuada, P.: Symbolic models for nonlinear control systems: alternating approximate bisimulations. SIAM J. Control. Optim. **48**(2), 719–733 (2009)
21. Reissig, G., Weber, A., Rungger, M.: Feedback refinement relations for the synthesis of symbolic controllers. IEEE Trans. Autom. Control **62**(4), 1781–1796 (2016)
22. Van der Schaft, A.: Equivalence of dynamical systems by bisimulation. IEEE Trans. Autom. Control **49**(12), 2160–2172 (2004)
23. Tabuada, P.: Verification and Control of Hybrid Systems: A Symbolic Approach. Springer, Cham (2009). https://doi.org/10.1007/978-1-4419-0224-5
24. Tabuada, P., Pappas, G.J.: Finite bisimulations of controllable linear systems. In: 42nd IEEE International Conference on Decision and Control (IEEE Cat. No. 03CH37475), vol. 1, pp. 634–639. IEEE (2003)
25. Tajvar, P., Meyer, P.J., Tumova, J.: Closed-loop incremental stability for efficient symbolic control of non-linear systems. IFAC-PapersOnLine **54**(5), 121–126 (2021). https://doi.org/10.1016/j.ifacol.2021.08.485. https://www.sciencedirect.com/science/article/pii/S240589632101260X, 7th IFAC Conference on Analysis and Design of Hybrid Systems ADHS 2021
26. Yang, L., Zhang, H., Jeannin, J.B., Ozay, N.: Efficient backward reachability using the minkowski difference of constrained zonotopes. Trans. Comput.-Aided Des. Integ. Circ. Syst. **41**(11), 3969–3980 (2022). https://doi.org/10.1109/TCAD.2022.3197971

Markov Decision Processes with Sure Parity and Multiple Reachability Objectives

Raphaël Berthon[✉], Joost-Pieter Katoen, and Tobias Winkler

RWTH Aachen University, 52062 Aachen, Germany
{berthon,katoen,tobias.winkler}@cs.rwth-aachen.de

Abstract. This paper considers the problem of finding strategies that satisfy a mixture of sure and threshold objectives in Markov decision processes. We focus on a single ω-regular objective expressed as parity that must be surely met while satisfying n reachability objectives towards sink states with some probability thresholds too. We consider three variants of the problem: (a) strict and (b) non-strict thresholds on all reachability objectives, and (c) maximizing the thresholds with respect to a lexicographic order. We show that (a) and (c) can be reduced to solving parity games, and (b) can be solved in EXPTIME. Strategy complexities as well as algorithms are provided for all cases.

Keywords: MDPs · Parity · Reachability · Multi-objective

1 Introduction

Markov decision processes (MDPs) [6,35] are prominent models for strategic planning and decision making in face of stochastic uncertainty. An important, yet intricate, problem is to determine if and how a combination of *multiple* properties, or *objectives*, is realizable in a given MDP. As objectives may be *conflicting*, it does not suffice to analyze each of them independently [4,20,37]. Instead, *trade-offs* between the objectives have to be taken into account. In this paper, we combine objectives of different nature: *Sure objectives* must be fulfilled on *all* possible executions of the MDP, even on those with probability 0. Thus, sure objectives do not depend on the exact transition probabilities; in fact, they can be analyzed by replacing the probabilities with an adversary. *Threshold objectives*, on the other hand, have to be satisfied with some probability of at least (or greater than) a given constant. Various combinations of sure and threshold objectives have been investigated in prior work [1,7,8,13,22,23]. Here, we focus

---

Berthon and Katoen are supported by the DFG Grant Nr. 520530521 "POMPOM", Katoen and Winkler are supported by the EU's Horizon 2020 research and innovation programme under the Marie Skłodowska-Curie grant agreement Nr. 101008233 "MISSION", Winkler is additionally supported by the DFG RTG 2236 "UnRAVeL".

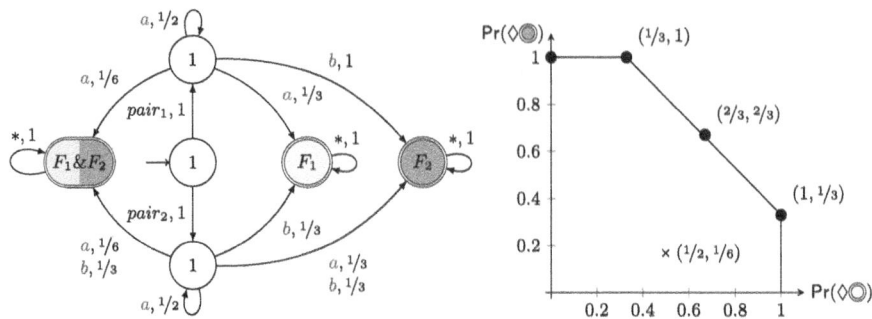

Fig. 1. An MDP with $Act = \{a, b, pair_1, pair_2, *\}$, target sets F_1, F_2, and parity condition ρ assigning 0 to the sinks and 1 to the non-sinks. *Right:* The Pareto frontier. (Color figure online)

on MDPs with a single sure ω-regular objective expressed as a *parity* condition together with n *reachability* threshold objectives towards sink states.

Running Example. In a game show a contestant plays a gamble to win either a bike, a surfboard, or both prizes. The gamble is as follows: The contestant must choose one out of two pairs of 6-sided dice. Each pair consists of a green and a red die. The four dice are all different; each of their faces shows either the bike, the board, both bike and board, or the symbol ↻ ("repeat"):

$pair_1$ red: $3 \times ↻, 1 \times$ both, $2 \times$ bike $green$: $6 \times$ board

$pair_2$ red: $3 \times ↻, 1 \times$ both, $2 \times$ board $green$: $2 \times$ both, $2 \times$ bike, $2 \times$ board

After committing to a pair of dice, the contestant rolls one die from their pair. The green die immediately ends the gamble with the resulting prize(s). The red die either ends the gamble or, in case of ↻, allows the contestant to roll again (the same die or the other one). However, since the show is broadcast on live TV, there is an additional rule: The gamble may not be prolonged indefinitely, i.e., the contestant may try the red die at most an arbitrary, but *a priori* fixed number of times. Clearly, optimal strategies depend on how much the contestant prefers one prize over the other. The MDP in Fig. 1 models this gamble. Prizes are encoded as reachability of sinks ($F_1 \triangleq$ bike, $F_2 \triangleq$ board); the additional constraint is a sure parity condition.

This Paper. We study the following three problems: Given a finite MDP, decide if it is possible to satisfy a sure parity objective and, at the same time, n sink reachability objectives with (a) all *strict*, or (b) all *non-strict* probability thresholds. In addition, we consider the problem (c) of checking existence of a *lexicographically* maximal achievable threshold vector w.r.t. a given linear order on the reachability objectives. In all cases, we are also interested in computing witnessing strategies if they exist. These problems are challenging, both computationally and conceptually. Already for two reachability objectives (without any sure objective) the set of achievable thresholds —the *Pareto frontier*— is a convex polytope with superpolynomially many vertices in general [25]. The problems

Table 1. Some existing and our new (in bold) results on multi-objective MDPs.

Sure objective	Probabilistic objective(s)	Complexity	Memory	Reference
1 sure parity	-	NP ∩ coNP	finite	[24]
1 sure parity	1 threshold parity	NP ∩ coNP	infinite	[8]
1 sure parity	n almost-sure parity	NP ∩ coNP	infinite	[7]
n sure parity	n almost-sure parity	$\mathsf{P}^{\mathsf{NP}}(=\Delta_2^{\mathsf{P}})$	infinite	[7]
1 sure parity	1 threshold reach	NP ∩ coNP	finite	[8]
-	n threshold sink reach	PTIME	finite	[25]
1 sure parity	n **strict threshold sink reach** (a)	NP ∩ coNP	finite	[Theorem 1]
1 sure parity	n **threshold sink reach** (c)	EXPTIME	finite	[Theorem 3]
-	lexicographic Streett	PTIME	finite	[17]
1 sure parity	**lexicographic sink reach** (b)	NP ∩ coNP	finite	[Theorem 2]

we study are more general than this and add further subtleties: While it is easy to show that the thresholds in the *interior* of the Pareto frontier are satisfiable with *any* sure parity objective (Sect. 3), identifying exactly which points on the *boundary* of the frontier are achievable is quite involved (Sect. 5).

Contributions. Our three main results are summarized in bold in Table 1. (a) Checking if a sure parity objective and n *strict* sink reachability thresholds are achievable simultaneously is in NP ∩ coNP (Sect. 3). This is done via a reduction to parity games, admitting a quasi-polynomial algorithm [14].

(c) We propose an algorithm that finds a strategy ensuring a sure parity objective while also maximizing the probability of reaching n sinks w.r.t. a *lexicographic* order (Sect. 4). It relies on a concept we call *projection*, a notion also used in prior work [4,17,25]. Our algorithm solves polynomially many parity games in sequence, hence the problem is (again) in NP ∩ coNP.

(b) We present an algorithm for finding a strategy satisfying a sure parity objective and n sink reachability objectives with *non-strict* thresholds (Sect. 5). Our algorithm alternates between computing Pareto frontiers, making projections, and pruning states not satisfying the sure parity objective; to our knowledge, this idea is new. Its time complexity is exponential in the size of the MDP, as it relies on computing exact Pareto frontiers.

We also treat strategy complexity for each case. Our results are a further step towards a solution for general combinations of sure and probabilistic objectives. An extended version of the results with detailed proofs can be found in [9].

Related Work. Previous research [7] on mixtures of sure and probabilistic objectives focused on *qualitative* thresholds, i.e., >0 and =1. Here, we also allow *quantitative* thresholds strictly between 0 and 1. We rely on some results of [8] that studied combining one sure parity and *one threshold parity* objective. This problem was shown to be in NP ∩ coNP, via a reduction to parity games with weights that can be solved in quasi-polynomial time [38]. The main difference to [8] is that we consider *multiple reachability* threshold objectives. The setting of one sure parity and one almost-sure parity has been further studied in [22] where it was shown that the restriction to finite memory strategies is still in NP ∩ coNP for MDPs, and coNP-complete for stochastic games.

The seminal paper [25] shows that computing the Pareto frontier for a mixture of either reachability or ω-regular objectives can be reduced to solving linear programs. An efficient technique, value iteration, is exploited by tools such as PRISM [33], Storm [30], and MultiGain [12]. The work [19] considers MDPs with two different kinds of stochastic mean-payoff objectives, and supports computing the Pareto frontier. *Percentile queries*, multiple threshold constraints that must each be satisfied with some probability, were studied in [36]. In [10,29], multiple reachability conditions associated to the expected or accumulated cost to reach a target are considered.

Lexicographic optimization is a widely employed principle in multi-objective decision making [34,40]. The idea is that a strategy should prioritize a primary objective while still doing best possible for a secondary objective, etc. The work of [17] imposes a lexicographic order on multiple, possibly conflicting, reachability, safety and ω-regular objectives. Reinforcement learning with lexicographic ω-regular conditions is studied in [28]. To the best of our knowledge, lexicographic optimization in MDPs together with a sure condition has not been studied yet.

Other approaches have been considered. Combinations of parity and mean-payoff [1], and parity and weighted games [38] have been studied in prior work. An alternative way to combining objectives is *strategy logic* (SL) [16], an extension of CTL that can express formulas involving the change of strategies. A probabilistic SL has been defined in [2].

2 Preliminaries

We write $\mathbb{N} = \{1, 2, 3, \ldots\}$ and $\mathbb{N}_0 = \mathbb{N} \cup \{0\}$. For $n \in \mathbb{N}$, we let $[n] = \{1, \ldots, n\}$, and $[n]_0 = [n] \cup \{0\}$. Vectors $\mathbf{v} \in \mathbb{R}^n$ are written in bold. For $\mathbf{v} \in [0,1]^n$ and $i \in [n]$, we denote by v_i the i-th component of $\mathbf{v}$. Given $\mathbf{v}, \mathbf{u} \in \mathbb{R}^n$, their *dot product* is defined as $\mathbf{v} \cdot \mathbf{u} = \sum_{i \in [n]} v_i \cdot u_i$. The symbol $\mathbf{e}_i \in \{0,1\}^n$ is the unit vector where the i-th component is 1 and all others are 0. The componentwise order on $\mathbb{R}^n$ is denoted with $\leq$. Given a finite set A, a *(probability) distribution* on A is a function $f \colon A \to [0,1]$ such that $\sum_{a \in A} f(a) = 1$. $\mathcal{D}(A)$ denotes the set of all distributions on A. We define the *support* $\mathsf{supp}(f) = \{a \in A \mid f(a) > 0\}$.

2.1 MDPs, Strategies, and Objectives

A *Markov decision process* (MDP) is a tuple $\Gamma = (S, Act, \mathbf{P})$ where $S \neq \emptyset$ is a countable set of *states*, $Act \neq \emptyset$ is a finite set of *actions*, and $\mathbf{P} \colon S \times Act \times S \to [0,1]$ is a *transition probability function* satisfying $\sum_{s' \in S} \mathbf{P}(s, a, s') \in \{0, 1\}$ for all $s \in S, a \in Act$. If the sum is 1 for a state-action pair s, a, then a is *enabled* at s. We write $Act(s)$ for the set of all actions enabled at s, and require that $Act(s) \neq \emptyset$. An MDP Γ is called *finite* if S is finite. A state $s \in S$ is called a *sink* if for all $a \in Act(s)$ we have $\mathbf{P}(s, a, s) = 1$. For technical convenience we assume

$|Act(s)| = 1$ for all sinks $s \in S$. Note that we may consider the same MDP with different initial states since the latter is not fixed in our definition. See Fig. 1 for a finite example MDP.

A (discrete-time) *Markov chain* (MC) is an MDP with $|Act| = 1$. We omit Act from the definition of MC and just write $\mathcal{M} = (S, \mathbf{P})$. We also identify $\mathbf{P}$ with a function of type $S \times S \to [0,1]$.

A *strategy* for an MDP Γ is a state machine $\sigma = (Q, q_\iota, \delta, o)$ where Q is a countable set of memory modes, $q_\iota \in Q$ is the initial mode, $\delta: Q \times S \to Q$ is a transition function, and $o: Q \times S \to \mathcal{D}(Act)$ is an output function with $\mathsf{supp}(o(q,s)) \subseteq Act(s)$ for all $q \in Q, s \in S$. σ is called *finite-memory* if $|Q| < \infty$, *memoryless* if $|Q| = 1$, and *deterministic* if $|\mathsf{supp}(o(q,s))| = 1$ for all $q \in Q, s \in S$.

Given an MDP $\Gamma = (S, Act, \mathbf{P})$ with strategy $\sigma = (Q, q_\iota, \delta, o)$, we define the *induced MC* $\Gamma[\sigma] = (S \times Q, \mathbf{P}^\sigma)$ where $\mathbf{P}^\sigma((s,q),(s',q')) = \mathbf{P}(s, o(q,s), s')$ if $\delta(q,s) = q'$, and otherwise $\mathbf{P}^\sigma((s,q),(s',q')) = 0$. In the following, we only consider finite MDPs, but when considering an infinite-memory strategy, the resulting MC is countably infinite. In the context of algorithms, we always assume that the probabilities in the given MDPs, strategies, and probability thresholds are *rational* numbers encoded as numerator-denominator pairs in binary.

Given an MC $\mathcal{M} = (S, \mathbf{P})$ with a distinguished initial state $s \in S$, we consider the σ-algebra $\mathcal{F}$ generated by the *cylinder sets* $\{\pi S^\omega \mid \pi \in S^*\}$ and the associated probability measure $\mathsf{Pr}_s^\mathcal{M}: \mathcal{F} \to [0,1]$ which is uniquely defined by requiring that for all $\pi = s_0 \ldots s_k \in S^+$, $k \geq 0$, we have $\mathsf{Pr}_s^\mathcal{M}(\pi S^\omega) = \prod_{i=0}^{k-1} \mathbf{P}(s_i, s_{i+1})$ if $s_0 = s$, and $\mathsf{Pr}_s^\mathcal{M}(\pi S^\omega) = 0$ if $s_0 \neq s$. See, e.g. [5, Chapter 10] for more details. The sets in $\mathcal{F}$ are called *measurable*. Further, we define $\mathsf{Paths}_s^\mathcal{M} = \{s_0 s_1 \ldots \in S^\omega \mid s_0 = s \wedge \forall i \geq 0: \mathbf{P}(s_i, s_{i+1}) > 0\}$.

An *objective* for an MDP Γ is a measurable[1] set of paths $\Omega \subseteq S^\omega$. A *reachability* objective for Γ is of the form $\Diamond F = \{\pi \in S^\omega \mid \exists k \geq 0: \pi(k) \in F\}$ where $F \subseteq S$. *Bounded reachability* objectives have the form $\Diamond^{\leq B} F = \{\pi \in S^\omega \mid \exists k \in [B]_0: \pi(k) \in F\}$, for some $B \in \mathbb{N}_0$. A *parity* objective for Γ is defined via a priority function $\rho: S \to [k]_0$, where $k \in \mathbb{N}_0$. For $\pi \in S^\omega$, let $\inf(\pi) = \{s \in S \mid \forall i \geq 0: \exists j \geq i: \pi(j) = s\}$ be the set of states visited infinitely often on π. Then the parity objective defined by ρ is $\{\pi \in S^\omega \mid \max \rho(\inf(\pi))$ is even$\}$. In the following, we identify the function ρ with the objective it defines.

2.2 Multi-objectives and Pareto Frontiers

A *multi-objective (MO) formula* for MDP Γ is a syntactic object $\varphi = \bigwedge_{i=1}^n atom_i$ with $atom_i \in \{\mathsf{S}(\Omega), \mathsf{Pr}_{\sim p}(\Omega) \mid \Omega \text{ an objective for } \Gamma, \sim \in \{>, \geq\}, p \in [0,1]\}$. An MC $\mathcal{M}$ with a distinguished initial state s satisfies a *threshold* constraint $\mathsf{Pr}_{\sim p}(\Omega)$ if $\mathsf{Pr}_s^\mathcal{M}(\Omega) \sim p$, a *sure* constraint $\mathsf{S}(\Omega)$ if $\mathsf{Paths}_s^\mathcal{M} \subseteq \Omega$, and it satisfies the formula φ (in symbols: $s \models_\mathcal{M} \varphi$) if it satisfies $atom_i$ for all $i \in [n]$. For an MDP Γ with strategy σ we write $s, \sigma \models_\Gamma \varphi$ if $s \models_{\Gamma[\sigma]} \varphi$, and $s \models_\Gamma \varphi$ if $s, \sigma \models_\Gamma \varphi$ for *some* strategy σ for Γ. In this paper, we only consider formulas of the form

---

[1] Measurability is actually only important for probabilistic objectives, not for sure objectives. However, all concrete objectives considered in this paper are measurable.

$S(\rho) \wedge \Pr_{\sim p_1}(\Diamond F_1) \wedge \ldots \wedge \Pr_{\sim p_n}(\Diamond F_n)$, i.e., conjunctions of one sure parity and $n \geq 1$ reachability objectives, either all with strict or non-strict thresholds.

We now define Pareto frontiers. Given an MO formula φ for MDP Γ containing n threshold constraints $\Pr_{\sim p_1}(\Omega_1), \ldots, \Pr_{\sim p_n}(\Omega_n)$ (in this order), we write $\varphi(p_1, \ldots, p_n) = \varphi(\mathbf{p})$ to emphasize the dependency of φ on the threshold vector $\mathbf{p} \in [0, 1]^n$. We also write $\varphi(\mathbf{x})$ without further qualifying $\mathbf{x}$ to indicate that the thresholds are *variables* $\mathbf{x} = (x_1, \ldots, x_n)$. We define the set of *achievable threshold vectors* as $Ach(\Gamma, s, \varphi(\mathbf{x})) = \{\mathbf{p} \in [0,1]^n \mid s \models_\Gamma \varphi(\mathbf{p})\} \subseteq [0,1]^n$. Note that $Ach(\Gamma, s, \varphi(\mathbf{x}))$ is *downward-closed* as $\sim \in \{\geq, >\}$, and *convex* since a convex combination $c \cdot \mathbf{p} + (1-c) \cdot \mathbf{p}'$, $c \in (0,1)$, of achievable threshold vectors $\mathbf{p}, \mathbf{p}'$ is achieved by a strategy that plays the strategy for $\mathbf{p}$ with probability c and the one for $\mathbf{p}'$ with probability $1-c$, see, e.g. [25]. Given a set of vectors X, we define $gen(X)$, the subspace generated by X, as the intersection of all subspaces of $\mathbb{R}^n$ containing X. It is the smallest subspace containing X. For the next definition recall that the *boundary* ∂X of a set $X \subseteq [0,1]^n$ is defined as $\overline{X} \setminus \text{int } X$, where $\overline{X}$ is the closure of X and int X is the interior, i.e., the largest open subset of X.

Definition 1 (Pareto frontier). *Let $\Gamma = (S, Act, \mathbf{P})$ be an MDP, $\varphi(\mathbf{x})$ an MO formula for Γ, and $s \in S$. We define $Pareto(\Gamma, s, \varphi(\mathbf{x})) = \partial Ach(\Gamma, s, \varphi(\mathbf{x}))$.*

The above definition is similar to the one from [3]; other authors define the Pareto frontier in a slightly different way, e.g., as the $\leq$-maxima of $Ach(\Gamma, s, \varphi(\mathbf{x}))$ [26].

The Pareto frontier is the boundary of a convex polytope of dimension at most n [25]. Such a polytope P has faces of lower dimension, from 0 (a vertex) to $n-1$. These faces are defined as follows: given a hyperplane H intersecting P, the polytope $H \cap P$ is a face of P iff P lies fully on one of the two closed half-spaces defined by H. When considering the polytope associated to a Pareto frontier (and by generalization the Pareto frontier itself), we can freely separate points between those strictly in the interior, those on the border, and those in the exterior. In what follows, we only consider faces defined by intersection with hyperplanes whose normal vectors only have non-negative components. More on convex polytopes can be found in [27].

Example 1. Consider the MDP Γ in Fig. 1 on page 2 and the formula $\varphi(\mathbf{x}) = \Pr_{\geq x_1}(\Diamond F_1) \wedge \Pr_{\geq x_2}(\Diamond F_2)$ (we ignore parity). Let s be the marked initial state.

- $\mathbf{p}_1 = (1, 1/3)$ is achievable from s by choosing $pair_1$, and then playing a repeatedly to reach F_1 with $2/3$, and both F_1 and F_2 with probability $1/3$.
- $\mathbf{p}_2 = (1/3, 1)$ is achievable from s by choosing $pair_2$, and then a repeatedly to reach F_2 with $2/3$ and both F_1 and F_2 with probability $1/3$.
- As mentioned earlier, a convex combination of two achievable points is achievable by following one of the two strategies with suitable probabilities. However, in this specific example, the vector $\mathbf{p}_3 = (2/3, 2/3) = 0.5 \cdot \mathbf{p}_1 + 0.5 \cdot \mathbf{p}_2$ is achievable with a *deterministic* strategy as well: First choose $pair_2$ in s and then b to reach state F_1 with $1/3$, state F_2 with $1/3$, and both F_1 and F_2 with probability $1/3$. These points will be relevant in Sects. 4 and 5.

– The above strategies are all *Pareto-optimal*, but there are also sub-optimal strategies, e.g., choosing $pair_1$ in s and then b leads to reaching F_1 and F_2 with probability 0 and 1, respectively. This is sub-optimal as $(0,1) \leq \mathbf{p}_2$.

We consider *clean* MDPs throughout the rest of the paper:

Definition 2 (Clean MDP). *Let $\Gamma = (S, Act, \mathbf{P})$ be an MDP. Γ is clean ...*

– *... w.r.t. a parity objective $\rho: S \to [k]_0$ if for all $s \in S$, we have $s \models_\Gamma \mathsf{S}(\rho)$, i.e., ρ is surely satisfiable from every state s.*
– *... w.r.t. target sets $F_1, \ldots, F_n \subseteq S$ if for all $s \in S$, we have $s \models_\Gamma \mathtt{Pr}_{\geq 1}(\Diamond F)$, where $F = \bigcup_{i=1}^n F_i$, and every state in F is a sink.*

Example 2. The MDP from Fig. 1 is clean w.r.t. ρ because from every state, there is a strategy that surely reaches a sink with priority 0. For instance, from the topmost state, the rightmost sink is reachable by playing action b. The MDP is also clean w.r.t. F_1, F_2 because from every state there exists a strategy reaching $F = F_1 \cup F_2$ with probability one, and because F contains sink states only.

Some remarks about clean MDPs are in order: (i)One can *clean* an MDP w.r.t. parity by identifying and removing states that violate $\mathsf{S}(\rho)$. The latter can be done by solving the 2-player deterministic parity game obtained by replacing the randomness in the MDP by an antagonistic player. Note that deciding the winner in a parity game (and hence checking if $s \models_\Gamma \mathsf{S}(\rho)$ holds for state s) is in NP ∩ coNP [14], and even in UP ∩ coUP [31], but is not known to be in PTIME. (ii) Reachability towards *sinks* only is a more severe restriction. We make it because simultaneous almost-sure reachability of n general target sets in an MDP is already PSPACE-complete [36].Intuitively, this is because strategies have to remember which targets were already seen. Contrarily, sink reachability often admits more practical complexities as shown in Sects. 3 and 4 (also see, e.g., [17] and [39]) and is still of practical interest. We leave a solution for general reachability for future work, since it would likely further improvements on the techniques we introduce.

3 Sure Parity and n Strict Reachability Thresholds

We study MO formulas of the form $\mathsf{S}(\rho) \wedge \bigwedge_{i=1}^n \mathtt{Pr}_{>p_i}(\Diamond F_i)$ in this section, i.e., with *strict* thresholds only. Non-strict thresholds are more involved, see Sect. 5. We start by stating the main result of this section. Note that it is formulated for MDPs that are clean w.r.t. the parity objective ρ and the target sets F_i. The assumption of being clean w.r.t. parity can be dropped, but this incurs the additional complexity of solving a parity game (see Definition 2 and subsequent remarks), and hence leads to an NP ∩ coNP complexity bound on the associated decision problem.

Theorem 1. *Let Γ be a clean MDP w.r.t. parity objective ρ and target sets $F_1, \ldots, F_n \subseteq S$. Further, let $\mathbf{p} \in [0,1]^n$, and let $s \in S$ be a state. Then:*

1. *The decision problem $\exists \sigma \colon s, \sigma \models_\Gamma \mathsf{S}(\rho) \wedge \bigwedge_{i=1}^{n} \Pr_{>p_i}(\Diamond F_i)$ is in* PTIME.
2. *A witness strategy σ using at most $2^{\mathsf{poly}(|\Gamma|+nD)}$ memory, where D is the bit-complexity of the rational numbers in $\mathbf{p}$, can be effectively constructed for the* YES-*instances.*

Proof (sketch). Using Corollary 3.5 of [25], we can test if $\varphi(\mathbf{p}) = \bigwedge_{i=1}^n \Pr_{>p_i}(\Diamond F_i)$ is achievable (note that we have dropped the parity objective). If it is not, then the answer is clearly NO. Otherwise, there exists a memoryless but possibly randomized strategy achieving $\varphi(\mathbf{p})$. As the inequalities in $\varphi(\mathbf{p})$ are strict, it can be shown that the reachability thresholds can be guaranteed after playing the strategy for some *finite* but exponential number of steps. As parity objectives are prefix-independent, we can then simply switch to a memoryless deterministic winning strategy for parity. For the latter argument to work, it is crucial that the MDP is clean w.r.t. the parity objective ρ, i.e., that it is possible to satisfy $\mathsf{S}(\rho)$ from every state of the MDP.

Example 3. Reconsider the MDP Γ from Fig. 1 with initial state s, yellow target F_1 and blue target F_2. To surely satisfy ρ, we must visit non-sink states only finitely often (this is a co-Büchi condition). We show that $s \models_\Gamma \mathsf{S}(\rho) \wedge \Pr_{>1/2}(\Diamond F_1) \wedge \Pr_{>1/6}(\Diamond F_2)$, achieving a value strictly greater than $\mathbf{p}_4 = (1/2, 1/6)$ which is strictly inside the Pareto frontier. To achieve this objective, we take the following strategy: we first play action $pair_1$, then a twice. By doing so, we have probability $1/4$ of reaching the leftmost state (contained in both F_1 and F_2), and probability $1/2$ of reaching the state only fulfilling F_1. If we now play action b, we satisfy condition ρ surely. We end up reaching F_1 with probability $3/4 > 1/2$ and reaching F_2 with probability $1/2 > 1/6$. Note that in general, thresholds that are achievable with strict inequalities are located strictly inside the Pareto frontier.

4 Sure Parity and Lexicographic Reachability

We are now interested in surely satisfying a parity objective while maximizing the probability of reaching n target sets in lexicographic order. Towards this goal we define the notion of *projection* in Definition 3, a concept also used extensively in Sect. 5. We then propose an algorithm using projection and prove it correct.

Recall that the *lexicographic order* on $[0,1]^n$ is the total order defined as $\mathbf{x} <_{lex} \mathbf{y}$ iff there is $k \in [n]$ such that (i) $x_k < y_k$ and (ii) $x_i = y_i$ for all $i \in [k-1]$. In the following, the order of our target sets $F_1, \ldots, F_n$ is relevant: For all $i, j \in [n]$, F_i appears before F_j iff F_i is more important than F_j.

One of the difficulties is that when considering the set of achievable points, the lexicographic supremum may not be achievable, i.e., the lexicographic maximum may not exist. We now formally give the main result of this section.

Theorem 2. *Let MDP Γ be clean w.r.t. parity objective ρ and target sets $F_1, \ldots, F_n \subseteq S$, and let $s \in S$ be a state. Then:*

- *It is decidable if $\mathbf{p}^* = \max_{lex}\{\mathbf{p} \in [0,1]^n \mid s \models_\Gamma \mathsf{S}(\rho) \wedge \bigwedge_{i=1}^n \Pr_{\geq p_i}(\Diamond F_i)\}$ exists by solving $\mathcal{O}(\mathsf{poly}(n))$ many parity games (hence the problem is in $\mathsf{NP} \cap \mathsf{coNP}$ [11]).*
- *A witnessing strategy using at most $2|\Gamma||\rho|$ memory can be effectively constructed for the YES-instances.*

Our approach considers every target set F_i one by one, following the lexicographic order. The general idea is to successively remove all transitions that do not achieve the maximal probability to reach F_i. Thus, after having pruned transitions w.r.t. the first i target sets, any strategy that maximizes the probability to reach the set $F = F_1 \cup \ldots \cup F_n$ also maximizes the probabilities of reaching $F_1, \ldots, F_i$ lexicographically [18]. In order to find the maximal probability to achieve a single objective, we adapt the notion of *projection* from [25,26]. The main difference is that we keep reachability objectives, instead of converting them into reward objectives, enabling us to use existing results [8] on combinations of sure and almost-sure objectives.

We define the MDP $\Gamma^{\pi\mathbf{v}}$, the projection of MDP Γ on a non-zero vector $\mathbf{v} \in [0,1]^n$ where we can freely assume $\|\mathbf{v}\|_1 = 1$. Intuitively, to obtain $\Gamma^{\pi\mathbf{v}}$, we consider a k-dimensional face of the Pareto frontier of $\bigwedge_{i=1}^n \Pr_{\geq p_i}(\Diamond F_i)$, maximal in the direction $\mathbf{v}$. This is thus an intersection with a hyperplane, and defines a face of dimension k. We remove all available actions that are used in none of the strategies achieving this face of dimension k, i.e. we remove all non-optimal actions when trying to maximize in the direction $\mathbf{v}$. Our purpose is to obtain a new MDP, in which every strategy that almost-surely reaches a final state in F also maximizes the probability to reach these states weighted with the direction $\mathbf{v}$. We remark that in this new MDP, the parity condition ρ may not be surely satisfied from every state; we will thus need to address this condition later.

Definition 3 (Projection). *Let $\Gamma = (S, Act, \mathbf{P})$ be clean w.r.t. $F_1, \ldots, F_n \subseteq S$. The projection $\Gamma^{\pi\mathbf{v}}$ of Γ in direction $\mathbf{v} \geq \mathbf{0}$ with $\|\mathbf{v}\|_1 = 1$, is defined in two steps: (1) Let $\Gamma' = (S', Act, \mathbf{P}')$ be an MDP where*

- $S' = S \cup \{\bot\}$ *where $\bot$ is a fresh sink state, and*
- $\mathbf{P}'$ *is defined similar to $\mathbf{P}$ with the following modifications (let $F = \bigcup_{i=1}^n F_i$):*
  - *For all $s \in S \setminus F$, $a \in Act(s)$, and $s' \in F$, we set*
    $\mathbf{P}'(s, a, s') = \mathbf{P}(s, a, s') \cdot \sum_{i:s' \in F_i} v_i$ *and* $\mathbf{P}'(s, a, \bot) = 1 - \sum_{s'' \in S} \mathbf{P}'(s, a, s'')$.
  - $\mathbf{P}'(\bot, a, \bot) = 1$, *where $a \in Act$ is arbitrary.*

(2) For each state $s \in S'$, let $y_s = \max_\sigma \Pr_s^{\Gamma'[\sigma]}(\Diamond F)$ be the maximum probability to reach F from s in Γ'. The MDP $\Gamma^{\pi\mathbf{v}}$ is then obtained from Γ' by removing all actions $a \in Act(s)$ that do not satisfy $y_s = \sum_{s' \in S'} \mathbf{P}'(s, a, s') \cdot y_{s'}$.

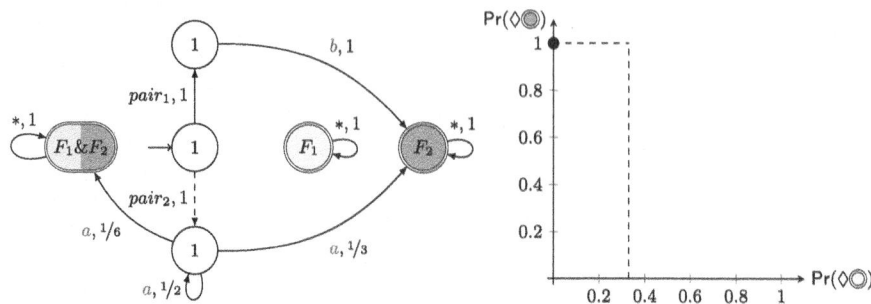

Fig. 2. The MDP from Fig. 1 projected on $\mathbf{v} = (0, 1)$. *Right:* The set of achievable points w.r.t. $\mathtt{S}(\rho) \land \Pr_{\geq x_1}(\Diamond F_1) \land \Pr_{\geq x_2}(\Diamond F_2)$ is $[0, 1/3) \times [0, 1) \cup \{(0, 1)\}$; the lexicographic maximum for order F_2, F_1 is thus $\mathbf{p}^* = (1, 0)$.

Algorithm 1. Sure parity and lexicographic reachability

Input: MDP Γ – clean w.r.t. parity objective ρ and $F_1, \ldots, F_n \subseteq S$, a state $s \in S$
Output: If $\mathbf{p}^* = \max_{lex}\{\mathbf{p} \in [0,1]^n \mid s \models_\Gamma \mathtt{S}(\rho) \land \bigwedge_{i=1}^n \Pr_{\geq p_i}(\Diamond F_i)\}$ exists, then the output is a witness strategy σ, otherwise the output is `false`.

1: $\Gamma_0 \leftarrow \Gamma$
2: $F \leftarrow \bigcup_{i=1}^n F_i$
3: **for** i from 1 to n **do**
4:     Compute $\Gamma_{i-1}^{\pi \mathbf{e}_i}$ ▷ See Definition 3.
5:     $\Gamma_i \leftarrow$ result of pruning all states not satisfying $\mathtt{S}(\rho) \land \Pr_{=1}(\Diamond F)$ in $\Gamma_{i-1}^{\pi \mathbf{e}_i}$.
6: **end for**
7: **if** s is not a state of Γ_n **then return false**
8: **else return** σ such that $s, \sigma \models_{\Gamma_n} \mathtt{S}(\rho) \land \Pr_{=1}(\Diamond F)$ ▷ By Theorem 2.
9: **end if**

Example 4. The MDP in Fig. 2 results from projecting the MDP from Fig. 1 on $\mathbf{v} = (0,1)$ ($\bot$ is not reachable). Only the actions reaching the blue target F_2 with maximal probability remain.

Given $\Gamma, F_1, \ldots, F_n$ and $\mathbf{v}$, it is clear from Definition 3 that we can construct the projection $\Gamma^{\pi \mathbf{v}}$ in polynomial time. Note that strategies in $\Gamma^{\pi \mathbf{v}}$ are still valid in Γ, but the converse is not necessarily the case as projection removes actions.

Lemma 1 *(Key property of projection). Let $\Gamma = (S, Act, \mathbf{P})$ be clean w.r.t. $F_1, \ldots, F_n, \bot \subseteq S$, and let $\mathbf{v} \geq \mathbf{0}$, $\|\mathbf{v}\|_1 = 1$. Then for all strategies σ of $\Gamma^{\pi \mathbf{v}}$, the following holds: $s, \sigma \models_{\Gamma^{\pi \mathbf{v}}} \Pr_{=1}(\Diamond F)$ iff there exists $\mathbf{x} \in [0,1]^n$ such that (i) $\mathbf{x} \cdot \mathbf{v}$ is maximal among the achievable $\mathbf{x}$, and (ii) $s, \sigma \models_\Gamma \bigwedge_{i=1}^n \Pr_{\geq x_i}(\Diamond F_i)$.*

5 Sure Parity and n Non-strict Reachability Thresholds

Finally, we consider the case of one sure parity condition and multiple *non-strict* threshold reachability objectives, i.e., formulas like $\mathtt{S}(\rho) \bigwedge_{i=1}^n \Pr_{\geq p_i}(\Diamond F_i)$. We do not impose a lexicographic ordering on the target sets. Our main result is:

Theorem 3. *Let MDP Γ be clean w.r.t. parity objective ρ and target sets $F_1, \ldots, F_n \subseteq S$. Further, let $\mathbf{p} \in [0,1]^n$, and let $s \in S$ be a state. Then:*

1. *The decision problem $\exists \sigma \colon s, \sigma \models_\Gamma \mathsf{S}(\rho) \wedge \bigwedge_{i=1}^n \mathtt{Pr}_{\geq p_i}(\Diamond F_i)$ is in EXPTIME.*
2. *A witness strategy σ using at most $2^{\mathsf{poly}(|\Gamma|+nD)}$ memory, where D is the bit-complexity of the rational numbers in $\mathbf{p}$, can be effectively constructed for the YES-instances.*

We solved the case where $\mathbf{p}$ is strictly inside the Pareto frontier in Sect. 3. It remains to show how to achieve $\mathsf{S}(\rho) \wedge \bigwedge_{i=1}^n \mathtt{Pr}_{\geq p_i}(\Diamond F_i)$ when $\mathbf{p}$ is exactly on the frontier. We first consider the case where $\mathbf{p}$ is a vertex of the Pareto frontier, that we will then use as a base case for an arbitrary point $\mathbf{p}$ of the frontier. We sketch the proof in the remainder of the section.

Since the Pareto frontier does not depend on the sure objective, to determine whether $\mathbf{p}$ is exactly on the Pareto frontier, it suffices to check if $s \models_\Gamma \bigwedge_{i=1}^n \mathtt{Pr}_{\geq p_i}(\Diamond F_i)$ and $s \not\models_\Gamma \bigwedge_{i=1}^n \mathtt{Pr}_{> p_i}(\Diamond F_i)$. The first formula checks if $\mathbf{p}$ is achievable, the second checks whether it is on the boundary. Hence the main difficulty is to decide whether adding a sure parity condition keeps the achievability of a point. We illustrate this in the following example.

Example 5. In the MDP of Fig. 1, playing $pair_1$ then a forever gives a total probability of 1 of reaching F_1 and probability $1/3$ of reaching F_2. This strategy does not surely satisfy the parity condition though, since there exists a path that visits the uppermost state, labelled 1, forever. Playing $pair_1$ and then b once surely reaches F_2. It is thus possible to satisfy $\mathsf{S}(\rho) \wedge \mathtt{Pr}_{=1}(\Diamond F_2)$, but not $\mathsf{S}(\rho) \wedge \mathtt{Pr}_{=1}(\Diamond F_1) \wedge \mathtt{Pr}_{\geq 1/3}(\Diamond F_2)$. Still, for every $\varepsilon > 0$, we can achieve $\mathsf{S}(\rho) \wedge \mathtt{Pr}_{\geq 1-\varepsilon}(\Diamond F_1) \wedge \mathtt{Pr}_{\geq 1/3-\varepsilon}(\Diamond F_2)$ by Theorem 1, using a finite-memory strategy.

5.1 Vertex of the Pareto Frontier

We first consider the easier case where we want to achieve a point which is a vertex of the Pareto frontier. We assume $\mathbf{p}$ to be a vertex of the Pareto frontier. Our proof relies on projection (Definition 3). Indeed, since $\mathbf{p}$ is a vertex, there exists some vector $\mathbf{v}$ such that $\mathbf{p}$ is the *unique* point of the Pareto frontier maximizing $\mathbf{p} \cdot \mathbf{v}$. We obtain the following lemma.

Lemma 2. *Suppose that the MDP Γ is clean w.r.t. parity objective ρ and target sets $F_1, \ldots, F_n \subseteq S$. Further, let $\mathbf{p} \in [0,1]^n$, and let $s \in S$ be a state. If $\mathbf{p}$ is a vertex of the Pareto frontier of $\bigwedge_{i=1}^n \mathtt{Pr}_{\geq p_i}(\Diamond F_i)$ from s, then we can decide if $s \models_\Gamma \mathsf{S}(\rho) \wedge \bigwedge_{i=1}^n \mathtt{Pr}_{\geq p_i}(\Diamond F_i)$ and if so give a finite-memory strategy.*

5.2 Arbitrary Point of the Pareto Frontier

We now consider any arbitrary point $\mathbf{p}$ of the Pareto frontier. Since $\mathbf{p}$ may be contained in a k-dimensional face of the frontier (with $k > 0$; $k = 0$ means

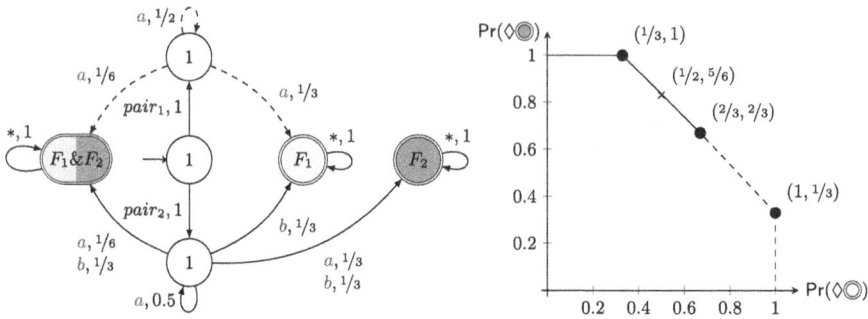

Fig. 3. The MDP of Fig. 1 after projection on $\mathbf{v} = (\,1\,,\,\boxed{1}\,)$.

that **p** is a vertex, see Sect. 5.1), projecting on this face will not be sufficient to obtain **p**. Similar to Algorithm 1, we iterate projections and state removal, thereby reducing the dimension of the Pareto frontier until either reducing **p** to a vertex, or entering a situation where we cannot project anymore. We remark that the latter only happens in specific cases. To properly define these cases, given a Pareto frontier P, we consider the smallest vector space containing P. We show that we cannot project any more when **p** is an interior point of P for this subspace, denoted $\mathbf{p} \in \text{int}_{\text{gen}(P)}(P)$.

Example 6. After projecting the MDP of Fig. 1 on $\mathbf{v} = (1,1)$, we obtain the MDP in Fig. 3 (including both the dashed and the solid transitions). Since from the uppermost state, no strategy surely satisfies the parity condition, $\mathtt{S}(\rho) \wedge \mathtt{Pr}_{=1}(\Diamond F)$ does not hold. We thus prune the dashed transitions, and the Pareto frontier is now restricted between $(1/3, 1)$ and $(2/3, 2/3)$, see Fig. 3 (right). To achieve e.g. $\mathbf{p} = (1/2, 5/6)$, which is strictly inside this line segment, it suffices to play $pair_2$, then a once, and then finally b once.

To obtain our result, we get the following lemma. After projecting on a given vector, and removing any state refuting $\mathtt{S}(\rho) \wedge \mathtt{Pr}_{=1}(\Diamond F)$ we obtain some polytope P; any point of P that is a topologically interior point in the smallest vector space containing P is achievable. Formally:

Lemma 3. *Let the MDP Γ be clean w.r.t. parity objective ρ and target sets $F_1, \ldots, F_n \subseteq S$. Further, let $\mathbf{v} \in [0,1]^n$, and $s \in S$. Let Γ_ρ be obtained by taking the MDP $\Gamma^{\pi \mathbf{v}}$ and pruning all states that refute $\mathtt{S}(\rho) \wedge \mathtt{Pr}_{=1}(\Diamond F)$. Let $B \subseteq [0,1]^n$ be the set of $\leq$-maximal points of the Pareto frontier of Γ_ρ from s. Then: For every $\mathbf{x} \in \text{int}_{\text{gen}(B)}(B)$, we have $s \models_{\Gamma_\rho} \mathtt{S}(\rho) \wedge \bigwedge_{i=1}^n \mathtt{Pr}_{\geq x_i}(\Diamond F_i)$, and we can compute a strategy that achieves this.*

The proof of this lemma is quite involved. Figure 4 provides some intuition on the proof. If **x** is inside an m-dimensional surface B, we can find $m+1$ elements of B such that **x** is within their convex hull. These are y^1, y^2, y^3 in Fig. 4. For every y^j, we can find a strategy satisfying $\bigwedge_{i \in [n]} \mathtt{Pr}_{\geq y_i^j}(\Diamond F_i)$. By playing such a strategy for sufficiently many steps, then switching to a strategy satisfying

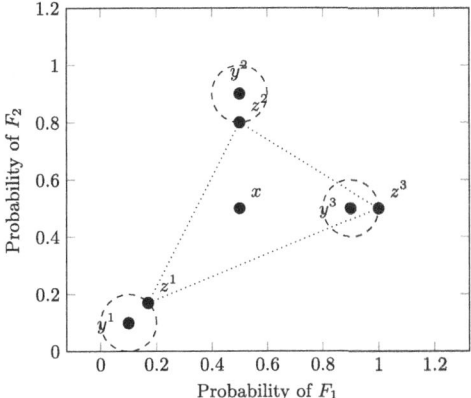

Fig. 4. To obtain x on the two-dimension plane, we first take three points y^1, y^2, y^3, then find ε small enough for x to be within the convex hull of any ε-approximation z^1, z^2, z^3 of y^1, y^2, y^3.

$\mathsf{S}(\rho) \wedge \Pr_{=1}(\lozenge F)$, we can ε-approximate the y^j while staying inside B. Points achieved by such approximation are denoted z^1, z^2, z^3 in Fig. 4. It then remains to show that if ε is small enough, $\mathbf{x}$ is within the convex hull of z^1, z^2, z^3 and thus can be achieved.

We can now give our result stating that we can verify the achievability of an arbitrary point of the Pareto frontier.

Lemma 4. *Suppose that the MDP Γ is clean w.r.t. parity objective ρ and target sets $F_1, \ldots, F_n \subseteq S$. Further, let $s \in S$ be a state, and $\mathbf{x} = (x_i)_{i \in [n]} \in [0,1]^n$ on the border of the Pareto frontier of $\bigwedge_{i \in [n]} \Pr_{\geq x_i}(\lozenge F_i)$*

- *Checking whether $s \models_\Gamma \mathsf{S}(\rho) \wedge \bigwedge_{i \in [n]} \Pr_{\geq x_i}(\lozenge F_i)$ is decidable;*
- *if YES a witnessing strategy with at most $2|\Gamma||\rho|$ memory can be effectively computed.*

Proof. We show that Algorithm 2 answers Lemma 4. The main differences with Algorithm 1 is that we must find the vector on which we project, and at the end of the loop of Algorithm 2 we have to split between the cases where $\mathbf{x}$ is a vertex or not.

During iteration i, we first find a vector orthogonal to the face of P_i that $\mathbf{x}$ belongs to. To do so, we may need to fully compute the Pareto frontier of Γ_{i-1}. In line 4 we project the current MDP Γ_{i-1} on vector $\mathbf{v}_i$, obtaining $\Gamma_{i-1}^{\pi \mathbf{v}_i}$. By Lemma 1, a strategy satisfies $\Pr_{=1}(\lozenge F)$ in $\Gamma_{i-1}^{\pi \mathbf{e}_i}$ iff it maximizes p such that $\Pr_{\geq p}(\lozenge F_i)$ in Γ_{i-1}. We then prune all states that do not satisfy the conjunction with the parity condition $\mathsf{S}(\rho) \wedge \Pr_{=1}(\lozenge F)$.

If $\mathbf{x}$ is an interior point, it follows by Lemma 3 that we can find a strategy σ such that $s, \sigma \models_{\Gamma_i} \mathsf{S}(\rho) \wedge \bigwedge_{i=1}^n \Pr_{\geq x_i}(\lozenge F_i)$. Otherwise, we start the loop again.

Algorithm 2. Sure parity and n non-strict reachability threshold objectives

Input: MDP Γ – clean w.r.t. ρ and $F_1, \ldots, F_n \subseteq S$, a state $s \in S$, a vector $\mathbf{x} \in [0,1]^n$
Output: A strategy σ such that $s, \sigma \models_\Gamma \mathsf{S}(\rho) \wedge \bigwedge_{i=1}^n \Pr_{\geq x_i}(\Diamond F_i)$ if it exists, else **false**.
1: Set $\Gamma_0 = \Gamma$ and $i = 1$.
2: **while** $\mathbf{x}$ is not a vertex of P_i, the Pareto frontier of Γ_i **do**
3:     Get $\mathbf{v}_i$ a vector orthogonal to the smallest face of P_i that $\mathbf{x}$ belongs to.
4:     Compute $\Gamma_i^{\pi \mathbf{v}_i}$ from s.   ▷ By Lemma 1.
5:     Set Γ_i by taking $\Gamma_{i-1}^{\pi \mathbf{v}_i}$ and pruning all states that do not satisfy $\mathsf{S}(\rho) \wedge \Pr_{\geq 1}(\Diamond F)$.
6:     If $\mathbf{x}$ is an interior point of P_i return σ s.t. $s, \sigma \models_{\Gamma_i} \mathsf{S}(\rho) \wedge \bigwedge_{i=1}^n \Pr_{\geq x_i}(\Diamond F_i)$ ▷
       By Lemma 3.
7:     $i := i + 1$
8: **end while**
9: Check if there exists σ such that $s \models_{\Gamma_i} \mathsf{S}(\rho) \wedge \bigwedge_{i=1}^n \Pr_{\geq x_i}(\Diamond F_i)$. ▷ By Lemma 2.
10: If such a σ does not exist then return **false**, else return σ.

Since every time we project, it is onto a face of the Pareto frontier of dimension smaller than the current Pareto frontier, we can only take the loop at most n times. After this we go to line 9, and since $\mathbf{x}$ is now a vertex of the Pareto frontier, we can use Lemma 2 to find whether there exists a strategy such that $s \models_{\Gamma_i} \mathsf{S}(\rho) \wedge \bigwedge_{i=1}^n \Pr_{\geq x_i}(\Diamond F_i)$.

If the output is **false**, we remark that since projection in line 4 keeps all states belonging to strategies such that $\Pr_{\geq p}(\Diamond F_i)$ in Γ_{i-1} (by Lemma 1), it keeps all states belonging to strategies such that $\bigwedge_{i=1}^n \Pr_{\geq x_i}(\Diamond F_i)$. Step 5 may prune states used in strategies such that $\bigwedge_{i=1}^n \Pr_{\geq x_i}(\Diamond F_i)$, but then by definition these strategies did not satisfy $\mathsf{S}(\rho)$. Hence, since none of the pruned states belonged to a strategies such that $s \models_{\Gamma_i} \mathsf{S}(\rho) \wedge \bigwedge_{i=1}^n \Pr_{\geq x_i}(\Diamond F_i)$, the algorithm is correct.

We show that Algorithm 2 solves at most a polynomial number of parity game of size polynomial in $|\Gamma|$.

Indeed, to obtain $|\Gamma_{i-1}^{\pi \mathbf{v}_i}|$, we have to compute the Pareto frontier of $|\Gamma_{i-1}|$. This new MDP only has one more state than Γ and at most $|\Gamma|$ additional transitions (that may lead from states originally in Γ to the new state $\bot$). We then remove once all states that do not satisfy $\mathsf{S}(\rho) \wedge \Pr_{=1}(\Diamond F)$, which can be done by solving a polynomial number of parity games. Every time we do this step, we project onto a face of the Pareto frontier of dimension smaller than the current Pareto frontier and this can only happen at most n times. Thus we end up solving a number polynomial in n of parity game of size polynomial in Γ, and compute Pareto frontiers for MDPs with n objectives at most n times.

We output **true** iff we find a strategy σ that is a solution of $s, \sigma \models_{\Gamma_i} \mathsf{S}(\rho) \wedge \Pr_{=1}(\Diamond F)$ iff it is a solution of $s \models_\Gamma \mathsf{S}(\rho) \wedge \bigwedge_{i \in [n]} \Pr_{\geq x_i}(\Diamond F_i)$, and we can find strategies satisfying the left hand formula that use $2|\Gamma||\rho|$ memory, as proved in Theorem 2.

Example 7. For the MDP from Fig. 1, with initial state s, and where F_1 is the yellow target and F_2 is the blue target, we check if $s \models_\Gamma \mathsf{S}(\rho) \wedge \Pr_{\geq 2/3}(\Diamond F_1) \wedge \Pr_{\geq 2/3}(\Diamond F_2)$. Point $(2/3, 2/3)$ is on the Pareto frontier but not a vertex of it, and following Algorithm 2, a vector orthogonal to it is $(1,1)$. After projection on $(1,1)$, we obtain the MDP in Fig. 3 (left), with both full and dashed transitions. As in Example 6, no strategy satisfies surely the parity condition in the uppermost state, we prune the dashed transitions, restricting the Pareto frontier to between $(1/3, 1)$ and $(2/3, 2/3)$. Now $(2/3, 2/3)$ is a vertex of the new Pareto frontier, Lemma 2 tells us to project on vector $(2, 1)$, and so we prune transition a from the lowermost state, only keeping transition b. Since we can satisfy the parity condition from the lowermost state, we obtain that the strategy playing b twice satisfies $\mathsf{S}(\rho) \wedge \Pr_{\geq 2/3}(\lozenge F_1) \wedge \Pr_{\geq 2/3}(\lozenge F_2)$.

6 Conclusion and Future Work

Combining sure parity and n reachability threshold objectives can be done via a reduction to parity games in the case of strict thresholds and when maximizing the threshold lexicographically, and in exponential time with non-strict thresholds. Finite-memory strategies suffice in all cases. One direction for future work is to implement our algorithms in the probabilistic model checker Storm [30]. Further open problems include the case where targets are not sinks, and the study of one sure parity and n parity threshold objectives. However, the exact memory required for one sure and *one* almost-sure parity is already unknown. It seems worthwhile to investigate if 1-bit Markov strategies suffice, as they do in countable MDPs with parity objectives [32]. In [1,8], the solution of sure parity and almost-sure reachability in MDPs relies on a reduction to a game with a *fair* opponent. Results in [15] concern *stochastic games* with a fair opponent, and may thus help extending the results from [1,8] to stochastic games. Another possible extension is to consider combinations of multiple objectives in *partially observable* MDPs (POMDPs), as in [21].

References

1. Almagor, S., Kupferman, O., Velner, Y.: Minimizing expected cost under hard Boolean constraints, with applications to quantitative synthesis. In: CONCUR. LIPIcs, vol. 59, pp. 9:1–9:15. Schloss Dagstuhl - Leibniz-Zentrum für Informatik (2016)
2. Aminof, B., Kwiatkowska, M., Maubert, B., Murano, A., Rubin, S.: Probabilistic strategy logic. In: IJCAI, pp. 32–38. ijcai.org (2019)
3. Ashok, P., Chatterjee, K., Kretínský, J., Weininger, M., Winkler, T.: Approximating values of generalized-reachability stochastic games. In: LICS 2020: 35th Annual ACM/IEEE Symposium on Logic in Computer Science, Saarbrücken, Germany, 8–11 July 2020, pp. 102–115. ACM (2020). https://doi.org/10.1145/3373718.3394761
4. Baier, C., Dubslaff, C., Klüppelholz, S.: Trade-off analysis meets probabilistic model checking. In: CSL-LICS, pp. 1:1–1:10. ACM (2014)
5. Baier, C., Katoen, J.-P.: Principles of Model Checking. MIT Press, Cambridge (2008)
6. Bellman, R.: A Markovian decision process. J. Math. Mech. 679–684 (1957)

7. Berthon, R., Guha, S., Raskin, J.-F.: Mixing probabilistic and non-probabilistic objectives in Markov decision processes. In: LICS 2020: 35th Annual ACM/IEEE Symposium on Logic in Computer Science, Saarbrücken, Germany, 8–11 July 2020, pp. 195–208. ACM (2020). https://doi.org/10.1145/3373718.3394805
8. Berthon, R., Randour, M., Raskin, J.-F.: Threshold constraints with guarantees for parity objectives in Markov decision processes. In: ICALP. LIPIcs, vol. 80, pp. 121:1–121:15. Schloss Dagstuhl - Leibniz-Zentrum für Informatik (2017). https://doi.org/10.4230/LIPIcs.ICALP.2017.121
9. Berthon, R., Katoen, J.-P., Winkler, T.: Markov Decision Processes with Sure Parity and Multiple Reachability Objectives (2024). arXiv:2408.01212
10. Bouyer, P., González, M., Markey, N., Randour, M.: Multi-weighted Markov decision processes with reachability objectives. In: GandALF. EPTCS, vol. 277, pp. 250–264 (2018)
11. Brassard, G.: A note on the complexity of cryptography (corresp.). IEEE Trans. Inf. Theory **25**(2), 232–233 (1979)
12. Brázdil, T., Chatterjee, K., Forejt, V., Kučera, A.: MULTIGAIN: a controller synthesis tool for MDPs with multiple mean-payoff objectives. In: Baier, C., Tinelli, C. (eds.) TACAS 2015. LNCS, vol. 9035, pp. 181–187. Springer, Heidelberg (2015). https://doi.org/10.1007/978-3-662-46681-0_12
13. Bruyère, V., Filiot, E., Randour, M., Raskin, J.-F.: Meet your expectations with guarantees: beyond worst-case synthesis in quantitative games. Inf. Comput. **254**, 259–295 (2017). https://doi.org/10.1016/j.ic.2016.10.011
14. Calude, C.S., Jain, S., Khoussainov, B., Li, W., Stephan, F.: Deciding parity games in quasipolynomial time. In: Proceedings of the 49th Annual ACM SIGACT Symposium on Theory of Computing, STOC 2017, pp. 252–263. ACM (2017). https://doi.org/10.1145/3055399
15. Castro, P.F., D'Argenio, P.R., Demasi, R., Putruele, L.: Playing against fair adversaries in stochastic games with total rewards. In: Shoham, S., Vizel, Y. (eds.) CAV 2022. LNCS, vol. 13372, pp. 48–69. Springer, Cham (2022). https://doi.org/10.1007/978-3-031-13188-2_3
16. Chatterjee, K., Henzinger, T.A., Piterman, N.: Strategy logic. Inf. Comput. **208**(6), 677–693 (2010)
17. Chatterjee, K., Katoen, J.-P., Mohr, S., Weininger, M., Winkler, T.: Stochastic games with lexicographic objectives. Formal Methods Syst. Des. (2023). https://doi.org/10.1007/s10703-023-00411-4
18. Chatterjee, K., Katoen, J.-P., Weininger, M., Winkler, T.: Stochastic games with lexicographic reachability-safety objectives. In: Lahiri, S.K., Wang, C. (eds.) CAV 2020. LNCS, vol. 12225, pp. 398–420. Springer, Cham (2020). https://doi.org/10.1007/978-3-030-53291-8_21
19. Chatterjee, K., Kretínská, Z., Kretínský, J.: Unifying two views on multiple mean-payoff objectives in Markov decision processes. Log. Methods Comput. Sci. **13**(2) (2017). https://doi.org/10.23638/LMCS-13(2:15)2017
20. Chatterjee, K., Majumdar, R., Henzinger, T.A.: Markov decision processes with multiple objectives. In: Durand, B., Thomas, W. (eds.) STACS 2006. LNCS, vol. 3884, pp. 325–336. Springer, Heidelberg (2006). https://doi.org/10.1007/11672142_26
21. Chatterjee, K., Novotný, P., Pérez, G.A., Raskin, J.-F., Zikelic, D.: Optimizing expectation with guarantees in POMDPs. In: AAAI, pp. 3725–3732. AAAI Press (2017)

22. Chatterjee, K., Piterman, N.: Combinations of qualitative winning for stochastic parity games. In: 30th International Conference on Concurrency Theory, CONCUR 2019. LIPIcs, vol. 140, pp. 6:1–6:17. Schloss Dagstuhl - Leibniz-Zentrum für Informatik (2019). http://www.dagstuhl.de/dagpub/978-3-95977-121-4
23. Clemente, L., Raskin, J.-F.: Multidimensional beyond worst-case and almost-sure problems for mean-payoff objectives. In: 30th Annual ACM/IEEE Symposium on Logic in Computer Science, LICS 2015, Kyoto, Japan, 6–10 July 2015, pp. 257–268. IEEE Computer Society (2015). https://doi.org/10.1109/LICS.2015.33
24. Allen Emerson, E., Jutla, C.S., Prasad Sistla, A.: On model checking for the μ-calculus and its fragments. Theor. Comput. Sci. **258**(1–2), 491–522 (2001). https://doi.org/10.1016/S0304-3975(00)00034-7
25. Etessami, K., Kwiatkowska, M.Z., Vardi, M.Y., Yannakakis, M.: Multi-objective model checking of Markov decision processes. Logical Methods Comput. Sci. **4**(4) (2008)
26. Forejt, V., Kwiatkowska, M., Parker, D.: Pareto curves for probabilistic model checking. In: Chakraborty, S., Mukund, M. (eds.) ATVA 2012. LNCS, pp. 317–332. Springer, Heidelberg (2012). https://doi.org/10.1007/978-3-642-33386-6_25
27. Grünbaum, B., Klee, V., Perles, M.A., Shephard, G.C.: Convex Polytopes, vol. 16. Springer, Cham (1967)
28. Hahn, E.M., Perez, M., Schewe, S., Somenzi, F., Trivedi, A., Wojtczak, D.: Good-for-MDPs automata for probabilistic analysis and reinforcement learning. In: TACAS 2020. LNCS, vol. 12078, pp. 306–323. Springer, Cham (2020). https://doi.org/10.1007/978-3-030-45190-5_17
29. Hartmanns, A., Junges, S., Katoen, J.-P., Quatmann, T.: Multi-cost bounded tradeoff analysis in MDP. J. Autom. Reason. **64**(7), 1483–1522 (2020)
30. Hensel, C., Junges, S., Katoen, J.-P., Quatmann, T., Volk, M.: The probabilistic model checker Storm. Int. J. Softw. Tools Technol. Transf. **24**(4), 589–610 (2022)
31. Jurdzinski, M.: Deciding the winner in parity games is in UP ∩ co-UP. Inf. Process. Lett. **68**(3), 119–124 (1998). https://doi.org/10.1016/S0020-0190(98)00150-1
32. Kiefer, S., Mayr, R., Shirmohammadi, M., Totzke, P.: Strategy complexity of parity objectives in countable MDPs. In: CONCUR. LIPIcs, vol. 171, pp. 39:1–39:17. Schloss Dagstuhl - Leibniz-Zentrum für Informatik (2020)
33. Kwiatkowska, M., Norman, G., Parker, D.: PRISM 4.0: verification of probabilistic real-time systems. In: Gopalakrishnan, G., Qadeer, S. (eds.) CAV 2011. LNCS, vol. 6806, pp. 585–591. Springer, Heidelberg (2011). https://doi.org/10.1007/978-3-642-22110-1_47
34. Miura, S., Wray, K.H., Zilberstein, S.: Heuristic search for SSPs with lexicographic preferences over multiple costs. In: SOCS, pp. 127–135. AAAI Press (2022)
35. Puterman, M.L.: Markov Decision Processes: Discrete Stochastic Dynamic Programming. Wiley Series in Probability and Statistics, Wiley, Hoboken (1994)
36. Randour, M., Raskin, J.-F., Sankur, O.: Percentile queries in multi-dimensional Markov decision processes. Formal Methods Syst. Des. **50**(2), 207–248 (2017). https://doi.org/10.1007/s10703-016-0262-7
37. Roijers, D.M., Vamplew, P., Whiteson, S., Dazeley, R.: A survey of multi-objective sequential decision-making. J. Artif. Intell. Res. **48**, 67–113 (2013). https://doi.org/10.1613/jair.3987
38. Schewe, S., Weinert, A., Zimmermann, M.: Parity games with weights. Log. Methods Comput. Sci. **15**(3) (2019). https://doi.org/10.23638/LMCS-15(3:20)2019

39. Winkler, T., Weininger, M.: Stochastic games with disjunctions of multiple objectives. In: GandALF. EPTCS, vol. 346, pp. 83–100 (2021). https://doi.org/10.4204/EPTCS.346.6
40. Wray, K.H., Zilberstein, S., Mouaddib, A.-I.: Multi-objective MDPs with conditional lexicographic reward preferences. In: AAAI, pp. 3418–3424. AAAI Press (2015)

Modelling Dynamical Systems: Learning ODEs with No Internal ODE Resolution

Johanne Cohen[1(✉)], Emmanuel Goutierre[1], Hayg Guler[2], Fatios Kapotos[3], Sida-Bastien Li[3], Michèle Sébag[1], and Bowen Zhu[3]

[1] LISN, CNRS, Université Paris-Saclay, Gif-sur-Yvette, France
{johanne.cohen,emmanuel.goutierre, michele.sebag}@universite-paris-saclay.fr
[2] IJCLAB, CNRS, Université Paris-Saclay, Gif-sur-Yvette, France
hayg.guler@ijclab.in2p3.fr
[3] CentraleSupelec, Université Paris-Saclay, Gif-sur-Yvette, France
{fatios.kapotos,sida-bastien.li,bowen.zhu}@student-cs.fr

Abstract. The quest for accurate modelling and simulation of dynamical systems is the Holy Grail of computational physics and numerical engineering. In deep learning, main approaches proposed in the literature include prediction by time series and modelling by Ordinary Differential Equations (ODEs). The usual methods for learning optimal parameters then consist of formulating the question as a reachability problem and then optimizing some suitable cost function for this reachability problem. However, these two approaches fail to model specific complex dynamical systems. The presented work considers the case of modelling and predicting the behaviour of beams in particle accelerators. The difficulty lies in the associated dynamic, which is highly versatile and possibly discontinuous. In order to extend the scope of dynamical system modelling to meet the particle accelerator modelling challenge, we present a new approach that can cover this context called implicit neural ODE (INode). It uses the modelling of discontinuous behaviour through integral operators; these operators are used to pre-process the data to get a more classical regression problem. Finally, the global model of the dynamical system is formulated as the solution of an ODE, which contains the solution of the regression problem. INode thus enables the learning of a data-driven ODE while removing the computationally heavy ODE resolution from the training loop.

The formal analysis of the approach establishes its consistency and convergence properties under moderate assumptions.

1 Introduction

Machine learning techniques have profoundly altered the field of dynamic system modelling and numerical engineering, leveraging the data and/or prior knowledge to meet the prediction challenge. Depending on the available knowledge, the sought model can be specified to satisfy physical constraints and/or used to generate new data consistent with the physics of the problem.

Two main approaches have been presented in the literature in the context of deep learning. A first approach and a long-studied direction tackle the prediction of time series forecasts. A second approach aims at identifying the ordinary differential equation (ODE) underlying the observed data [5,19,22,30].[1]

From an abstract point of view, these (deep learning) methods consist of formulating the question as a reachability problem and then optimizing some suitable cost function on this reachability problem. Such parameters are obtained using some descent gradient method. This leads to particular methods, such as the adjoint method. These methods fit on methods trying to optimize the reachability of some desired goal from some given initial data, in some incremental manner, guided by the local optimization of some cost function: we provide a concise review in Sect. 3.1 and how this approach relates to some reachability question.

Our motivating application is modelling particle accelerators, where particles are generated and submitted to diverse electromagnetic fields to deliver a beam with a prescribed behaviour. Efficient simulators have been designed to predict the beam behavior with the desired level of accuracy. Their limitation is that they are too computationally heavy to support, e.g., the number of experiments required to calibrate the accelerator and achieve the prescribed behavior with the desired accuracy. Several approaches, centered on learning surrogate simulators and inspired from time series forecasting, have been presented in the literature [2,9,23,29,33,34].

However, it turns out from our experiments that the above existing approaches cannot model the highly versatile behaviour of a particle accelerator in the general case. The behaviour of the beam is controlled by the several dozen electromagnetic fields involved in the accelerator chambers, referred to as *particle accelerator control settings* in what follows, and control settings when there is no confusion to be feared. Depending on the control settings in particular, a significant fraction of the particles may be ejected and the behaviour of the beam becomes discontinuous.

Consequently, the existing methodology must be adapted to cover these highly versatile and possibly discontinuous behaviours. One of our key ideas is not to work directly over the inputs but on their transformation by a smooth, deterministic operator. This aims to reduce the versatility of the data involved. An example of a considered operator is the integral of the inputs. Notice that this is inspired by ideas coming from analogue computations [3,4,31], where, for example, integration was considered as an operation far more stable than derivative. This is especially true when the observable trajectory u experiences a discontinuity at some time t^*, implying that the derivative is infinite at this point, unlike the integral.

---

[1] A third emerging direction involves generative modelling, where the generated examples are filtered using the prior knowledge to support the identification of the sought system. This approach is outside the scope of the paper and will not be considered in the remainder.

Indeed, a new methodology, enabling the modelling of dynamic systems with highly diverse and possibly discontinuous behaviour, is presented in this article and referred to *Implicit Neural ODE* (INODE). INODE works in four stages.

1. Firstly, the observed behavioral data of the dynamical system is transformed using integral operators; a new, continuous-by-design trajectory is computed, called *implicit trajectory*. Notice that we use the vocabulary sequential data and trajectory interchangeably in the following.
2. Secondly, the implicit trajectory is used to define a classical regression problem aimed to predict the observed trajectory from the implicit trajectory.
3. Thirdly, the implicit trajectory is by design solution of an ODE involving the original trajectory. We then consider the implicit ODE where the original trajectory is replaced by its approximation, learned in Step 2. The implicit ODE is solved and yields an approximate implicit trajectory.
4. Lastly, the original trajectory is estimated from the approximate implicit trajectory using the model learned in Step 2.

The main contributions of the approach are as follows. Firstly, INODE aims to characterize parametrized ODEs (one ODE for each control settings), while prominent NODE approaches aim to characterize *the* ODE best fitting the data. Secondly, the theoretical analysis of the approach shows that, under moderate assumptions, i) the solution of the implicit ODE is unique (stage 3. above); ii) the original trajectory can be approximated with arbitrary precision in the large sample limit (universal approximation property). Thirdly, and importantly, the ODE resolution is contained in Stage 3., thus outside of the learning loop (Stage 2. above), with a significant gain in computational complexity.

2 Related Work

Neural ODEs. Ordinary Differential Equations is a widespread tool to model the evolution of systems over time [12]. The theory of dynamical systems, Ordinary Differential Equations (ODEs), Partial Differential Equations (PDEs) and their applications to various contexts is a common topic in mathematics, physics, and, more generally, in every applied science. Their use and adequateness in the context of deep learning have been pointed out recently [27]. In particular, neural ODEs [5] have demonstrated their performance in various contexts, particularly with respect to traditional approaches in deep-learning problems. Neural ODEs are particularly suitable for contexts with temporal dependencies, providing extensions of tools such as Recurrent Neural Networks (RNNs) and Long Short-Term Memory (LSTM) networks [13], and hence refining traditional modeling methods (see for an overview [8]). Research and applications involving Neural Ordinary Differential Equations (Neural ODEs) are typically explored within two main frameworks: Continuous-depth deep learning models [5] and Physics-Informed Neural Networks (PINNs) [25]. PINNs leverage insights from the continuous nature of physics underlying the differential equations, thereby aiding learnability by incorporating pre-existing knowledge. By integrating physics-informed principles with machine learning, as exemplified in PINNs, the learning process is anchored in established physical laws that the neural networks must adhere to. This approach assumes prior knowledge about the observed data, ensuring a more robust and interpretable learning process. Achieving such promising results involves a suite of techniques, including adaptive sampling methods [7], specialized neural network architectures [28], and dedicated optimization tools [14]. Applying them to high-dimensional problems remains a challenge.

The work considered here is more closely related to the continuous-depth deep learning model approach and its variants [5,27]. Neural ODEs are often associated with problems related to learning time series [5,17,26]. In particular, an approach is to consider the learning of suitable latent variables using models such as *Latent ODEs*. Some continuous-time extensions of (classical discrete-time) transformers have been proposed. It has been demonstrated that many variants of transformers can be extended to deal with irregular time series modeling [6]. Here, we learn directly from some hyperparameters and from the initial data but we do not work on time series. Learning a Neural ODE is classically based on techniques such as the adjoint method, and this requires in practise some numerical methods. Methods to reduce the practical costs have been proposed: [16] uses higher-order derivatives of solution trajectories to improve the process. Some methods have been proposed in constitutive equations by hard-wiring some constraints from physics to help the learning process in this specific context of mechanics [21].

The Context of Particle Accelerators. Building a surrogate particle accelerator is a difficult task because the associated learning process requires to deal with discontinuous inputs (trajectories) to be learned. This is a difficulty in ODE-based approaches, as differentiability implies continuity. On the other hand, differentiation is well-known from an engineering point of view as a very numerically unstable operation. Our approach is based on not working with functions and their derivatives but working with their integrals.

Concerning various approaches and practical models to deal with similar contexts, we can mention the following: investigations with discontinuous dynamics resulted in the formulation of Neural Jump Stochastic Differential Equations, [15] enabling the modelling of systems experiencing sudden shifts. This innovation expands the utility of Neural ODE frameworks to represent diverse real-world scenarios more accurately. In addressing practical challenges such as irregularly-sampled data, the articles [19,20,26] showcased the adaptability of Neural ODEs to real-world data collection scenarios, enhancing their utility in various applications.

The scope of Neural ODEs has been broadened by introducing a framework for modeling systems ruled by integro-differential equations [35]. This seems to be pivotal in areas where understanding cumulative effects over time is essential. This progress diversifies the Neural ODE toolkit, accommodating a broader spectrum of dynamic behaviours.

3 Problem Description

In supervised learning, we deal with a set of inputs $\mathcal{X} \subset \mathbb{R}^n$ and corresponding outputs $\mathcal{Y} \subset \mathbb{R}^m$, both subsets of Euclidean spaces. The aim is to accurately approximate a *ground truth* function, depicted as a mapping $F : \mathcal{X} \to \mathcal{Y}$.

Here, F is typically considered as the solution of a reachability problem: it is assumed that there is some underlying Ordinary Differential Equation (ODE) $\frac{du}{dt}(t, \mathbf{c}) = f(\mathsf{u}(t, \mathbf{c}), t, \mathbf{c})$ so that F is the function that maps initial condition $\mathsf{u}(0, \mathbf{c}) = x$ to its value $\mathsf{u}(T_1, \mathbf{c})$ at some given fixed time T_1 (or possibly at a set of given times $(T_i)_i$, but for simplicity we consider here first the case of only one measurement $T = T_1$). In other words, F is the flow of some underlying ODE, or if one prefers, the set of points reached from an initial condition x at some given time T_1.

To achieve this, we want to approximate the underlying *ground truth* function f by a family of approximating functions f_θ parametrized by $\theta \in \Theta$. The set of parameters, Θ, typically resides within another subset of an Euclidean space.

In a generic view, this problem then comes back to tackle an optimization problem generally expressed as $\inf_{\theta \in \Theta} \mathbb{E}_{x \sim \mathcal{X}}[\ell(F_\theta(x), F(x))]$, wherein $\ell : \mathcal{Y} \times \mathcal{Y} \to \mathbb{R}$ serves as a *loss function* and the input distribution is represented by a probability measure on $\mathcal{X}$. In this work, the definition of the loss function is the regression loss $\ell(x, y) = ||x - y||^2$.

If one prefers, assume that F is some fixed function, T_1 is some fixed parameter. We search for the optimal θ minimizing the loss function, measuring the (square of the) distance between $F(x)$ and $F_\theta(x)$, where F is the flow associated to f, and F_θ is the flow associated to f_θ (both at time $T = T_1$). As usual, in the context of deep learning, the difficulty is that when we state that F and T_1 are fixed, we only know F on some input/output points, but we do not know explicitly fully function F. We search for the best θ, minimizing the loss function on the available data.

In the more general case of many time/parameters measurements (i.e.; possibly a set of given times $(T_i)_i$), we consider an *observable trajectory* as a function $\mathsf{u} : [t_0, T] \times \mathbb{R}^p \to \mathbb{R}^n$. We may then have T distinct from a single point T_1, but all the $(T_i)_i$: namely, $\sup_i T_i < T$. We always assume a trajectory is L^∞ (integrable and bounded), representing the state trajectory of a system, where p denotes the dimension of the state space. In practice, this trajectory is influenced by a set of p control parameters $\mathbf{c} \in \mathbb{R}^m$ which dictate the system's dynamics. Our objective is to learn an approximation for u based on known initial conditions $\mathsf{u}(t_0, \mathbf{c})$ and the control parameters $\mathbf{c}$, with the available data.

3.1 Background in Our Context

This can be related to Recurrent Neural Networks (RNNs). In RNNs, the system's evolution from one time step to the next is described by the equation:

$$\mathsf{u}(t_{i+1}, \mathbf{c}) = \mathsf{u}(t_i, \mathbf{c}) + f_\theta(\mathsf{u}(t_i, \mathbf{c})) \qquad (1)$$

Here, f is a function parameterized by θ, representing the system's behaviour dynamics. θ corresponds to the weights.

The concept of NODE [5] can be interpreted as a continuous generalization of the classical recurrent neural network by adding additional layers and progressively decreasing the step size. NODE's novelty lies in considering the function u as the solution of an ODE:

$$\frac{du}{dt}(t, c) = f(u(t, c), t, c) \tag{2}$$

A neural network f_θ is employed to approximate the unknown f function. The estimate function $\hat{u}$ is the solution to the initial value problem defined by f_θ, and $u(t_0, c)$.

$$\frac{d\hat{u}}{dt}(t, c) = f_\theta(\hat{u}(t, c), t, c) \tag{3}$$

The state estimate at any time can thus be predicted by solving the ODE:

$$\hat{u}(t, c) = \text{ODESolve}(f_\theta, u(t_0, c), t_0, t) \tag{4}$$

ODESolve($f_\theta, u(t_0, c), t_0, t$) serves as a black-box that solves Eq. (3), starting with the value $u(t_0, c)$ at time t_0. In summary, the NODE algorithm can be formulated in our context as follows:

1. Initialize a neural network f_θ (weights are randomly initialized)
2. For a control parameter c, and its associated trajectory $u(t, c)$
   (a) Compute $\hat{u}(t, c)$ such that $\frac{d\hat{u}}{dt}(t, c) = f_\theta((\hat{u}(t, c), t, c)$ using an ODE solver.
   (b) Compute the loss $L(\hat{u}(T, c)) = \ell(\hat{u}(T, c), u(T, c)) = ||\hat{u}(T, c) - u(T, c)||^2$.
   (c) Compute the gradient of the loss $\ell(\hat{u}(T, c), u(T, c))$ using the adjoint method (see [5] or below).
   (d) Update the network weights based on the gradient.

To calculate the gradient of the loss function $L(\hat{u})$, the adjoint method is classicaly utilized [5]: The adjoint state $a(t)$ is defined as the gradient of L with respect to the state u, i.e., $a(t, c) = \nabla_{u(t,c)} L$. The key insight of the adjoint method is that the dynamics of $a(t, c)$ is given by another ODE:

$$-\frac{da(t, c)}{dt} = a(t, c)^T \frac{\partial f_\theta(u(t, c), t, c)}{\partial u(t, c)}.$$

This ODE is solved backwards in time, from t_1 to t_0, with the initial condition $a(t_1) = \nabla_{u(t_1)} L$, obtained from the derivative of the loss with respect to the output of the ODE solver. The gradient $\nabla_\theta L$ is then computed by integrating

$$a(t)^T \frac{\partial f_\theta(u(t, c), t, c)}{\partial \theta}$$

over time from t_0 to t_1. Despite the efficient memory consumption of such a method, the full differential equation must be solved at each training step, which becomes particularly challenging when dealing with complex underlying neural networks or extended sequences.

This method provides an efficient way to find (possibly locally) optimal parameters θ, solving the associated reachability problem corresponding to the considered supervised learning problem.

4 Our Approach

We propose a new model, INODE, inspired by NODE, aimed at learning the observable trajectory u. Our approach involves incorporating additional information, referred to as *integrated trajectory* v, required to be a solution of a differential equation (controlled by u). In other words, v is designed to fulfill the differential equation:

$$\frac{d\mathsf{v}}{dt}(t, \mathbf{c}) = g(\mathsf{u}(t, \mathbf{c}), \mathsf{v}(t, \mathbf{c}), t) \tag{5}$$

where function g called *driving function*, defined as $g : \mathbb{R}^n \times \mathbb{R}^m \times [t_0, T] \to \mathbb{R}^m$ if it satisfied the three properties:

1. **Lipschitz Continuity in State Space**: for every $\mathbf{x_v} \in \mathbb{R}^m$ and $t \in [t_0, T]$, the function $g(\cdot, \mathbf{x_v}, t)$ is k_u-Lipschitz. This means there exists a constant $k_u \in \mathbb{R}_+$ such that for all $(\mathbf{x_{u,1}}, \mathbf{x_{u,2}}) \in \mathbb{R}^n \times \mathbb{R}^n$, $\|g(\mathbf{x_{u,2}}, \mathbf{x_v}, t) - g(\mathbf{x_{u,1}}, \mathbf{x_v}, t)\| \leq k_u \|\mathbf{x_{u,2}} - \mathbf{x_{u,1}}\|$.
2. **Lipschitz Continuity in Integral State Space**: for every $\mathbf{x_u} \in \mathbb{R}^n$ and $t \in [t_0, T]$, the function $g(\mathbf{x_u}, \cdot, t)$ is k_v-Lipschitz. Thus, for all $(\mathbf{x_{v,1}}, \mathbf{x_{v,2}}) \in \mathbb{R}^m \times \mathbb{R}^m$, $\|g(\mathsf{u}, \mathbf{x_{v,2}}, t) - g(\mathsf{u}, \mathbf{x_{v,1}}, t)\| \leq k_v \|\mathbf{x_{v,2}} - \mathbf{x_{v,1}}\|$.
3. **Continuity over the Time**: the function g is continuous with respect to time, meaning that for any $\forall (\mathbf{x_u}, \mathbf{x_v}) \in \mathbb{R}^n \times \mathbb{R}^m$ the mapping $t \to g(\mathbf{x_u}, \mathbf{x_v}, t)$ is continuous over the interval $[t_0, T]$.

Note that for a given integrated trajectory v, multiple potential suitable functions g may exist, but we only require to know one. Inspired by analogue computational methods [31], our focus will primarily be on the integral operator, rendering the integrated trajectory as $\mathsf{v}(t, \mathbf{c}) = \int_{t=t_0}^{t} \mathsf{u}(s, \mathbf{c}) \, ds$. In this case, v corresponds to the integral.

Theorem 1 (Driving Function Existence). *Let* $\mathsf{u} : [t_0, T] \times \mathbb{R}^p \to \mathbb{R}^n$ *be a observable trajectory, let g be a driving function.*

For any initial condition $\mathsf{v}_0(\mathbf{c})$, *there exists a unique absolutely continuous function* $\mathsf{v} : [t_0, T] \times \mathbb{R}^p$, *such that for all* $\mathbf{c} \in \mathbb{R}^p$, *for all* $t \in [t_0, T]$, $\mathsf{v}(t, \mathbf{c}) = \mathsf{v}_0(\mathbf{c}) + \int_{t_0}^{t} g(\mathsf{u}(\tau, \mathbf{c}), \mathsf{v}(\tau, \mathbf{c}), \tau) \, d\tau$.

In other words, there exists a unique function denoted as $\mathcal{F}$ and referred to as the *integral operator*, which maps observable trajectory u to integrated trajectory v: $\mathsf{v} = \mathcal{F}(\mathsf{u})$.

Assuming initial knowledge of the integrated trajectory's value $\mathsf{v}(t,\mathbf{c}) = \mathcal{F}(\mathsf{u})(t,\mathbf{c})$, we then proceed to train a neural network f_θ to learn the observable trajectory $\mathsf{u}(t,\mathbf{c})$ from integrated trajectory $\mathsf{v}(t,\mathbf{c})$:

$$\mathsf{u}(t,\mathbf{c}) = f_\theta(\mathsf{v}(t,\mathbf{c}), t, \mathbf{c}). \tag{6}$$

This neural network f_θ can be seen as defining a new ODE g_θ as follows:

$$\frac{d\mathsf{v}}{dt}(t,\mathbf{c}) = g\left(f_\theta(\mathsf{v}(t,\mathbf{c}),t), \mathsf{v}(t), t, \mathbf{c}\right) \\ = g_\theta(\mathsf{v}(t,\mathbf{c}), t, \mathbf{c}) \tag{7}$$

This ODE will be learnt in two ways. Either we solve the regression problem defined by Eq. (6). Or we use the adjoint method to solve Eq. (7). Once we have learned g_θ, we might not have direct access to integrated trajectory v. However, at any time step, integrated trajectory v can be estimated by solving the ODE:

$$\widehat{\mathsf{v}}(t,\mathbf{c}) = \mathrm{ODESolve}(g_\theta, \mathsf{v}(t_0), t_0, t, \mathbf{c}) \tag{8}$$

Under the same assumptions as [5], g_θ fulfills the conditions of the Picard-Lindelöf theorem since most neural networks architecture are Lipschitz continuous functions, ensuring the unicity of the solution. For a formal proof and statement, refer to Theorem 2.

Theorem 2. *Let g be a driving function and f be a smooth estimator. For any initial condition $\mathsf{v}_0(\mathbf{c})$, there exists a unique function v such that for all $t \in [t_0, T]$, for all $\mathbf{c} \in \mathbb{R}^p$,*

$$\frac{\partial \mathsf{v}}{\partial t}(t,\mathbf{c}) = g\left(f\left(\mathsf{v}(t,\mathbf{c}),t,\mathbf{c}\right), \mathsf{v}(t,\mathbf{c}), t\right) \text{ and } \mathsf{v}(t_0,\mathbf{c}) = \mathsf{v}_0(\mathbf{c})$$

In other words, there exists a unique integrated trajectory that is the solution to this new ODE defined in Eq. (7). Then, we can estimate observable trajectory u by applying the following equation:

$$\widehat{\mathsf{u}}(t,\mathbf{c}) = f_\theta(\mathsf{v}(t), t, \mathbf{c}) \tag{9}$$

A key point of our approach is that we have access to additional information $\mathsf{v} = \mathcal{F}(\mathsf{u})$ during the training. This additional information is calculated from the datasets in a preprocessing step (Step 1 below).

In summary, the INODE algorithm can be described as follows:

1. **Step 1: Operator preprocessing:** Compute $\mathsf{v} = \mathcal{F}(\mathsf{u})$ for training data.
2. **Step 2: Learning the ODE**
    (a) Initialize a neural network f_θ and consequently g_θ (Refer to Eq. (7) for understanding the relationship between the two networks)
    (b) For a control parameter $\mathbf{c}$, and its associated trajectory $\mathsf{u}(t,\mathbf{c})$
        i. Compute $\widehat{\mathsf{v}}(t,\mathbf{c})$ such that $\frac{d\widehat{\mathsf{v}}}{dt}(t,\mathbf{c}) = g_\theta(\widehat{\mathsf{v}}(t,\mathbf{c}), t, \mathbf{c})$ using an ODE solver.

ii. Compute the loss $L(\hat{v}(t,c)) = \ell(\hat{v}(t,c), v t, c))$.
  iii. Compute the gradient of the loss $\ell(\hat{v}(t,c), v(t,c))$ using the adjoint method.
  iv. Update the network weights based on the gradient.
3. **Step 3: Evaluation**
   (a) Compute $\hat{v}$ by solving the ODE defined by g_θ (see Eq. (8))
   (b) Estimate u as $\hat{u} = f_\theta(v)$ (see Eq. (9))

The INODE algorithm acts as a universal approximator for any observable trajectory since this observable trajectory is integrable and essentially bounded. In other words, it allows us to approximate the discrepancy between u and $f_\theta(v)$ to an arbitrary degree, given the value of v (see Theorem 3): given the function v, there exists a neural network f_θ that acts as a universal approximator for $u(t, c)$ (see. Step 2).

Following the notation from [18], we denote by $\mathcal{NN}^\rho_{m,n,k}$ the set of multilayer perceptrons that have m neurons in the input layer, n neurons in the output layer and an arbitrary number of hidden layer each containing at least k neurons each, using the activation function ρ.

Theorem 3 (Universal Approximation knowing the integrated trajectory). *Let u be a observable trajectory and g be a driving function. Let v be the integrated trajectory of u guaranteed by Theorem 1: $v = \mathcal{F}_{g,v_0}(u)$.*

For all $\varepsilon > 0$, for all $k > 0$, there exists a neural network

$$f_\theta \in \mathcal{NN}^{\text{ReLU}}_{m+p+1, n, \max(m+p+2, n)}$$

such that f_θ is a smooth estimator with a uniform Lipschitz constant bounded by k, and for all $t, c \in [t_0, T] \times \mathbb{R}^p$, $\|f_\theta(v(t,c), t, c) - u(t,c)\| < \varepsilon$.

Then, having such an approximation, we can infer another approximation for observable trajectory, this time without prior knowledge of the value of integrated trajectory (see Theorem 4).

Theorem 4 (Universal Approximation). *Let u be a observable trajectory and g be a driving function. Let v be the integrated trajectory of u corresponding to solution guaranteed by Theorem 1: $v = \mathcal{F}_{g,v_0}(u)$.*

For all $\varepsilon > 0$, for all constant $k > 0$, there exists $\varepsilon_1 > 0$ such that all neural networks $f_\theta \in \mathcal{NN}^{\text{ReLU}}_{m+p+1, n, \max(m+p+2, n)}$ that satisfies the two following properties

1. *f_θ is a smooth estimator with a uniform Lipschitz constant bounded by k,*
2. *and for all $t, c \in [t_0, T] \times \mathbb{R}^p$, $\|f_\theta(v(t,c), t, c) - u(t,c)\| < \varepsilon_1$.*

   verify: $\|f_\theta(\mathbf{G}_{g,v_0}(f_\theta)(t,c), t, c) - u(t,c)\| < \varepsilon$ where $\mathbf{G}_{g,v_0}(f_\theta)(t,c) = \int_{t_0}^t g(f_\theta(v(\tau,c), \tau, c), v(\tau,c), \tau) \, d\tau$.

In other words, Theorem 4 says that if f_θ is a universal approximator for $\mathsf{u}t, \mathbf{c}$, then the trajectory computed by Step 3 defines a universal approximator for $\mathsf{u}t, \mathbf{c}$.

We have outlined our method using a generic operator between u and v, requiring only this operator to be differentiable. For example, we can outline three specific operators:

- the integral operator that we call INODE,
- the exponential smoothing (EXP-INODE), inspired by the exponential smoothing operator widely recognized in the literature,
- and a combination of the strategies of the first two approaches (Comb-INODE).

INODE : *Case when* v *corresponds to the integral of* u : $\mathsf{v}(t, \mathbf{c}) = \int_{t=t_0}^{t} \mathsf{u}(s, \mathbf{c}) \, \mathrm{d}s$.

As previously mentioned, integration is known to be a numerically more stable operation than differentiation, particularly in the context of analog computations [31]. It is especially true when the signal u experiences a discontinuity at some time t^*, where the derivative would be infinite, unlike the integral. This concept can be illustrated with the Heaviside function H whose derivative, in the sense of distributions, is the Dirac delta distribution δ, whereas its integral is the ramp function R (also called ReLU function). When the observable trajectory u is the Heaviside function H, the integrated trajectory $\mathsf{v}(t, \mathbf{c})$ is continuous (see Fig. 1). Unlike NODE-based methods that must approximate the Dirac delta function, our method is designed to learn a mapping from the ramp function to the Heaviside function, which can be expressed in the following way: $H(t) = g(R(t), t) = \frac{R(t)}{t}$.

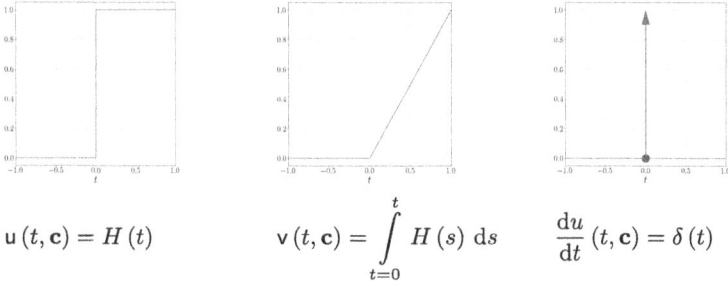

$\mathsf{u}(t, \mathbf{c}) = H(t)$ $\qquad \mathsf{v}(t, \mathbf{c}) = \int_{t=0}^{t} H(s) \, \mathrm{d}s \qquad \frac{\mathrm{d}\mathsf{u}}{\mathrm{d}t}(t, \mathbf{c}) = \delta(t)$

Fig. 1. Function u whose its integral is more stable than its derivative. The red point in the middle figure signifies the discontinuity.

Note that Step 3.(a) of INODE, which involves computing an integral, can be efficiently performed using simple numerical methods such as the Trapezoidal rule.

EXP-INODE: *Case when* v *corresponds to the Exponential Smoothing of* u:

$$v(t, c) = u(t_0, c)e^{-\lambda(t-t_0)} + \int_{t=t_0}^{t} \lambda e^{-\lambda(t-s)} u(s, c) \, ds \tag{10}$$

where λ is a positive constant. The exponential smoothing (or exponential moving average) results from averaging the past signal, but applying forgetting weight exponentially decreases with time. This operator is also chosen for its smoothing capability, acting as a low-pass filter inspired by signal processing theory.

Note that v fulfills an (ODE): $\frac{dv}{dt}(t, c) = \lambda(u(t, c) - v(t, c))$. This equation confirms that Step 3.(a) of Method INODE can be implemented.

Similar to the previous case, calculating $v(t, c)$ can be achieved by relying on the dataset because it involves a composition of an integral and an exponential function.

Comb-INODE: *Case when* $v(t, c)$ *corresponds to a combinations of the two previous cases:* Finally, the combination of these two approaches—integrating and exponential smoothing—suggests a hybrid strategy where both operators are employed to harness their respective advantages.

$$v(t, c) = \begin{bmatrix} v_1(t, c)) \\ v_2(t, c)) \end{bmatrix} = \begin{bmatrix} \int_{t_0}^{t} u(s, c)) \, ds \\ u(t_0, c)e^{-\lambda(t-t_0)} + \int_{t=t_0}^{t} \lambda e^{-\lambda(t-s)} u(s) \, ds \end{bmatrix} \tag{11}$$

Now, an integrated trajectory corresponds to $u = f_\theta(v) = f_\theta\left(\begin{bmatrix} v_1 \\ v_2 \end{bmatrix}\right)$. We then have a system of ODEs that we can also solve:

$$\frac{dv(t, c)}{dt} = \begin{bmatrix} \frac{dv_1(t,c)}{dt} \\ \frac{dv_2(t,c)}{dt} \end{bmatrix} = \begin{bmatrix} f_\theta(v(t, c)) \\ \lambda(f_\theta(v(t, c)) - v_2(t, c)) \end{bmatrix}.$$

5 Experiments Results

We aim to create a surrogate for a part of the ThomX particle simulator [32]: This accelerator began operation in 2021 and is currently in the commissioning phase). We aim to focus on the linear accelerator section (corresponding to the part where the electron beam accelerates from zero velocity to the speed of light).

Dataset. The overall Linac dataset [11] includes 4000 simulations, computed by Astra [10] and uniformly divided into a training set (80%), a validation set (10%) and a test set (10%). The overall dimension of the control settings (c) is 36 [24]. It includes detailed simulation outputs of electron beams passing through a linear accelerator, capturing a series of parameters that describe beam dynamics.

Each trajectory i is represented by 4000 points, denoted as $(t_j, \mathsf{u} t_j, \mathbf{c}_j)_{1 \leq j \leq 4000}$ where t_j are the time instances and $\mathsf{u}(t_j, \mathbf{c}_i)$ are the corresponding values of the trajectory under specific parameters $\mathbf{c}_i$. The time for each point t_j is uniformly drawn from the interval $[0, 9.393]$.

Experiments Set-Up. We evaluate our three variant approaches alongside the baselines (LSTM [13] and NODE [5]). Additionally, we explore two methodologies to resolve Eq. (7): One approach employs a regression technique, and the second refines the outcome of the first using the adjoint method. We will specify the use of the adjoint method by incorporating the term *"with finetuning"*.

All experiments, except LSTM, employ a Fully Connected Neural Network (FCNN). The NODE is trained using the loss function defined in the seminal paper [5]. The selection of hyperparameters, such as the optimizer, specifics of the ODE solver, sampling points, method of normalization, learning rate, and criteria for loss, is conducted through extensive testing using Optuna [1], a hyperparameter optimization framework.

Performance Measurement. To compare the performance of the different models, we compare the predicted trajectory position with the trajectory from the dataset (referred to as ground truth). We will use the coefficient of determination R^2 to compare these two trajectories. The R^2 value is defined as:

$$R^2 = 1 - \frac{\sum_{i=1}^{n}(y_i - \hat{y}_i)^2}{\sum_{i=1}^{n}(y_i - \bar{y})^2}$$

where y_i are the ground truth values, $\hat{y}_i$ are the predicted values, and $\bar{y}$ is the mean of the actual values. The R^2 value indicates how well the predicted trajectory matches the ground truth, with a 1 indicating a perfect match.

Experiments Results. This section aims to evaluate the capability of INODE and its variations to accurately represent a LINAC's behaviour and compare their effectiveness with conventional time series forecasting models.

Table 1. Comparison of Global Metrics Across Different Methods

Method	LSTM	NODE	INODE	EXP-INODE	Comb-INODE
R^2	0.9588	0.9448	0.9911	0.9920	0.9944

The results presented in Table 1 highlight the superior performance of the INODE models over traditional time series prediction models such as LSTM and NODE.

We aim to visualize the predicted beam properties compared to the ground truth trajectory given by the Astra [10] beam tracking simulator. We focus on three significant parameters to characterize the beam:

1. *Emittance* measures the spread of a particle beam in phase space, representing the beam's quality and tendency to diverge. It describes the size and angular divergence of the beam, with lower emittance values indicating a higher quality beam that is more focused and less prone to spreading out.
2. *Average horizontal position-angle correlation in the beam transverse coordinates* (denoted by XX'_{avr}) is the covariance between the horizontal position of particles within the beam and their horizontal propagation angles.
3. *Horizontal angular spread in the beam transverse coordinates* (denoted by X'_{ms}) is a measure of the spread of particle angles in the horizontal plane, quantified as the root mean square (RMS) of these angles. It indicates how much the particles within the beam deviate from the average direction in the horizontal axis.

Figures 2, 3, 4, 5 and 6 offer a visual comparison of model performance across varying levels of sequence discontinuity in the dataset. Although the NODE model achieves satisfactory R^2 scores, its main strength lies in predicting the general scale of the sequences rather than handling discontinuities. In contrast, even the base INODE model accurately captures these discontinuities or exponential smoothing.

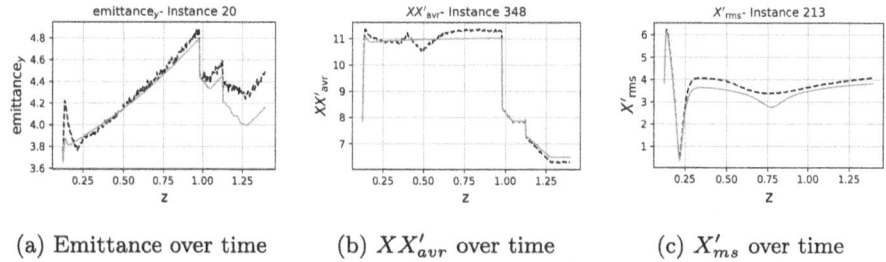

(a) Emittance over time   (b) XX'_{avr} over time   (c) X'_{ms} over time

Fig. 2. Performance of LSTM. The orange curve represents the ground truth trajectory, while the blue curve represents the trajectory predicted by the model. (Color figure online)

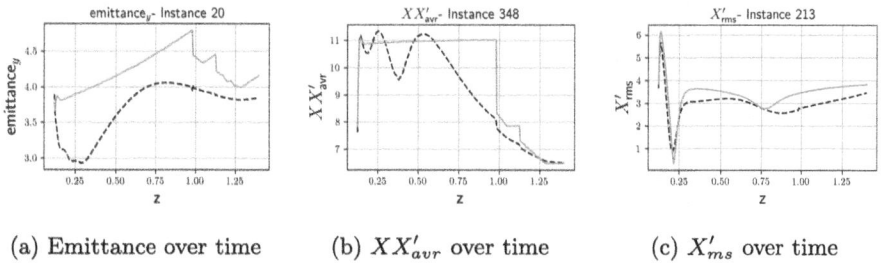

(a) Emittance over time   (b) XX'_{avr} over time   (c) X'_{ms} over time

Fig. 3. Performance of NODE. The orange curve represents the ground truth trajectory, while the blue curve represents the trajectory predicted by the model. (Color figure online)

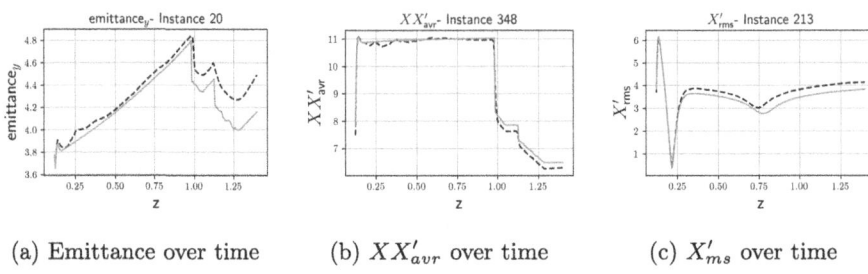

(a) Emittance over time (b) XX'_{avr} over time (c) X'_{ms} over time

Fig. 4. Performance of INODE. The orange curve represents the ground truth trajectory, while the blue curve represents the trajectory predicted by the model. (Color figure online)

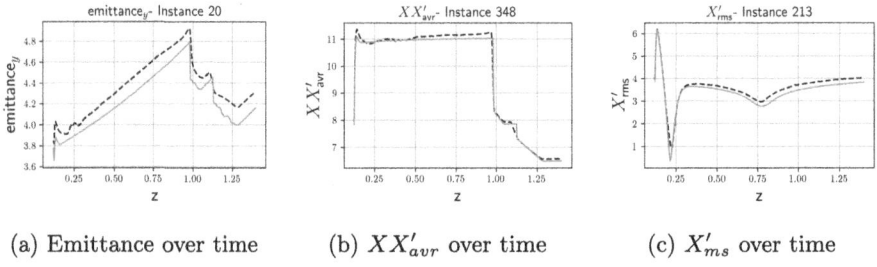

(a) Emittance over time (b) XX'_{avr} over time (c) X'_{ms} over time

Fig. 5. Performance of EXP-INODE. The orange curve represents the ground truth trajectory, while the blue curve represents the trajectory predicted by the model. (Color figure online)

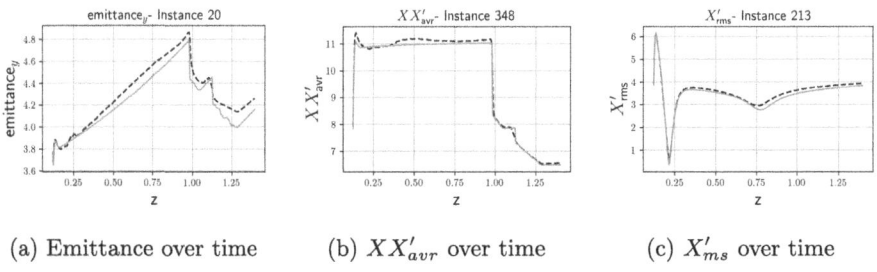

(a) Emittance over time (b) XX'_{avr} over time (c) X'_{ms} over time

Fig. 6. Performance of Comb-INODE. The orange curve represents the ground truth trajectory, while the blue curve represents the trajectory predicted by the model. (Color figure online)

6 Conclusion

This article reviews the basic approach based on Neural Ordinary Differential equations for time series prediction and ODE modelling. It consists of searching from some underlying ordinary differential equation and searching for some (possibly local) optimal of a cost function over the associated reachability problem.

This is done by using a gradient descent and requires the computation of the gradient of this cost function. The adjoint method is a classical method for evaluating this gradient, which solves another differential equation. This can be seen as a continuous time version of the backpropagation algorithm [5].

This study introduces the Implicit Neural ODE (INode) framework, extending the approach in dynamical systems modelling, specifically tailored for applications involving discontinuous behaviours, such as those observed in particle accelerators. Unlike traditional methods, INode leverages integral operators to transform the input data into a more manageable form, bypassing the discontinuities' complexities. Our results demonstrate that INode not only addresses the challenges of modelling systems with abrupt behavioural changes.

Through rigorous theoretical analysis, we have established the consistency and convergence properties of the INode framework, validating its effectiveness under moderate assumptions. Moreover, our approach's ability to learn data-driven ODEs without direct interaction with the heavy computational process of ODE resolution, particularly when discontinuous are involved, marks a significant advancement over existing techniques, offering improved efficiency and robustness.

Acknowledgments. The authors would like to thank Olivier Bournez for his insightful and fruitful comments and constructive discussions on neural networks.

References

1. Akiba, T., Sano, S., Yanase, T., Ohta, T., Koyama, M.: Optuna: a next-generation hyperparameter optimization framework. In: Proceedings of the 25th ACM SIGKDD International Conference on Knowledge Discovery and Data Mining (2019)
2. Biedron, S., et al.: Snowmass21 accelerator modeling community white paper (2022)
3. Bournez, O., Campagnolo, M.L.: A survey on continuous time computations. In: Cooper, S.B., Löwe, B., Sorbi, A. (eds.) New Computational Paradigms. Springer, New York (2008). https://doi.org/10.1007/978-0-387-68546-5_17
4. Bournez, O., Pouly, A.: A survey on analog models of computation. In: Brattka, V., Hertling, P. (eds.) Handbook of Computability and Complexity in Analysis. TAC, pp. 173–226. Springer, Cham (2021). https://doi.org/10.1007/978-3-030-59234-9_6
5. Chen, R.T.Q., Rubanova, Y., Bettencourt, J., Duvenaud, D.K.: Neural ordinary differential equations. In: Bengio, S., Wallach, H., Larochelle, H., Grauman, K., Cesa-Bianchi, N., Garnett, R. (eds.) Advances in Neural Information Processing Systems, vol. 31. Curran Associates, Inc. (2018)
6. Chen, Y., Ren, K., Wang, Y., Fang, Y., Sun, W., Li, D.: ContiFormer: continuous-time transformer for irregular time series modeling. In: Advances in Neural Information Processing Systems, vol. 36 (2024)
7. Daw, A., Bu, J., Wang, S., Perdikaris, P., Karpatne, A.: Mitigating propagation failures in physics-informed neural networks using retain-resample-release (R3) sampling. arXiv preprint arXiv:2207.02338 (2022)

8. DiPietro, R., Hager, G.D.: Deep learning: RNNs and LSTM. In: Handbook of Medical Image Computing and Computer Assisted Intervention, pp. 503–519. Elsevier (2020)
9. Fedeli, L., et al.: Pushing the frontier in the design of laser-based electron accelerators with groundbreaking mesh-refined particle-in-cell simulations on exascale-class supercomputers. In: SC22: International Conference for High Performance Computing, Networking, Storage and Analysis, pp. 1–12. IEEE (2022)
10. Flöttmann, K.: ASTRA: a space charge tracking algorithm, manual, 2017. Technical report (2017)
11. Goutierre, E.: Linac dataset from Thomx (2024). https://zenodo.org/records/11084340
12. Hirsch, M.W., Smale, S., Devaney, R.L.: Differential Equations, Dynamical Systems, and an Introduction to Chaos. Academic Press (2012)
13. Hochreiter, S., Schmidhuber, J.: Long short-term memory. Neural Comput. **9**(8), 1735–1780 (1997)
14. Hu, Z., Shukla, K., Em Karniadakis, G., Kawaguchi, K.: Tackling the curse of dimensionality with physics-informed neural networks. arXiv preprint arXiv:2307.12306 (2023)
15. Jia, J., Benson, A.R.: Neural jump stochastic differential equations. In: Wallach, H., Larochelle, H., Beygelzimer, A., d'Alché-Buc, F., Fox, E., Garnett, R. (eds.) Advances in Neural Information Processing Systems, vol. 32. Curran Associates, Inc. (2019)
16. Kelly, J., Bettencourt, J., Johnson, M.J., Duvenaud, D.K.: Learning differential equations that are easy to solve. In: Advances in Neural Information Processing Systems, vol. 33, pp. 4370–4380 (2020)
17. Kidger, P.: On neural differential equations. arXiv preprint arXiv:2202.02435 (2022)
18. Kidger, P., Lyons, T.: Universal approximation with deep narrow networks. In: Abernethy, J., Agarwal, S. (eds.) Proceedings of Thirty Third Conference on Learning Theory. Proceedings of Machine Learning Research, vol. 125, pp. 2306–2327. PMLR (2020)
19. Kidger, P., Morrill, J., Foster, J., Lyons, T.: Neural controlled differential equations for irregular time series. In: Advances in Neural Information Processing Systems, vol. 33, pp. 6696–6707. Curran Associates, Inc. (2020)
20. Lechner, M., Hasani, R.: Learning long-term dependencies in irregularly-sampled time series
21. Masi, F., Stefanou, I.: Evolution TANN and the identification of internal variables and evolution equations in solid mechanics. J. Mech. Phys. Solids **174**, 105245 (2023)
22. Massaroli, S., Poli, M., Park, J., Yamashita, A., Asama, H.: Dissecting neural odes. In: Advances in Neural Information Processing Systems, vol. 33, pp. 3952–3963 (2020)
23. Miller, J.A., et al.: A survey of deep learning and foundation models for time series forecasting. arXiv preprint arXiv:2401.13912 (2024)
24. Purwar, H., et al.: Random error propagation on electron beam dynamics for a 50 MeV S-band linac. J. Phys. Commun. **7**(2), 025002 (2023)
25. Raissi, M., Perdikaris, P., Em Karniadakis, G.: Physics informed deep learning (Part I): data-driven solutions of nonlinear partial differential equations (2017)
26. Rubanova, Y., Chen, R.T.Q., Duvenaud, D.K.: Latent ordinary differential equations for irregularly-sampled time series. In: Advances in Neural Information Processing Systems, vol. 32. Curran Associates, Inc

27. Ruthotto, L.: Differential equations for continuous-time deep learning. arXiv preprint arXiv:2401.03965 (2024)
28. Sitzmann, V., Martel, J., Bergman, A., Lindell, D., Wetzstein, G.: Implicit neural representations with periodic activation functions. In: Advances in Neural Information Processing Systems, vol. 33, pp. 7462–7473 (2020)
29. Sun, K., Chen, X., Zhao, X., Qi, X., Wang, Z., He, Y.: Surrogate model of particle accelerators using encoder-decoder neural networks with physical regularization. Int. J. Mod. Phys. A **38**, 2350145–2928 (2023)
30. Tzen, B., Raginsky, M.: Neural stochastic differential equations: deep latent gaussian models in the diffusion limit (2019)
31. Ulmann, B.: Analog and Hybrid Computer Programming. Walter de Gruyter GmbH & Co KG (2023)
32. Variola, A., Haissinski, J., Loulergue, A., Zomer, F., et al.: Thomx technical design report (2014)
33. Vay, J.-L., et al.: Modeling of advanced accelerator concepts. J. Instrum. **16**(10), T10003 (2021)
34. Xiao, L., Ge, L., Li, Z., Ng, C.-K.: Advances in multiphysics modeling for parallel finite-element code suite ACE3P. IEEE J. Multiscale Multiphys. Comput. Tech. **4**, 298–306 (2019)
35. Zappala, E., et al: Neural integro-differential equations, vol. 37, no. 9, pp. 11104–11112 (2023)

Author Index

A
Adamson, Duncan 73

B
Bartocci, Ezio 3
Benso, Andrea 88
Berthon, Raphaël 203
Bruse, Florian 38

C
Cohen, Johanne 221

D
D'Alessandro, Flavio 88

F
Fleischmann, Pamela 73
Frehse, Goran 187

G
Ganty, Pierre 21
Goutierre, Emmanuel 221
Guler, Hayg 221

H
Hirvensalo, Mika 167
Huch, Annika 73

K
Kapotos, Fatios 221
Katoen, Joost-Pieter 203
Kawamura, Akitoshi 167
Kincaid, Zachary 154
Kojima, Misaki 54
Kučera, Antonín 9

L
Lange, Martin 38
Li, Sida-Bastien 221

M
Manini, Nicolas 21

N
Nishida, Naoki 54

P
Papi, Paolo 88
Pimpalkhare, Nikhil 154
Potapov, Igor 167
Priser, Gwendal 187

R
Ranzato, Francesco 21
Ryzhikov, Andrew 104

S
Sébag, Michèle 221
Stérin, Tristan 120

T
Tveretina, Olga 141

V
Vanneaux, Elena 187

W
Winkler, Tobias 203
Woods, Damien 120

Y
Yuyama, Takao 167

Z
Zhu, Bowen 221

© The Editor(s) (if applicable) and The Author(s), under exclusive license to Springer Nature Switzerland AG 2024
L. Kovács and A. Sokolova (Eds.): RP 2024, LNCS 15050, p. 239, 2024.
https://doi.org/10.1007/978-3-031-72621-7

SPRINGER NATURE

GPSR Compliance

The European Union's (EU) General Product Safety Regulation (GPSR) is a set of rules that requires consumer products to be safe and our obligations to ensure this.

If you have any concerns about our products, you can contact us on ProductSafety@springernature.com

In case Publisher is established outside the EU, the EU authorized representative is:

Springer Nature Customer Service Center GmbH
Europaplatz 3
69115 Heidelberg, Germany

The manufacturer's authorised representative in the EU is Springer Nature Customer Service Centre GmbH, Europaplatz 3, 69115 Heidelberg, Germany. If you have any concerns regarding our products, please contact ProductSafety@springernature.com

Printed and bound by CPI Group (UK) Ltd, Croydon, CR0 4YY

25/03/2026

02078189-0014